Transportation Research, Economics and Policy

Series editors

The objective of the series is to provide academics, students, policy analysts and practitioners with authoritative and accessible information to develop practical solutions, assist in policy formulation and inform debate on transportation issues and questions.
Topics in the series may include:

- transportation economics
- management of transportation by firms
- government bodies and regulators and the integration of transportation into cities, regions and countries

The unifying feature of each volume is the development or application of quantitative economic models and methods to examine current and emerging issues relating to transportation. Coverage includes the analysis of data, models of market structure, the evolution of institutions and emerging governance structures all in the context of transportation economics, policy and management. Both research monographs and edited volumes will be considered for publication in the series.

More information about this series at http://www.springer.com/series/6647

Mikhail Blinkin · Elena Koncheva
Editors

Transport Systems of Russian Cities

Ongoing Transformations

Springer

Editors
Mikhail Blinkin
National Research University Higher School of Economics
Moscow
Russia

Elena Koncheva
National Research University Higher School of Economics
Moscow
Russia

ISSN 1572-4387
Transportation Research, Economics and Policy
ISBN 978-3-319-47799-2 ISBN 978-3-319-47800-5 (eBook)
DOI 10.1007/978-3-319-47800-5

Library of Congress Control Number: 2016954019

Printed on acid-free paper

This Springer imprint is published by Springer Nature
The registered company is Springer International Publishing AG
The registered company address is: Gewerbestrasse 11, 6330 Cham, Switzerland

Acknowledgments

The authors are sincerely grateful to the colleagues, with whom they have spoken in recent years:

Vukan Vuchic, professor at the University of Pennsylvania and former consultant for the US Department of Transportation,

Donald Shoup, professor at the University of California, Los Angeles,

Jeffrey Kenworthy, professor at the Curtin University Sustainability Policy Institute, Western Australia,

Boris Kerner, professor at the University of Duisburg-Essen, Germany,

Johanna Zmud, Head of Transportation, Space, and Technology Program at RAND Corporation,

Jaime Lerner, former governor of the state of Paraná and mayor of Curitiba, Brazil,

Kenneth Livingston, former mayor of London,

Pierre Laconte, former president of ISOCARP (International Society of City and Regional Planners), Brussels,

Alain Flausch, secretary-general of UITP,

Yuri Bocharov, member of the Russian Academy of Architecture and Construction Sciences,

Alexander Mikhailov, professor at the Irkutsk National Research Technical University,

Sergei Tarkhov, leading research fellow at Lomonosov Moscow State University.

Special thanks to the rector of the NRU HSE Yaroslav Kuzminov, on whose initiative the Institute for Transport Economics and Transport Policy Studies was created.

Contents

Introduction

This book is a collaborative effort from a team of authors, most of whom work at the Institute of Transport Economics and Transport Policy, a research center organized in 2011 as part of the National Research University "Higher School of Economics" (HSE).

The research fields of our institute include problems of operation and development of all transport modes and sectors. However, our primary field of research consists of urban transport issues. My own long-term scientific and sociopolitical interests have largely determined this priority. I have published many papers, scientific and journalistic, devoted to "city–car–public transport" issues since the 1970s.

I found one of them mentioned a few years ago in "The New Yorker" magazine. The journalist introduced me to his readers with high praise: "Mikhail Blinkin, the author of a legendary paper titled "The Etiology and Pathogenesis of Moscow Traffic Jam" (Gessen 2010). It is not for me to say whether the aforementioned article has really become legendary, but it has certainly played a constructive role judging by the paradigm shift in Moscow transport policy in recent years.

I should also add that many of my employees, the authors of the various chapters of this book, are actively involved in professional and public discussions that arise around the transport problems in Moscow and in other major Russian cities. Furthermore, they also participate to some extent in the decision-making processes in the field of urban transport policy all over the country.

Our interest in urban transport issues can also be explained by the close cooperation of our institute with the UITP,[1] the secretary-general of which, Alain Flausch, has become an honorary doctor of the HSE on our initiative.

Moreover, in addition to research, consulting, and project activities, our institute curates a master's program "Urban Transportation Planning" at the Vysokovsky Graduate School of Urbanism—one of the educational sections of the HSE.

[1]International Association of Public Transport.

HSE has traditionally been a university open to broad international dialogue. Thus, the previously mentioned master's program, as well as the work of our institute, constantly involves foreign colleagues, not to mention undergraduate and graduate students from foreign universities, who regularly come for internships to the Graduate School of Urbanism.

The idea of this book was largely dictated precisely by our experiences while communicating with our foreign colleagues—scientists and planners, and students and graduate students from all over the world.

Based on the results of this discussion, we have somewhat revised our own ideas about the world's best transport policies. At the same time, we were surprised to find out that even those of our colleagues, who have repeatedly visited modern Russia and the former USSR, have very incomplete (and even distorted!) ideas about the transport systems and problems encountered by Russian cities. Furthermore, we frequently found them to be confused about the fundamental differences between these issues and the issues faced by cities in the Western world.

In particular, many of them continue to believe in the myths of the great virtues of Soviet public transport, which was supposedly destroyed because of "haphazard and hurried privatization of the 1990s."

Some of our foreign guests have passionately recommended that we borrow solutions from current European practice (e.g., the creation of bicycle lanes), without understanding the unique planning specifics of Russian cities. They also fail to comprehend the behavior of Russian drivers, who, unlike their European counterparts, represent the first generation of Russian citizens to be able to drive.

However, the most curious thing we found out was that our foreign colleagues were often in the position of the character of a Moliere comedy, who did not know that he had been speaking in prose his whole life.[2] Here is a typical example.

I was driving down the Garden Ring with my American friend, a respected international expert in the field of urban and transportation planning, and I told him that Moscow, unlike US cities, had a one-level street and road network.

"I don't understand what you're saying," my friend interrupted me.

"I mean," I answered my guest, "that since the Robert Moses era streets (streets, avenues, arterials) in your cities belong to one level, and roads (freeways, expressways) belong to a different one. You do not have combinations. If it's a street, then it's not a road, and if it's a road, then it's not a street."

"Can it be any other way?" my colleague answered, puzzled.

After such conversations, you begin to remember how you unwittingly tried many years ago to fit the strikingly different workings of the transport systems of foreign cities, which you had seen since the fall of the Iron Curtain, into the Procrustean bed of your own Soviet experience. Unfortunately, each of us is an involuntary hostage of our past experiences…

---

[2]Monsieur Jourdain: "Par ma foi! Il y a plus de quarante ans que je dis de la prose sans que j'en susse rien."

As a result, we became more and more confident in the usefulness of this text, addressed to foreign colleagues, which contains systematic and professionally presented information on how transport systems in Russian cities are constructed, how these systems have changed over the last 100 years, and how they were planned and operated during different epochs…

I will mention a few typical cases, which we believe to be especially important when dealing with foreign readers.

The Soviet system of development and land use, the Soviet-style development model, initially focused entirely on public transport. This system may seem similar to transit-oriented development (TOD) to our foreign colleagues; however, there are vast differences between these models (described in detail in the book).

The "clash of cities and cars" has been discussed by urbanists on both sides of the Atlantic for many years. We would like to add to the international pool of knowledge on the subject a very interesting case in which all the key planning and transportation norms are fundamentally different from those faced by our foreign colleagues.

I will begin my explanation of our situation by stating the fact that the severity of this "clash" depends, primarily, on the interaction between three parameters:

- population density (in the built-up area),
- share of built-up area in streets and roads (ε),
- motorization level.

The first two parameters have been shaped in Russian cities by the Soviet system of development and land use. Population density (*d*) in Russian cities is as high as in cities of the Asian type: 60–80 residents per hectare or more. The share of built-up area in streets and roads, on the other hand, is extremely low, less than 0.1. Let me remind you for comparison that in North American cities, the parameter ε usually equals 0.3–0.35, and even in the "compact" cities of Western Europe, it is about 0.2–0.25.

The third parameter, motorization level (*m*), was artificially capped at a maximum of 60 cars per 1000 inhabitants until 1990, and in the next 25 years, it increased exponentially to 300–400 or more cars per 1000 inhabitants. The number of square meters of asphalt per car is therefore extremely low compared to foreign cities. Currently, the motorization rate in Moscow is rather high (more than 400 cars per 1000 inhabitants), while the share of trips by public transport is more than 80 %.

Let us now turn to a standard criterion for any transport planner: total vehicle kilometers of travel (VKT). Applying it to Russian planning norms, we find that given an equal Euclidean distance, network distance is always larger for a Russian citizen than for his counterpart in North America or Western Europe.

The explanation for this phenomenon is that street and road network connectivity has traditionally been very low in Russian cities. This is due to many things, including the deficit of bridges over rivers and overpasses over railways, as well as the goal of old Soviet planning to be able to block exit from any residential area with "one armored personnel carrier."

Accordingly, the rerun coefficient (average value of network distance to Euclidean distance ratio, ς) in Russian cities is extremely high. This ratio is equal to $4/\pi$ ($\approx$1.27) for the perfect "Manhattan Grid" (1) and rarely exceeds 1.2–1.25 (Newell 1980) for well-designed foreign cities. In Moscow, unfortunately, $\varsigma \approx 1.7$[3]; so the amount of "extra" mileage is always at least 40 %.

$$\zeta = \iint \frac{|x| + |y|}{\sqrt{x^2 + y^2}} \mathrm{d}s \tag{1}$$

Let us return to the above-mentioned problem of stratification of the road network: the functional division of roads and streets. The value of such stratification is best revealed by employing the classic "two-fluid model" devised by Herman and Prigogine (1979). More precisely, this can be seen through the examination of the parameter n of this model, treated as an "indicator of the quality of traffic service in the network."

Williams J.C. notes in an article that "field studies have shown that n varies from 0.8 to 3.0, with a smaller value typically indicating better operating conditions in the network. In other words, n is a measure of the resistance of the network to degraded operation with increased demand. Higher values of n indicate networks that degrade faster as demand increases" (Williams 1997).

A lengthy digression was required for us to observe the most interesting feature of Moscow's road network. The "indicator of the quality of traffic service," n, is equal to 1.25 for high-density ($\varepsilon = 0.28$!) street networks in the center of the capital, but is several times higher (2.62 and 3.62, respectively) for the key Moscow highways—the Third Ring Road and the Moscow Ring Road, which are typical examples of Moscow's hybridization of streets and freeways.

These figures demonstrate the sad fact that Moscow's key multilane highways, which were constructed to match the parameters of an urban freeway while preserving the functions of urban streets (such as not being separated from developed areas), "degrade faster as demand increases" compared to ordinary urban streets.

The consequences arising from all these very unusual combinations of key planning and transport parameters are the subject of detailed analysis in many chapters of this book.

Let me draw your attention to another case. Formal institutions of transport systems management in Russian cities have passed through real "roller coasters" since 1991:

- from the total centralization and directive planning, characteristic of the Soviet period, to almost complete "laissez-faire" approach, observed in the 1990s;
- later, in the 2010s—a return of many central legislative and enforcement requirements, with neither ideological nor financial security.

[3]The assessment is based on the processing of more than 1 million tracks of mobile phones of Moscow car owners, conducted in 2015 by the Big Data research group at HSE under the direction of Andrei Zhulin.

In the majority of Russian cities, the major role in the public transport sector is played by paratransit, specifically jitneys, which are typical of Third World countries. Very few cities in Russia, such as Moscow and Kazan, have succeeded to some extent in getting rid of this archaic transport. In many other cities, local authorities balance the need to support the remaining Soviet public transport networks that completely duplicate their paratransit networks and confronting car owners and their demands to widen and build new roads. A number of Russian cities simply lack proper understanding of the essence of the problem and how to solve it.

We also believe that the implementation in Russian practice of road pricing and demand management ideas had some interesting consequences that could be useful to our international colleagues with respect to the planning and operation of urban transport systems. It appears that the collective subconscious of Russian citizens is governed by a few simple ideas of free riding and paternalism. In particular, the belief is that the city government can and must solve all of the city's transport problems with its limited budget, providing car owners with roads and parking lots, and other residents with high-quality public transport.

At the same time, standard road pricing and demand management measures are unpopular,[4] and willingness to pay for paid parking and toll roads (and, especially, for the paid entrance to some areas) is not shown even by the owners of cars such as the Bentley or the Porsche Cayenne.

Recently, in an international forum, I happened to discuss this problem with Kenneth Livingston, the former mayor of London, and the creator of the congestion charge zone (CCZ). Ken pessimistically noted that British car owners were not very happy about the introduction of the London congestion charge. However, after learning that most Muscovites do not need to pay for the residential storage of a car and that until 2013 parking was free even in the capital's "golden mile," and he agreed that Moscow's case would be harder to resolve than London's.

Let us add to all this the characteristics of the behavior of Russian drivers. The majority of Russian car owners are car owners of the first generation, that is, young or middle-aged people whose parents did not have (and could not have!) their own car. Two serious problems result from this circumstance for transport planners.

Firstly, Russian citizens are extremely sensitive to the standard and necessary steps the city's authorities take, such as increasing the cost of car ownership, reducing the attractiveness of everyday car trips, and redistributing roadway traffic into public transit systems.

Secondly, and perhaps most importantly, the prevalence of first-generation car owners (as well as poor road conditions and institutional defects common in Russia) results in abnormally high accident rates. Russia (along with China, India, Brazil, Cambodia, Egypt, Kenya, Mexico, Turkey, and Vietnam) is included in the group of "Risky States – 10": the ten countries with the highest level of transport-related risks (annual road fatalities per 10,000 vehicles). These 10 countries account for

[4]For example, the practice of canceling already entered dedicated lanes for buses in some cities.

more than half of the total number of road fatalities in the world. The indicator of transport risks in Russia is about five units, which is approximately 10 times (sic!) higher than in Western Europe. Therefore, the Russian transport planner is obliged to be very careful when implementing standard Western European measures in the field of pedestrian and bicycle infrastructure.

All these and many other related subjects are disclosed in great detail in the various chapters of this book. Due attention is also paid to the history and current practice of urban transport planning, the Russian experience of applying sociological methods to study the transport behavior of citizens, analysis of the state of municipal and private segments of public transport, and much more.

I hope that the book will be quite helpful to foreign colleagues from the sphere of urban and transportation planning, willing to expand their understanding of the specific nature and possible scenarios for the development of urban transport systems.

Mikhail Blinkin

References

Herman RP, Prigogine I (1979) A two-fluid approach to town traffic. Science 204(4389):148–151

Gessen K (2010) Stuck. The New Yorker. http://www.newyorker.com/magazine/2010/08/02/stuck-3. Accessed 31 Aug 2016

Newell G (1980) Traffic flow on transportation networks. MIT Press Cambridge, Mass

Williams JC (1997) Macroscopic Flow Models. In: Gartner N, Messer CJ, Rathi AK (eds) Traffic flow theory. Oak Ridge National Laboratory, Oak Ridge, p 1–31

Chapter 1
Transport Planning and Transport Modeling

Alexander Kulakov and Konstantin Trofimenko

Abstract This chapter focuses on the evolution of transport planning in Russia during the 20th century. Since transport planning is a socio-economic practice, it is not surprising that it has been drastically influenced by the ideologies that succeeded each other in Russia during the period. On the other hand, the evolution of transport planning also took place due to technical progress and change of the traditional paradigm of urban mobility. From this point of view, this chapter will be of interest not only for specialists in transportation studies, but also for researchers in the field of interconnection of humanitarian and technical sciences. In addition, the chapter covers and evaluates formal documents of transport planning adopted in Russia, as well as the history of the use of transport models.

1.1 Russian Transport Planning in the 20th Century: Historical Practices and Current Consequences

Historically, the practice of transport planning (and of urban planning in general) in Russia has mostly been a reflection of the socio-economic processes happening in the country during the 20th century.

Before World War I, urban planning norms were determined by The Construction Charter of the Russian Empire, the last edition of which dates back to 1912.

The sixth section of the Charter "On the construction of cities and on urban buildings and facilities" contained, in particular, a rule that "the city is built in accordance with plans approved with respect to the established order". At the same time, the Charter contained quite liberal regulations of the free land market. Article 178 of the Charter stated that; "urban inhabitants", i.e. the owners of "vast" land

A. Kulakov · K. Trofimenko (✉)
Institute for Transport Economics and Transport Policy Studies,
National Research University Higher School of Economics, Moscow, Russia
e-mail: ktrofimenko@hse.ru

M. Blinkin and E. Koncheva (eds.), *Transport Systems of Russian Cities*, Transportation Research, Economics and Policy,
DOI 10.1007/978-3-319-47800-5_1

parcels, had the right to sell their parcels entirely or partially, without any restrictions, except for "caution in case of fire".

The Charter also contained fairly reasonable standards for planning the proportions of the built-up area of the city and the parameters of urban streets.

Russian transport planners—well educated urban engineers and thinkers—attentively studied the best foreign practices. A good example of this is the monograph on urban railways by the engineer Girshon, published in St. Petersburg in 1900 and containing a detailed analysis of urban transport systems in the industrialized world.

At the end of the nineteenth century, these urban engineers, together with the representatives of the educated classes, believed the mass appearance of cars on roads to be inevitable and desirable. Thus, in September 1896 the Russian Minister of Railways M. Khilkov[1] signed a letter "On procedure and conditions of transport of goods and passengers in self-propelled carriages on the roads of the department of communications", which became one of the first governing documents in the history of world motorization. It is necessary to pay tribute to the wisdom of the minister, because "self-propelled carriage" used to be a technical exotic like the lunar rover is today. In fact, even the word "automobile" didn't exist in the Russian language (Goltz 2005) back then.

The first fruits of the national automotive industry—the first "road-vehicle-factories"—appeared at the turn of the XIX century. In the Imperial Russian Technical Society, as well as in the merchant clubs of both capitals, the prospects of mass motorization in conjunction with the creation of a network of highways were actively discussed. These discussions were quite substantive. In 1913, a group of the most influential businessmen in Russia came together to begin the formation of a "large scale" automotive industry in accordance with "Ford's model" in Russia. In 1914 the Road Fund was established as a fiscal institution charged with target financing of the road network development. This event occurred more than forty years before of the creation of the famous and extremely efficient US analogue—"The Highway Trust Fund".

[1]Prince Mikhail Khilkov (1834–1909) is a legendary figure in transport history. Here is a brief summary of his life and achievements. He was a graduate of the St. Petersburg Corps of Pages (1853), and later an officer in the Life Guards (1853–1859). He later worked as a railway lumberjack, stoker, machinist, chief of service of rolling stock and traction at the Transcontinental Railroad in the US (1864–1871), mechanic at the Liverpool locomotive repair plant (1872–1873). Then, he became minister of public works, communications, commerce and agriculture in Bulgaria (1882–1885). He was later the head of a number of Russian railways, and the organizer of railway construction in various parts of the country, including Siberia and Turkestan. Finally, in 1895 he became the minister of communications of the Russian Empire, a post he held until 1905. Khilkov was an outstanding theorist and practitioner of construction and operation of railways, concession agreements and tariff policy. The educator and reformer-Westerner was also a passionate motorist—he participated in the first Russian automobile races, including the automobile race on the Black Sea coastal highway (1903). He was also the initiator of the opening of the first urban and suburban bus routes in Russia.

The business, scientific and technical elite in Russia came to a consensus about the need to adapt the theory and practice of urban planning to the realities of high motorization and business in the future. A number of serious publications reflect this consensus, in particular, the monograph by Professor Dubelir (1910), in which the trend of "increasing intensity of the traffic of carriages and, in particular, of automobiles" is accepted as the most important urban development fact that must be taken into account in urban planning. Based on his observations of this trend, G. Dubelir gave the very reasonable recommendation of changing parameter ε—the share of land used for streets and roads in the built-up areas of the city, which will be further discussed in Chap. 2—from 0.2 to 0.4 depending on the planning parameters.

Here we should pay attention to the fundamentally different interpretations of the concepts of "urban planning" and "urban planner", which have been used in Russia during the last 100 years (Table 1.1).

In the pre-Soviet era, the norms in accordance with which the city had to be built were developed by urban engineers, called "gradoustroitel", as was required by The Construction Charter of the Russian Empire. The literal translation of the term is "an engineer working to improve a city". Urban engineers engaged in this profession, who knew foreign languages (as was customary at the time), called themselves "stadtplaners" or "urban planners".

The Soviet era witnessed the appearance of the term "gradostroitel", which literally means "a city builder", but officially has had the same translation ("urban planner"), both during the Soviet era and today.

The essence of the difference was studied in depth by the famous Russian urbanist Professor Glazychev (2011).

> "The term "gradostroitelstvo"", says Glazychev, "has no direct analogue in the world dictionary, it comes from the absolutist tradition dating back to the Assyrian king Sennacherib. It is a symbolic reflection of the phenomenon of a ruler building something new on an empty space or on a space that the ruler perceives as empty".

Let us add a poetic reference for clarity:

Here a new city shall be wrought,

Defiance to the haughty neighbor.[2]

It is clear that the ruler who decides to build a city—the only stakeholder and the decision maker in the situation—does not take into account anyone else's interests or coordinate his or her decisions with anybody.

"Gradoustroistvo" (urban planning), according to Glazychev, in contrast to the "gradostroitelstvo" "involves the formation of a framework in which the development of the city takes place when a lot of players are involved, including stakeholders willing to implement various solutions".

[2]Pushkin (1993) The Bronze Horseman. Translated by Walter Arndt.

Table 1.1 Urban planning concepts in Russia

Field		Profession	
Transliteration	Literal translation	Transliteration	Literal translation
Gradoustroistvo	Improvement of a city	Gradoustroitel	An engineer working to improve a city
Gradostroitelstvo	Construction of a city	Gradostroitel	A city builder

Source Created by author based on Glazychev (2011)

A historical essay (Zhukov et al. 2008) notes that Russian urban engineers have formed a "planning framework within which the activities of the private developers have not come into conflict with the rights of the owners of the "old" property, as well as the interests of the urban community as a whole". It is easy to see that this definition is almost identical to the contemporary definition of urban planning.

On the eve of the October Revolution, leading Russian engineers used this concept because of their good education and knowledge of the latest world practices. In the early years of Bolshevik rule (before the 1930s) issues of urban planning and development in the USSR were still dealt by urban and provincial engineers (Bocharov 2003).

Many young intellectuals (of left wing views) believed in the possibility of combining socialism with the ideas of rational and humanistic urban planning and, accordingly, continued their "seditious" research until the mid-1930s. Zilbertal[3] and Sheleyhovskiy[4] were among the most prominent representatives of this generation.

Zilbertal noted (Zilbertal 1932, 1937) that "… the issues of movement (judging by the context of the phrase, it is about transport planning) are not a purely mathematical problem, and really depend on the extent to which people value their time and convenience".

Sheleyhovskiy (1946), who worked in the Soviet design office, believed that cities are designed for people who are free to choose their place of residence and mode of transport "…unlike the inhabitants of a zoo". He insisted on the need for

[3]Abram Zilbertal (1890–1942) was an outstanding Russian engineer and scientist in the field of urban transport, settlement patterns and transport behavior of citizens. In the 1930s he was the chief engineer of the Leningrad tramway trust. His monographs (Zilbertal 1932, 1937) are cited in this chapter. These books, which have never been reprinted and have long become a bibliographic rarity, have remained relevant in modern times.

[4]Georgy Sheleyhovskiy (1892–1946) was an outstanding Russian engineer and scientist in the field of urban studies, settlement patterns, transport systems and urban ecology. His book (Sheleyhovskiy 1946), cited in this chapter, was banned from publication. Sheleyhovskiy was urged to stop writing on similar topics in order to avoid serious consequences. Copies of Sheleyhovskiy's manuscripts have been preserved by his students and colleagues. During the thaw of the 1960s, copies of these copies were distributed as "*samizdat*" (self-published) texts. The electronic version of the book, prepared in 2008 by the initiative and under the editorship of S. Vaksman is available online at http://waksman.ru/Russian/Vehi/shel/p.htm.

"…the synthesis of at least four essential planning categories—settlement pattern, transport, street network and the form of the city plan". Here are two of his quotes:

> …what consequences arise for the residents of a city in which the free or natural settlement pattern is violated? The answer is simple: the majority of the residents of the city will be uncomfortable;
>
> …we will not make any hypotheses about the pace of motorization in our cities … streets should be designed to accommodate for demand in 50 years. When dealing with such time periods we have no reason to impose any restrictions on the car as a transport mode. When designing a city, we must check its highways in terms of their ability to let through maximum possible traffic flow …. Let us define the maximum scope of what we can expect in the future to 1 car per household….

By the 1930s the Soviet Union had developed a rather effective transport planning methodology, which included studying population mobility and travel distance, surveying passengers on different transport modes, searching for ways to increase capacity of individual links and nodes of the transport system, and evaluating the various options from an economic point of view. There was also a search for patterns in transport demand in urban areas, as well as a mathematical apparatus to model them (Figs. 1.1, 1.2, 1.3 and 1.4).

It is important to note that early Soviet researchers were very well versed in the statistics of urban transport systems and the practice of transport planning in Western countries.

However, the westernized views of Russian urban engineers did not fit into the new Soviet reality.

The Soviet government did not make its decision about cities, roads and cars immediately, but when it did, it did so firmly and permanently. The date and circumstances of the adoption of these historic decisions is well-known: June 1931, the resolution of the plenum of the Central Committee on the report of L. Kaganovich on municipal services in Moscow.

The party strongly condemned the "bourgeois theory of urban planning" in its entirety. Academic M. Bocharov has compared this decision with the way the Soviet Union's later dealt with the sphere of genetics and cybernetics. Participants on both sides of the hot theoretical debate over "socialist cities" and the "socialist settlement pattern" were all repressed together. The very best urban engineers (in particular, S. Shestakov—the head of "Large Moscow" project) were repressed in the same way as people who barely merited the title "urbanist" (L. Sabsovich) and "unurbanists" (L. Ohitovich).

Some urban engineers, including P. Balinskiy, the most famous of that generation, managed to emigrate long before the described events took place.[5] A couple

[5]Pyotr Balinskiy (1861–1925)—graduate of the St. Petersburg Institute of Civil Engineers, the author of subway projects in St. Petersburg (1901) and Moscow (1902). The developer of the scheme of high-speed railways in both capitals, integrated with the mainline railway network. Balinskiy emigrated from Russia in 1920, and died in Paris 5 years later.

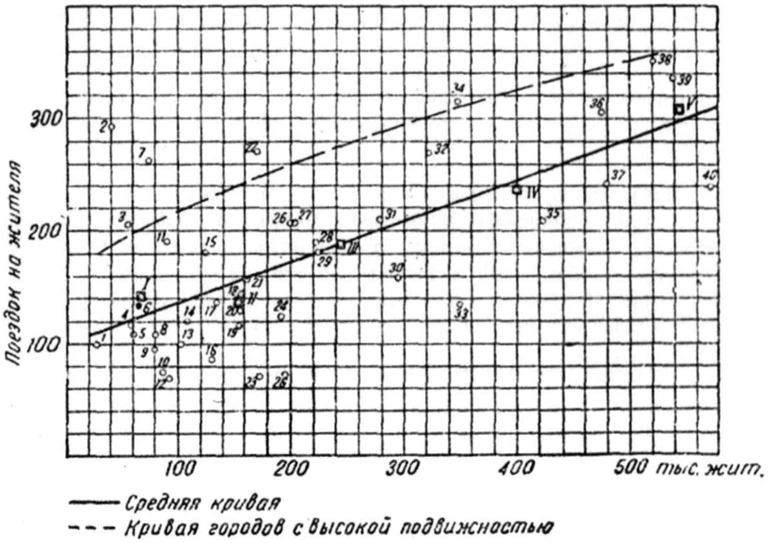

Fig. 1.1 Data on mobility in cities of the former USSR [*Source* Zilbertal (1937)]

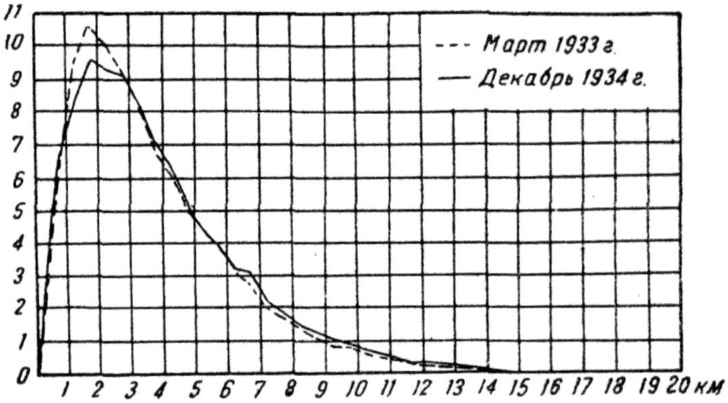

Fig. 1.2 Data on the distribution of passengers by trip length in Leningrad tramway [*Source* Zilbertal (1937)]

of them, such as professor Dubelir, managed to switch to purely engineering subjects just in time.

The bourgeois-liberal term "urban development" was replaced by "planned management of municipal services" and Soviet urban development, the main target of which was "propaganda to show the monumental achievements of the Soviet regime".

The "time and convenience" of the population (as well as their quality of life in general) was viewed as completely irrelevant, and was covered up with banal

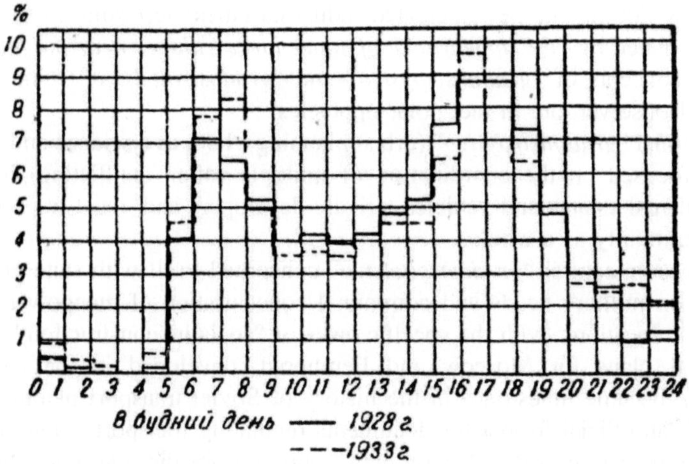

Fig. 1.3 Data on the distribution of passengers by hour in Berlin tramway [*Source* Zilbertal (1937)]

Fig. 1.4 Index of the relationship between the mobility level and the level of unemployment in the US [*Source* Zilbertal (1937)]

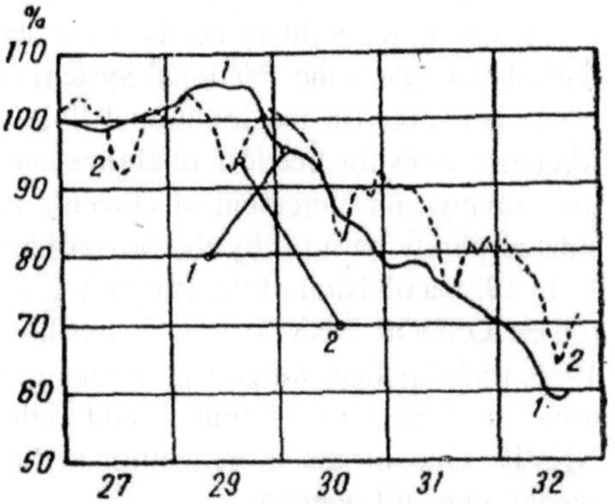

populism that used terms such as "model city" and "the growth of capital of the socialist type" to justify itself. The chief architect of the city was accountable to the Soviet authorities not for the quality of the "environment of the urbanized municipalities and local communities", but rather by achievements in the field of large-scale propaganda. The fact that architects, who replaced urban planners, were by definition not as skilled at planning calculations and mathematical formulas, was not a drawback, but instead, remarkably, turned into an advantage. Pictures, and not calculations fully met the Bolsheviks' interpretation of the idea of "… architectural design of the city, giving it proper beauty" (Kaganovich quote by Medvedev 2005).

Under these conditions it was quite natural that the "*basis for planning decisions was founded to accommodate processions and workers' demonstrations in front of tribunes of the authorities (everywhere) and military parades—in capitals*"

(Glazychev 2008). The newly created industrial cities, according to same author, were dominated by "*the appearance of super-large villages near the factories*".

Thus, concepts of "*gradoustroystvo*" and "*gradostroitelstvo*", synonymous to the casual observer, are in fact polar opposites.

Traditional "*gradoustroystvo*" (urban planning), both as a science and a practice, initially focused on the coordination of multiple (often conflicting) interests of owners of real estates and vehicles, private developers and, most importantly, the urban community as a whole.

Soviet architectural "*gradostroitelstvo*" concerned itself with state propaganda, focusing on military needs and communal concentration of manpower in certain geographic locations, with the specific tasks of "socialist construction".

General plans for Moscow and Leningrad, developed in the early 1930s (Fig. 1.5), became milestones in the history of Soviet transport planning. On the one hand, an official format for documents regulating transport planning issues—the transport section of the General plan—appeared for the first time. On the other hand, transportation planning (as well as urban planning in general) was declared to be purely secondary in favor of architectural and planning activities aimed, as already noted, to display the "monumental achievements of the Soviet regime".

Similar processes took place in the field of roads and vehicles. It is clear that this area is tied to the military needs of the state from top to bottom in every country; a typical example is the "National System of Interstate and Defense Highways" in the USA. However, in the traditional liberal paradigm roads and cars are, first and foremost, tools for freedom of choice and transport self-sufficiency of a household, time saving and increment of citizen's facilities. These critical aspects were considered purely harmful by the Soviet ideology.

In 1935 a division of the People's Commissariat of Internal Affairs of the USSR (GUSHOSDOR NKVD) was established, and was put in charge of highways. In 1936, traffic police, which was in charge of registration of vehicles, control of their use, "the regulation of traffic" and "education of drivers" was merged with the NKVD. One of the aforementioned Soviet institutions existed until 1953; the second one still operates.

Just as all other unauthorized manifestations of social or economic activity, transport self-sufficiency was seen as unwanted by communist leaders. It is no coincidence that construction under socialist rule began with a slogan typical of the Soviet mindset "take the horses away!".

Surveys of the real foreign transport planning experience were replaced by the same banal demagogy: "While in capitalist countries different modes of transport—tram, bus, subway—compete with one another, here, in a socialist economy, these transport modes will not compete among themselves. Instead, they will complement each other and give us transportation efficiency unprecedented anywhere in the world and impossible anywhere else".[6]

---

[6]The report of L. Kaganovich on the June session of Central Committee of the Communist Party of the Soviet Union in 1931.

Fig. 1.5 "Stalin's" general plan of Moscow (1935) became the standard in Soviet urban planning documentation for years ahead

Soviet transport planning in the middle of the 20th century was roughly in the same predicament as other applied disciplines. This point was very accurately reflected in a novel by classics of Soviet scientific fiction, the Strugatskiy brothers, describing life on a habitable planet, lost in deep space. State ideology there required considering the home planet a concave hollow world; however, for secret artillery calculations the military was allowed to take into account the true curvature of the surface.

Let us turn to the already mentioned parameter ε—the share of land used for streets and roads in the built-up areas of the city. In world practice, this parameter was initially oriented towards a high level of motorization, which, as already noted, contradicted the Soviet ideology. The motorization at the time was less than 60 cars per 1000 inhabitants and was planned to reach 180 cars per 1000 inhabitants "after the construction of communism". The plan relied mostly on public transport, thus there was no need to keep this parameter at a high level. However, according to the recommendations of the Academy of Architecture of the USSR (Levchenko 1947) referred to in Chap. 2, "the percentage of the territory occupied by streets" had to be changed from 15 % (in small towns) to 25 % of the total residential area.

Notice that $\varepsilon \approx 0.25$ is a typical figure adopted in the design practice of Western European cities, where administrative restrictions on the ownership and use of private cars did not exist. So, accounting for the "the true curvature of the surface" was not forbidden for experts from the Academy of Architecture of the USSR.

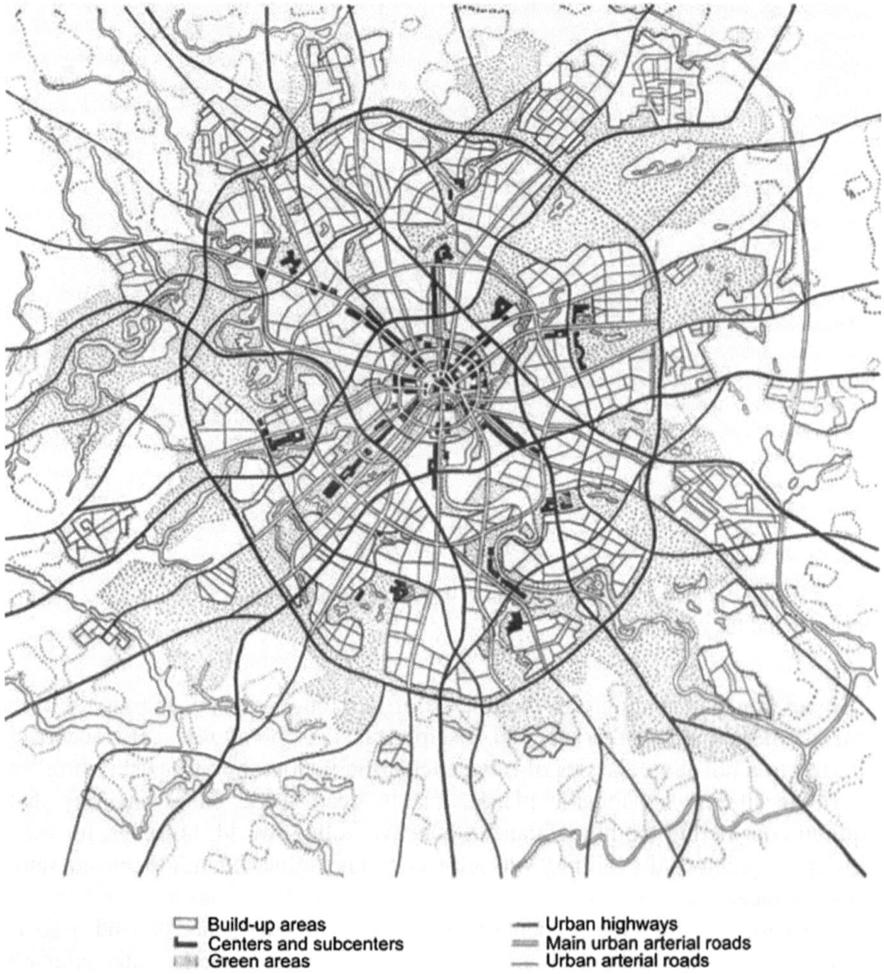

Fig.1.6 General plan of Moscow, 1971—a document, that has not lost its relevance to this day. In many cities of the Russian Federation large-scale solutions in the field of transport planning are blindly copied from documents (such as this one) developed in 1970s. In some cases this even turns out to be beneficial to the cities in question

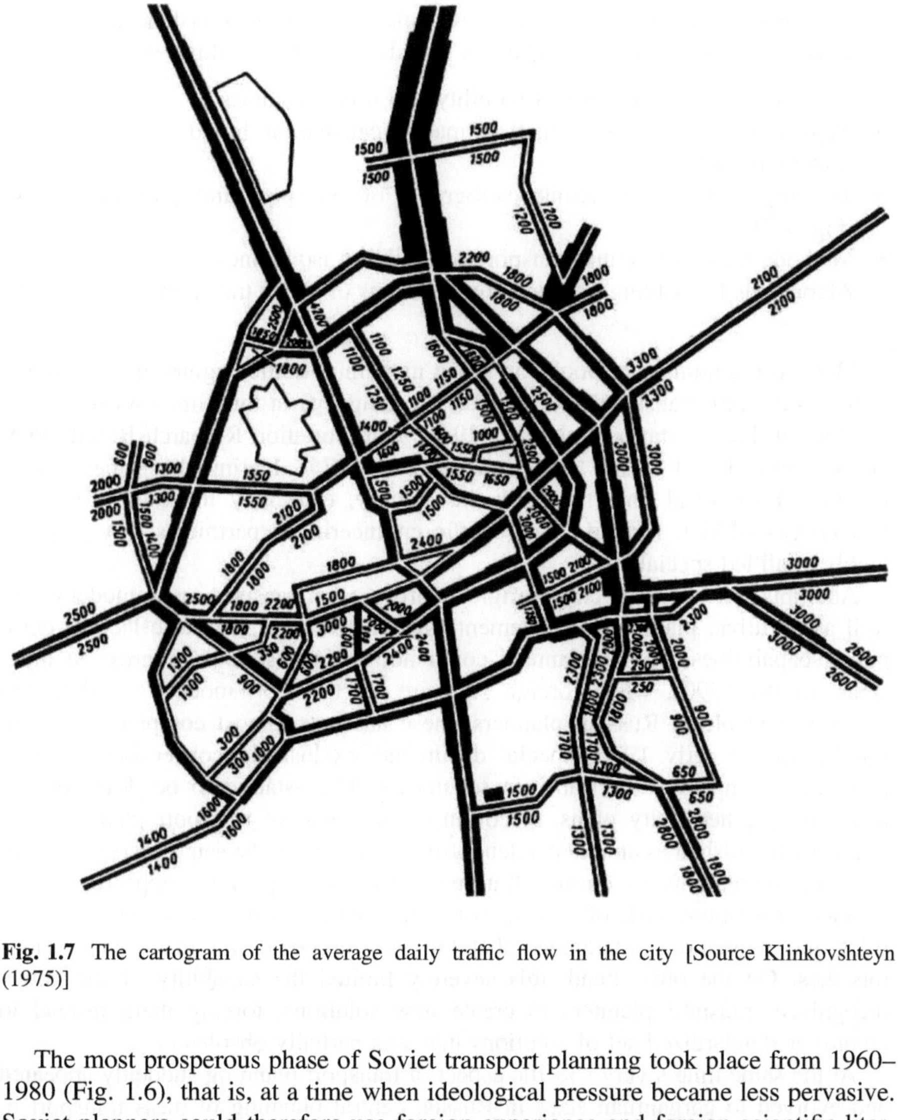

Fig. 1.7 The cartogram of the average daily traffic flow in the city [Source Klinkovshteyn (1975)]

The most prosperous phase of Soviet transport planning took place from 1960–1980 (Fig. 1.6), that is, at a time when ideological pressure became less pervasive. Soviet planners could therefore use foreign experience and foreign scientific literature without having to fear for their life and career. During this time several elite research and design schools that dealt with the field of transport planning were formed in Moscow, Leningrad, Kiev, Minsk, and Sverdlovsk. Even though ideological pressure significantly decreased, the Soviet practice of urban planning continued to be defined by the semi-compulsory settlement system (where people had little choice as to where they lived) and the total dominance of public transport. Thus, the achievements of Soviet transport planners from the 1960s to the 1980s were mainly of the engineering and mathematical kind. In this field a relatively effective set of engineering calculations was developed, including, in particular:

- A model for predicting urban mobility and traffic volumes;
- Algorithms for origin-destination matrix calculation based on gravity and entropy models;
- Techniques for constructing cartograms of transport and passenger flows (Fig. 1.7);
- Methods for constructing transport accessibility isochrones in urban areas;
- Algorithms for solving optimization problems of urban transport networks, etc.

Moreover, a number of books on urban modeling, traffic engineering and traffic flow theory were translated into the Russian language at this time (Wilson 1970; Institute of Transportation Engineers 1941; Transportation Research Board 1965; Drew 1968; Height 1963; Inose and Hamada 1975). During the same years a number of technical universities in the country, especially the State Technical University—MADI, formed strong traffic engineering departments that prepared highly qualified specialists.

Attempts were made to create original software for transport macromodeling, as well as an urban planning management system. However, due to a lack of computing capabilities, Soviet planners could not achieve serious progress in these areas. In the 1990s, when foreign software for transport modeling and design became available to Russian planners, these attempts almost completely ceased. Finally, in the early 1970s special documents exclusively concerning transport planning—comprehensive transport schemes (CTS)—started to be developed in addition to general city plans. Accordingly, the field of transport planning was finally distinguished as an independent profession. During the same years numerous rules and regulations governing all aspects of transport planning appeared. These detailed regulations had, of course, some advantages: formal adherence to these rules and regulations guaranteed that the planner could not make any serious mistakes. On the other hand, this severely limited the capability of the newly recognized transport planners to create new solutions, forcing them instead to employ a standardized set of solutions that was partially obsolete.

At the same time a very specific aspect of transport planning suddenly appeared and evolved to gargantuan sizes in Russia: central planning of mass transport of cargo using linear programming transportation. This interest was largely caused by the 1975 Nobel Prize in Economics, which was jointly awarded to the Soviet academician Leonid Kantorovich and American professor Tjalling Charles Koopmans for the creation of linear programming theory. The main difference between Soviet and American experiences was that, in America, the problem of optimal routing of deliveries between factories fundamentally assumed that they could have surpluses and that distributors could have varying demand. This was traditionally included in the scope of corporate governance, but was transferred to the city as a whole. It is clear that the possibility for such a formulation of the problem was impossible under total state ownership which included all the companies, suppliers, customers and carriers.

Computing centers dedicated to solving the problem of route optimization were organized in the largest Russian cities. They did so by assigning suppliers to consumers, and on this basis directive plans for mass transport of cargo were developed. Officially, this planning method created a significant positive economic effect. However, transport planners involved in the process understood that these "optimal plans", were, in reality, not very effective.

Nevertheless, Soviet transport planning in the seventies and eighties was at a high level in all major aspects of the field:

- methodology for mobility surveys and modeling;
- detailed regulation of different elements of transport planning;
- technologies for calculations relating to passengers and freight capacities of the various elements of the transport system;
- mathematical apparatus, etc.

It is safe to say that this level was considerably higher than the level of transport planning in cities of the former Soviet Union today, or in third-world countries.

However, the Soviet practice of transport planning had a fundamental methodological defect, since it was based on strictly controlled patterns of transport behavior of the population and commodity producers. This reflected the realities of a "socialist city", which had several main distinguishing features, such as:

Semi-compulsory settlement patterns. For example, a residential neighborhood for the AVIASTAR plant in Ulyanovsk was planned with the basic requirement that 100 % of the population must be connected to the workplace: everyone who lived in the neighborhood had to work at the plant, and all the employees of the plant had to live in that neighborhood. Given the fact that the plant and its symbiotic neighborhood were situated on the left bank of Volga river, and the rest of the city stood on the right bank, it is easy to understand the problems this planning caused a few years after construction.

Directive freight transport planning, mentioned above.

Artificial capping of the motorization rate, explained earlier, which lasted almost until the end of the 1980s. The motorization rate in 1938 was 6 cars per 1000 inhabitants, and 60 cars per 1000 inhabitants in 1991. For comparison, in the UK these figures in the same years were 68 and 436, respectively. These figures for the same years were 128 and 575 in Australia, and as high as 231 and 710 cars per 1000 inhabitants in the US.

While the motorization rate was low, the flaws of this urban planning system stayed invisible. Moreover, in a situation where a "citizen's own feet" (Zilbertal 1932) were the only alternative to public transportation for the majority of the population, there was no way for the average citizen to notice a lack of car related infrastructure. It should be noted that the effective replacement of individual car ownership with Soviet public transport, "the best in the world", was sold to the people as part of the official mythology.

In the foreign practice of urban and transport planning, at least since the 1960s, it has been assumed that individual and public transport are not substitutes but complements for each other (Jones 1981).

Moreover, objective comparisons of Soviet public transport with the best world analogues were clearly not in favor of the former. Since the easing of censorship of the press, which occurred in the late 1980s, objective data on the actual difference between "socialist" and "bourgeois" public transport appeared on the pages of specialized journals. The quality of public transport in Moscow, according to the survey conducted in 1985 (see Chap. 3), can be considered "very low" in comparison to any objective criteria or international transport systems. Obviously these quality control standards were never incorporated in Soviet transport planning regulations, but the gap between the planned parameters (good and optimistic) and the actual circumstances (usually very poor) was an everyday reality for the Soviet people.

By the end of the Soviet era, transport planning documents and practices that were quite decent in methodological and instrumental terms developed in the country. Unfortunately, these documents and practices were based on very artificial hypotheses about the transport behavior of citizens. In addition, the transport realities of Russian cities were far from the standards included in these documents.

1.2 Transport Planning in the Post-Soviet Period

The basic personal freedoms finally granted to the residents of Russian cities after the fall of the Soviet regime automatically led to a rapid increase in ownership of cars and small businesses. The trajectory of Russian motorization growth is lagging behind the phase their foreign counterparts were in 40 years ago (in the case of the UK), 55 years ago (in Australia) and even 80 years ago (in the US). However, the difference is much smaller compared to Japan, South Korea or Brazil, but, regardless, a continuous 20-year growth rate of 5–8 % per annum is impressive. By the end of 2015, there were 350 cars per 1000 people in Russia. Of course, in big cities, their concentration is much higher: about 400 cars per 1000 inhabitants (more than 550 cars per 1000 people in Vladivostok).

Mass motorization in the city turned out to be a very difficult challenge for the traditional national urban development paradigm, as well as for many government (municipal, civil) institutions, one way or another associated with this process.

When there was a low level of motorization traffic jams were exotic and only happened in the case of force majeure. The official methodological documents from the 1970s suggested measuring capacity of road sections by intentionally causing "short-term congestion using a road police employee". It is unlikely that transport planners could have ever predicted that in 40 years congestion would become an everyday reality on almost all the urban streets and suburban roads.

In most cases, the causes of these bottlenecks can partially be explained by extremely weak institutions governing the use of cars (movement, daytime parking, storage, etc.) and traffic management in general. However, they are caused, first and foremost, by inadequate planning of cities as a whole, and their street and road networks in particular.

In the 1990s and 2000s Russia faced the "clash of cars and cities" and degradation of traditional systems of mass public transport, which largely repeated similar processes that had taken place in the western world during the initial stages of motorization growth.

The graphs of growth and decline in urban land-based public transport ridership in US cities and Russia ones, presented in Chap. 10, reveal the degree of similarity of these processes. In both cases, the process of degradation of public transport took place under the same slogan: "Public transport is an option either for the poor or for a self-supporting business".

Accordingly, funding (in those cities where it was available!) was allocated only to the needs of road infrastructure, which in Russia mainly meant widening the streets and the constructing multilevel interchanges in the city center.

The "Vicious Circle", described by Professor V.R. Vuchic, accurately describes the processes which took place in the wealthiest Russian cities from 1990–2000 (Fig. 1.8). However, in poorer cities even these processes did not take place, as their transport systems did not develop in the slightest during these crucial last 15–20 years.

The process of transport planning in the 1990s also underwent significant degradation. Many skilled workers retired, emigrated, or changed their profession. Qualified teams of transport planners survived mainly, and in much smaller numbers, in planning organizations in Moscow and St. Petersburg.

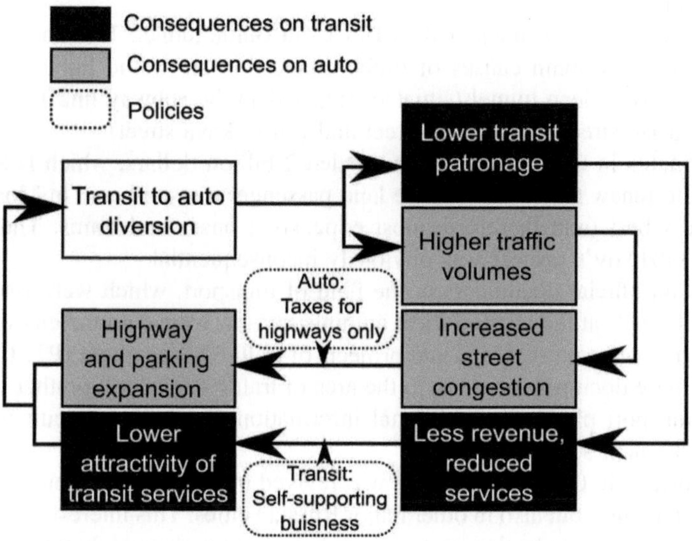

Fig. 1.8 The Vicious Circle of urban transportation [*Source* Reproduced by authors from Vuchic (1999)]

From 1990–2000 Russian cities developed next to no planning documentation. Comprehensive transport schemes were no longer mandatory documents. Even in cities where transport plans were created, the degree of their implementation was extremely low.

Trends in spatial development of Russian cities during the period in question are presented in Chap. 2. Let us add to this that the development of Russian cities was determined mainly by two factors:

- Short-term financial interests of developers somehow affiliated with the local authorities. These developers were given the opportunity to build new facilities with a complete disregard for the availability of sufficient transport capacity at the construction site: both in terms of public transport, and especially in terms of the road network;
- The aforementioned "absolutist tradition" dating back to Joseph Stalin, if not to the king Sennacherib. This "absolutist tradition", oddly enough, has only strengthened despite the new socio-political conditions. Mechanisms for independent expertise and citizen participation in the process of urban planning were still in their infancy. The mayor's power was no longer constrained by the framework of the General plan or urban development regulations, approved by the central government authorities.

The most striking examples of this occurred during the development of Moscow's transport system during the mayorship of Yuri Luzhkov (Fig. 1.9). Billions of dollars from the budget were invested in projects committed to widening old Soviet avenues and the construction of multilevel road junctions and tunnels. The effect these projects had on transport problems was mostly either negligible or nonexistent.

A typical example is the so-called Halabyan-Baltic tunnel. The mayor, sincerely believing that the main causes of traffic congestion are traffic lights, ordered the construction of a deep tunnel (situated deeper than the subway line!) between two ordinary urban streets (Alabyana street and Baltiyskaya street).

Investments in the project have exceeded 2 billion dollars, which is an amount sufficient to renew the entirety of the land passenger transport grid in Moscow with the world's best (and therefore, most expensive) buses and trams. The transport effect of Luzhkov's project was obviously inconsequential.

The main official documents in the field of transport, which were issued in the 1990s and 2000s at the level of local administrations, were comprehensive schemes of traffic management (CTMS) and projects of traffic management (PTM). In world practice, these documents belong to the area of traffic engineering rather than to the area of transport planning. Additional information about these documents is provided in the next section.

Real interest in transport planning was revived in the 2010s, not only in Moscow and St. Petersburg, but also in other major Russian cities. This interest was caused by the fact that traffic congestion moved from being a hypothetical threat to a daily and highly inconvenient reality. In these circumstances, even the most self-confident and

Fig. 1.9 General plan of Moscow during the mayorship of Yuri Luzhkov (1990–2005). The emphasis on the reconstruction of road junctions is clearly visible (*large* and *small circles*)

stubborn mayors began to turn to recommendations from world science and practice. Unfortunately, the appeal of "world science" in most cases resulted in an attempt to reproduce old Soviet practices instead of copying the fruits of foreign experience. This can be explained by a long pause in professional tradition, as well as by the majority of Russian planners being ignorant of the best foreign practices and theoretical developments.

Demand, as usual, has created supply: since the turn of the 21st century, transport planners from Western Europe and the United States have started to come to Russian cities to give lectures and classes. Some other English-speaking "professionals" have also come, despite having more in common with the heroes of the novel "The adventures of Huckleberry Finn" by Mark Twain,[7] than with actual urban and transport planners.

Obviously, the process of filling a 20-year gap in transport planning cannot happen in the blink of an eye. However, representative offices of competent foreign design companies have started to appear in Russia, planners of the older generation and their students have returned to the profession, graduates of foreign universities have started to appear in Russian design organizations. So while the process is far from completion, at the moment it is advancing rapidly.

Let us note the most obvious issues that reduce the effectiveness of transport planning projects, and complicate the prospects of their successful implementation.

1. "Blind" copying of Western practices. The implementation of the methods and practices of European transport planning often take place without due regard to Russian specifics (determined by the Soviet urban planning heritage), chaotic urban development during the post-Soviet era, weak development of civil institutions, etc. A typical example of this is the exaggerated attention paid to the development of cycling infrastructure, manifested in the majority of new transport planning documents. Let us note parenthetically that this emphasis in European transport planning emerged after Western European countries made drastic reductions of transport risks (see Chap. 8), making roads safe enough for bicycles. In Russia, the level of transport risk is at the same level as it was in Europe in the 1950s, which makes inviting cyclists onto urban streets a highly unintelligent decision.
2. The lack of data concerning the performance of urban transport systems and the travel behavior of citizens (which most European transport planners use as a cornerstone for their projects). In Russia, the raw data that a transport planner is used to getting from any city administration is sadly lacking. The transport planner is reduced to using the official accounts of municipal transport companies, accounting data from the traffic police, as well as information (usually outdated) from the municipal master plan concerning land use, development and population density. Furthermore, the information from different municipal authorities is usually not integrated, topographical information is inaccurate and hardly accessible, and data concerning the performance of private operators is totally absent. Moreover, surveys of transport and pedestrian flow are carried out rarely and are not standardized methodologically. Here is a typical scenario: a planner, based on Western European experience, aims to reduce the total VKT[8] of the urban road network, but immediately encounters a problem; no one has ever measured VKT in the city.

[7]See Chap. XXII: "At the court house! For 3 nights only! The World-Renowned Tragedians David Garrick the younger! And Edmund Kean the elder!".

[8]VKT (vehicle kilometers travelled)—an indicator used in transport planning and traffic safety assessment.

3. The discrepancy between the real and declared goals of the individual ordering or supervising the project. The problem of the system analyst having incomplete information about the true objectives of the decision maker is quite typical for all operational research segments. However, in Russian transportation planning this problem is even more prominent than usual. Russian transport planners can only guess at the real interests of urban officials (and of the key players in real estate development and construction business behind them). Here is a commonplace example: a planner, understanding the necessity of increasing road network connectivity, raises the question of building several relatively inexpensive underpasses and overpasses. This would connect several neighboring districts of the city otherwise divided by railways. The decision maker (urban official) insists on the implementation of a major project of reconstruction of the main thoroughfare of the city. A simple analysis shows that transport efficiency does not play any role in the final decision: for a private construction company closely connected to the administration, the second option (which allows focusing its facilities on a single site) is much more profitable in terms of comparative costs and margins.

1.3 The Russian System of Transport Planning Documents

This section focuses on well-established practices of modern transport planning in Russia, including the development of official transport planning documents.

Up until 2014 there was no such thing as "transport planning" in the Urban Planning Code of the Russian Federation. Transport planning was instead carried out in the form of projects given by administrations of cities and regions and paid for from their budgets.

The main documents at the city level include (Fig. 1.10):

- The transport section of the General plan;
- Comprehensive transport schemes (CTS);
- Comprehensive traffic management schemes (CTMS);
- The projects of traffic management (PTM).

Here is a brief analysis of the existing interpretations and the functioning practices of the listed documents.

CTS and the transport section of the General plan The composition and technology of development of the CTS is most accurately described in the document "Guidelines for the development of comprehensive transport schemes for large cities" enacted in 1982. The status of the CTS has been defined in these recommendations as follows:

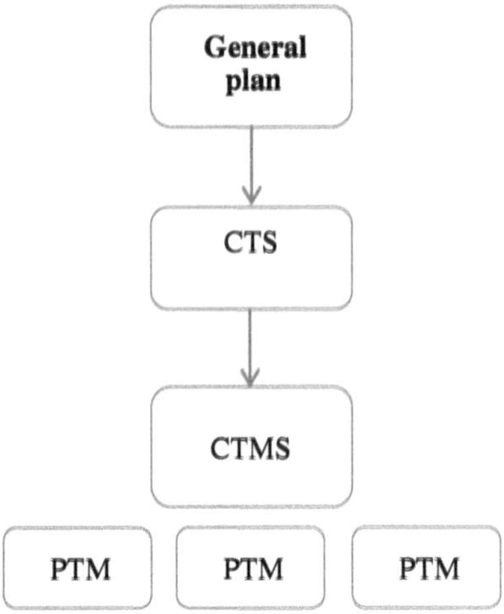

Fig. 1.10 "Classical" hierarchy of transport planning documents (*Source* Created by authors)

A comprehensive transport scheme is

- The main document for transport and engineering organizations to develop projects for the construction of passenger transport and road engineering structures as well as for the development and reconstruction of the existing urban transport network.
- Guidelines for the <cities and regional administrations[9]> pertaining to the development of economic plans concerning urban public transport.

In actual practice, a CTS contains all the information concerning questions of current performance and strategic development of urban transport systems, including (Fig. 1.11).

The network of streets and roads with separation of its arterial and local sections;

Public transport ridership and, accordingly, the target development indicators (including a material and technical base of enterprises of different transport modes), such as:

Heavy rail transport—suburban railways and subways;

Urban land-based public transport, including route networks and the required amount and typical structure of the vehicle stock.

Freight transport volumes and, accordingly, the target development indicators (including a material and technical base of enterprises of freight transport modes[10]);

Inter-city transport, including railway stations, airports, sea and river ports.

At the same time, CTS is considered to be a subsection and a large expansion of the transport section of the General Plan: "…if the development of the CTS reveals that certain sections of the general plan need to be updated, clear recommendations regarding what must be corrected and how must be provided".

[9]The original document mentions Soviet era government structures which are difficult to translate into foreign languages.

[10]As noted above, freight transport planning was supposed to be carried out with the intention of solving the "transport problem of linear programming".

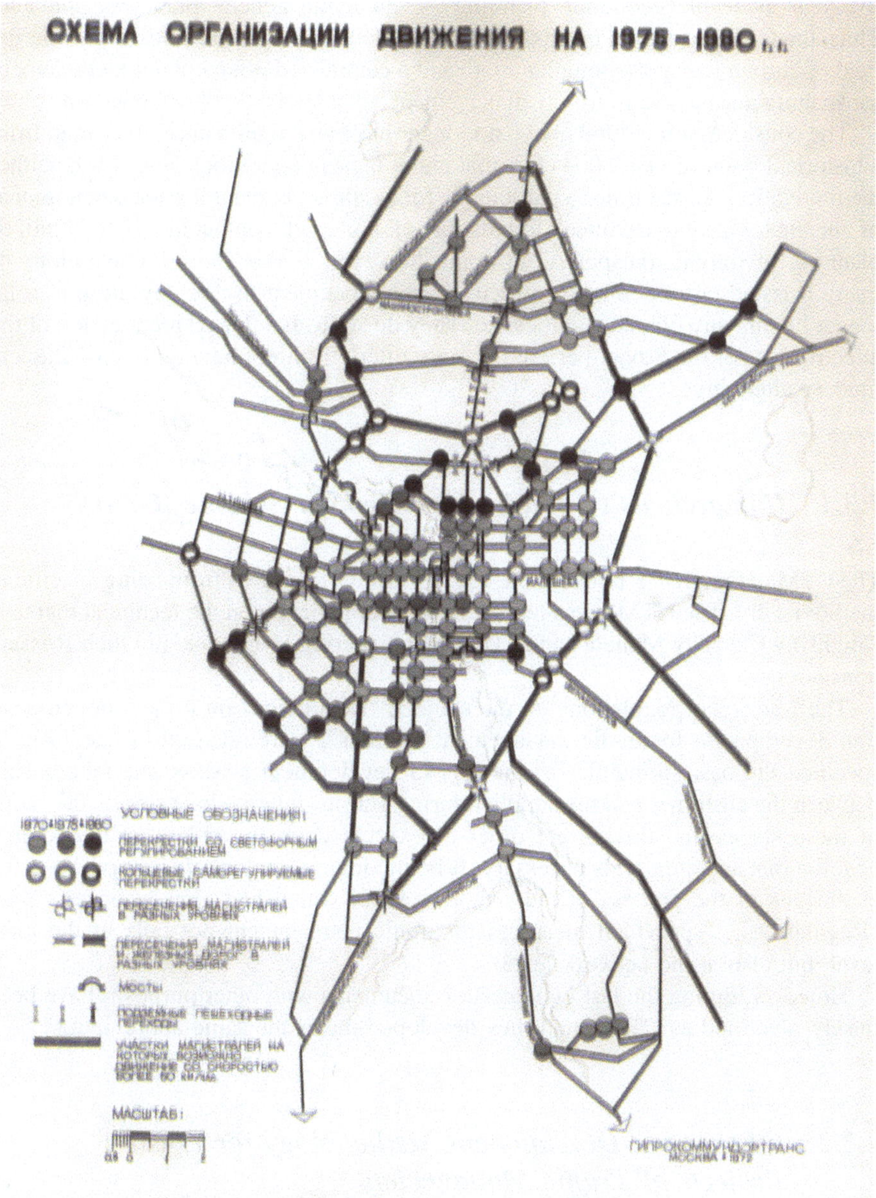

Fig. 1.11 The level of detail of CTS materials in the city of Yekaterinburg (*Source* The materials of A. Tsarikov)

In reality, while the development of a new version of the general plan historically took up to 40 years, the CTS was an up to date and detailed part of the transport section of the General plan. It could therefore be applied to the established five-year cycle of "economic planning", even if the general plan was outdated. Thus, the CTS was a kind of mix between an ideologically neutral instrument in the field of "urban transportation planning" and a centralized policy planning document for manufacturing and transport management, mandatory for the Soviet era.

The considerations found above no longer have any significance other than from a historical point of view. It is clear that under present conditions cities have neither the motivation nor the time to participate, for example, in planning the development of the material and technical foundation of transport companies, or centralized planning of freight transport. The point, however, is that the administrations of many Russian cities continue to focus on this document to this day, despite it no longer having any official significance. They do so from a flawed recollection of the effectiveness of the Soviet period, or, more often, from not having a more efficient modern alternative.

1.3.1 Comprehensive Traffic Management Scheme (CTMS)

The CTMS (Fig. 1.12), unlike the CTS, has always been free from being specific to the Soviet era. The CTMS was largely based ideologically on the technical manuals "Highway Capacity Manual" and "Traffic Engineering Handbook", in their Russian translations.

The last Soviet regulations on the subject, "Regulations on project documentation development for traffic management in cities", were released in late 1990. It specified the basic principles of the CTMS development process, the relationship between the customer and the supplier during various intermediate stages, the scope of these stages and the correct order to execute them in, and a list of required negotiations and approvals necessary. It is important to note that in response to the challenges of the process of mass motorization, which had already begun, these "Regulations…" provided measures for traffic management not only at the local level, but also at the network level.

However, during the last two decades documents with other purposes have been falsely identified as CTMS and thus developed under the same abbreviation.

1.3.2 CTMS as a Development Methodology for Local Projects of Traffic Management

This version of the CTMS provides a general idea of traffic management in the city and a guide for the development of local projects in the area of traffic management (PTMs). It is assumed that the location and parameters for all existing and planned facilities of the road network have already been defined in the General Plan (or in the CTS and/or Program for the integrated development of transport infrastructure).

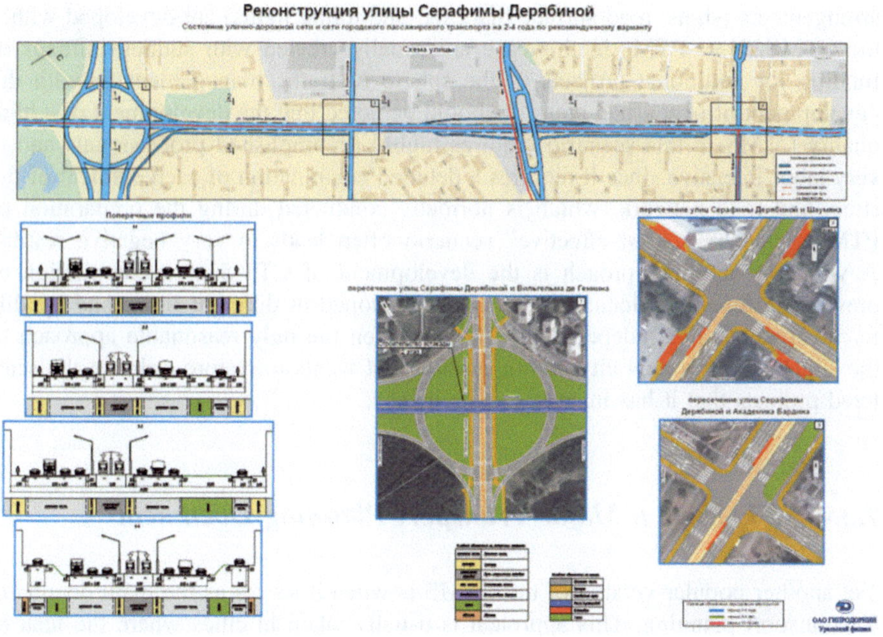

Fig. 1.12 The level of detail of the CTMS materials in the city of Yekaterinburg (*Source* The materials of A. Tsarikov)

Since everything else is already covered in various aforementioned sections CTMS should therefore only answer questions of traffic management and future traffic conditions on the network. Accordingly, the emphasis is on traffic forecasting, conceptual issues pertaining to speed limits, traffic management on all major sections and intersections, and the sequence and structure of the development of relevant PTMs. Under this definition of CTMS the details of engineering arrangements (signs, road surface marking, traffic lights) and their development are transferred to the level of local PTMs. This approach is commonly taken by many successful urban practices. However, this always assumes the existence of a common plan for city-wide traffic management, the parameters of which must be taken into account when relevant PTMs are being developed at different times and by different organizations.

1.3.3 CTMS as a Scheme of Traffic Management for the Entire City

This scenario is typical in small towns with small budgets, who, to save money, have just one CTMS for the entire city. In this case, the details of engineering arrangements (signs, road surface marking, and traffic lights) are developed within the CTMS. New PTMs in this case are usually created with sudden influxes of funding for additional projects in the sphere of traffic management or with the development of new urban areas. It should be noted that the development of a high quality CTMS in this scenario requires highly accomplished professionals and is very labor intensive since it requires a detailed examination of each element of the street and road network, which is normally conducted during the preparation of PTMs. Thus, this "cost effective" scenario often leads to very negative results. A variation of this approach is the development of CTMS as a compilation of previously developed local PTMs, often developed at different times and by different organizations independently. This is often the only reasonable approach to the situation, for a new city administration that wants to restore order in the scattered projects that it has inherited as its legacy.

1.3.4 CTMS as a Major Transport Planning Document

Yet another popular version of the CTMS is when it serves as the main document for transport planning. This approach is usually taken in cities where the idea of "car oriented mobility" wins due to a chronic lack of funding.

In this scenario, the city administration agrees that the majority of trips will be performed by cars, jitneys and surviving fragments of tram and local bus routes, and uses this as a basis for further development and management. The mayor is usually sincerely convinced that proper traffic management, especially the "optimization of traffic lights" using ITS from the capital, will solve all the major transport problems of the city.

Under this definition, the CTMS includes not only the activities associated with traffic management, but also the adjustment of the land public transport route network and performance indicators. Moreover, the CTMS in this scenario also includes the proposals for construction of roads and bridges. The authors of such projects openly argue that the CTMS should complement and adjust the General Plan.

Given the imbalance between the road network and the level of motorization described earlier, this approach is obviously a deadlock. This imbalance essentially cannot be overcome or even relieved by means of traffic management. As was already stated, this requires significant investment and radical reforms in the sector of public transport, as well as standard measures to counter the demand for daily car trips.

In fact, the CTMS in this interpretation is an attempt to postpone solving basic problems of the urban transport system.

The CTMS is usually developed for a period of up to 5 years and is based on the comparison of several design options.

From 2000–2010, it was suggested that the typical activities discussed during the development of the CTMS should include:

1. Introducing one-way traffic.
2. Establishing zoning restrictions for freight transport.
3. Creating a unified navigation system, with a primary focus on transit traffic flows.
4. Providing convenience and safety for pedestrians, and recognizing that this is one of the most crucial yet difficult sections of traffic management. When organizing pedestrian traffic attention must be paid to existing pedestrian routes, as well as to the locations of places of interest.
5. Road and bridge construction. Despite the fact that a section on improving traffic conditions is not included in the standard list of tasks when developing the CTMS, its inclusion in the CTMS has been conventional practice.
6. Optimization of traffic management at certain intersections, usually at intersections of key roads and streets of citywide importance.

An analysis of the CTMS conducted from 2000–2010 has shown that only a few of these documents are sufficiently informative and practically useful.

1.3.5 The Project of Traffic Management (PTM)

The most widespread document related to transport development in Russian cities has been and remains a local project of traffic management (PTM).

During the development of a PTM, attention is paid to traffic on individual sections of the street and road network of the city, i.e. at the local level (Fig. 1.13). The purpose of PTM development is to define specific traffic management methods that are to be used in parts of the street and road network, as well as the implementation of these methods using appropriate hardware and software for the traffic management at a local level.

The use of specific methods of traffic management and tools for their implementation is planned on the basis of standard criteria of capacity increase and integrated performance indicators of road safety for that particular link or node of the street and road network.

Moreover, the designer understands that optimization by increasing capacity for the local-level element of the street and road network is not always efficient at the network level. However, demonstrating positive changes at a key intersection of the city to the customer always increases the designer's chances for a timely payment for the completed project and for obtaining a new contract.

With this in mind, it becomes obvious that projects for traffic light control at individual intersections are not always coordinated with each other. Considering the poor level of service, typical of the street and road network of Russian cities, lack of coordination leads to extremely negative consequences. Local congestion expands to the neighboring elements of the road network, and can even lead to uncontrolled traffic.

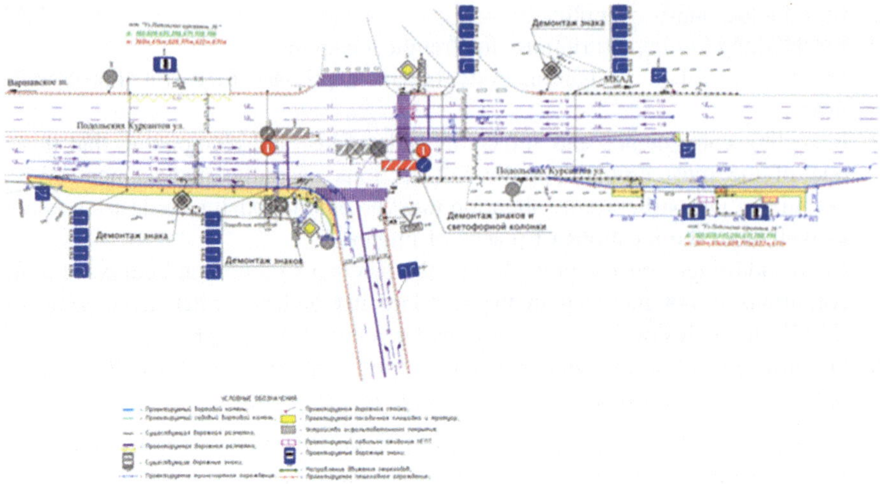

Fig. 1.13 The level of detail of PTM materials (*Source* The materials of A. Tsarikov)

Analysis of the current practice of PTM development demonstrates that a holistic system of organization of design projects in this area is sadly lacking.

Both municipal authorities and private institutions can act as customers in the development of PTMs. The former case is typical for elements of the street and road network in the central parts of cities. The latter is typical for newly developed areas; usually the sites of new shopping malls or new residential areas. In the first case, there is a competition to determine who will develop the PTM. In the second case, the choice is made by the private customer. In both cases, the developer of the PTM is usually not obliged to provide any qualifications, confirmed by state certification bodies or professional associations of design projects. At the same time, the PTM is usually given to whoever can provide it at the lowest cost and in the shortest time.

Representatives of a third party (traffic police and major municipal transport operators) usually participate in the negotiations of the terms for the development of PTMs, and give their approval of the completed project. The participation of traffic police in this process can be justified to some extent by the fact that an officer of this department is often the best specialist in traffic management if this is happening in a provincial town. The participation of municipal transport operators is a tribute to historical tradition: the share of such enterprises in the total volume of public transport operations during the Soviet period was enormous.

In recent years, this procedure has started to disappear. The process is gradually becoming unified: municipal transport authorities that fund PTMs are now either all-powerful (and no longer require the participation of a third party), or act as third parties with the power to veto PTMs, if they are ordered by private entities.

1.4 Reasons to Be Optimistic

The critical assessment of the state of affairs in Russian urban transport planning, provided above, has not gone unnoticed by legislative and executive authorities. In 2014, important amendments in the field of transport planning processes were made to the Urban Planning Code of the Russian Federation. In particular, a requirement was implemented that Russian cities develop "A Program for comprehensive transport infrastructure development", while the federal government was given the right to establish the guidelines for these programs.

The corresponding decree of the Government of the Russian Federation, adopted in 2015 and approving these guidelines does not contain any interesting innovations. The fact that an emphasis was made on the need for forecasting transport demand and prioritizing public transport at an official level for the first time is the only thing worthy of mention.

However, the need for the development and execution of programs for comprehensive transport infrastructure development in cities, which are more or less unified in terms of targets and suggested activities, is in and of itself a sign of progress.

A federal law, passed in 2015, "On organization of regular passenger road transport and urban electric land transport", required the development and approval of "documents concerning public transport planning" and "a register of public land-based transport routes" by municipal authorities. These requirements have enabled the resolution of demand from administrations of Russian cities to develop (update) the network of public transport routes, many of which have been preserved (in an edited form) from the middle of the 20th century.

A federal law "On the organization of the traffic management" is expected to be approved in the near future. The main aim of the law is to organize and standardize the process of developing CTMSs and PTMs.

In view of the above-mentioned legislative innovations, as well as based on the analysis of our own urban transport planning practices, it is possible to draw some conclusions.

Taking into account Russian realities, the conceptual triangle "Land Use and Development—Road Network—Public Transport", which has to be taken into consideration when developing (or updating) the General Plan, should be focused on the intersection "Road Network—Public Transport". Russian urban planners should therefore focus on, the potential, limits, and possibilities surrounding, the development of the urban transport system (Fig. 1.14).

Given the tight spatial constraints characteristic of Russian cities, complex development programs of transport infrastructure, as well as the CTS, must be focused primarily on the development of public transport. Moreover, the plans for public transport development should be harmonized with each other and apply to all operators, including municipal public transport and commercial operators.

The development of CTMS should be carried out while referring to the general goals of traffic management in the city, as well as acting as a set of guidelines for

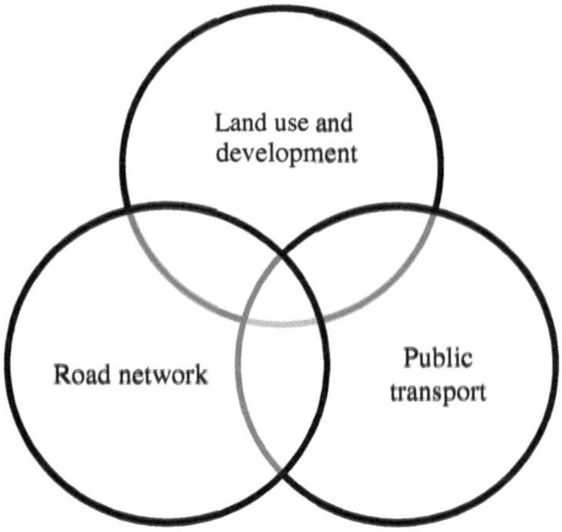

Fig. 1.14 Urban planning conceptual triangle (*Source* Created by authors)

the development of local projects of traffic management (PTM). In this respect the CTMS can be considered a part of the CTS.

"The Program for comprehensive transport infrastructure development" should be considered as the summarizing transport planning document, containing, inter alia, the list of projects for construction and reconstruction of transport infrastructure. At the same time the legal status of the program provides grounds for adjustments to the previously adopted General Plan.

All the transport planning documents should be reflected in the "big" city transport model, which has to be to systematically updated.

1.5 The Use of Transport Models

Transport modeling in both meanings of the term—traffic modeling and travel modeling (or classical urban transportation planning system model)—has quite a long history of development in Russia. However, it was only recently recognized as the most important tool for assessing the effectiveness of decisions in the field of urban and transportation policy.

The field of traffic modeling has been present in Russia since the very emergence of Russian translations of monographs by Haight (1963), Drew (1968), etc. A book by Gasnikov (2013) gives a fairly accurate description of the current state of this field in Russia.

To agree with our foreign colleagues (Transportation Research Board 2011), we have to admit that this discipline is increasingly moving away from traffic engineering and transportation planning and becoming a subsection of applied mathematics.

Nevertheless, a number of classic models from traffic flow theory continue to play an important role in the modern practice of transportation planning. We will confine ourselves to a single case, the choice of which is determined by the need to clarify an important statement made in the Foreword.

The classical model (Herman and Prigogine 1979) links the parameters T_r—running time, T—travel time, T_s—stop time (all of them per unit distance) using the following equation:

$$T_r = T_{min}^{1/n+1} \times T^{n/n+1}$$

where T_{min}—average minimum trip time per unit distance and

$$T = T_r + T_s.$$

The key to this model is the ergodic hypothesis, according to which $f_s = \frac{T_s}{T}$—the fraction of time a test vehicle circulating in a network is stopped is equal to the average fraction of vehicles stopped during the same period.

In an article, Williams (1997) states that "field studies have shown that n varies from 0.8 to 3.0, with a smaller value typically indicating better operating conditions in the network. In other words, n is a measure of the resistance of the network to degraded operation with increased demand. Higher values of n indicate networks that degrade faster as demand increases". If the network is arranged in a reasonable manner, and movement is organized competently, $n \leq 1$ (Hong et al. 2005; Nelsona and Sopasakis 1998; Lee et al. 2005). Accordingly, as demand increases, speed is slowly reduced. Moreover, a number of publications on the subject have revealed the fact that usually, $n \ll 1$ for urban freeways.

Here are the values of "the indicator of the quality of traffic service" for the street and road network in Moscow (Table 1.2).

These figures show that, with an adequate system of traffic management and traffic light control, street traffic in Moscow's city center has satisfactory elasticity of speed limits when faced with increased demand.

At the same time, Moscow's high-speed highways demonstrate abnormally high elasticity to demand growth ("degrade faster as demand increases"). According to the chart, the Third Transport Ring and the Moscow Ring Road (where there are no traffic lights!) are in a significantly worse situation, compared not only to foreign freeways and arterials, but even to the Garden Ring, where traffic light control is organized.

The abnormally high sensitivity of Moscow's main highways to demand growth is explained, first of all, by the high frequency of exits due to the inadequacy of Moscow's road network as well as the lack of alternative exits from the network to

Table 1.2 The indicator of the quality of traffic service for the street and road network in Moscow

Network fragment	*N*
The network as a whole	1.428
City center	1.253
Garden ring	1.729
Third transport ring	2.621
Moscow ring road	3.619

Source Calculations by M. Blinkin based on the field surveys conducted in 2000s by Boris Tkachenko the majority of urban areas. Most of the exits from the Third Transport Ring and the Moscow Ring Road operate in the "stochastic traffic light" mode, in which queues at freeway exit ramps cause the road to be blocked. This inevitably creates major bottlenecks on radial roads that emerge exactly at the place where reconstructed sections meet old Moscow streets.

Application of the "classical urban transportation planning system model" has become widespread in a professional and general sense in Russia since the 1980s. The original distributors of these ideas (and authors of pioneering publications on this subject) were the highly esteemed architects—Michail Posokhin and Alexey Gutnov, and influential experts in systems analysis and applied mathematics—Yury Popkov and Boris Shmulyan (Shmulyan et al. 1983).

For a long time "travel models" were used mainly for demonstrations. The developers of the models (who, since the nineties, were operators of foreign software) demonstrated their progress in the craft of modeling to customers. The leaders of design institutes or major road construction corporations, in turn, organized a sort of "amusement for the authorities", where they would show the best place to build a bridge or how best to align roads.

Nowadays, that "romantic period" of the use of "travel models" is long gone. Careful transport modeling is required now not only by the city government (represented by the chief urban planner), but also by large development companies, construction contractors, as well as by numerous "third parties", including traffic police, public organizations and environmental movements.

Almost all Russian research and design centers associated with transport planning use standard packages common in the world market of specialized software: PTV Vision, Aimsun, EMME, Cube, TransCAD, etc. There is also a sample of the original Russian software—TransNet, designed by Vladimir Shvetsov from the Institute for System Studies of RAS (though it hasn't received any widespread use).

The authors of this book represent one of the most respected Russian transport modeling centers, which is part of the Institute of Transport Economics and Transport Policy (HSE). Our experiences are related primarily to the operation of the PTV Vision system, and, in recent years—of the Aimsun system. Regular communication with foreign colleagues working with the same software leads us to the conclusion that our experience of adapting and upgrading the software, due to the necessity of taking into account Russian specifics, may be interesting to our reader.

To illustrate this, let us describe a typical example from our practice of using transport modeling for the analysis of the situation on the road network and proposing design solutions. The purpose of this analysis is to answer two basic questions:

- what are the characteristics of the existing and projected versions of the operation of the transport system (or its separate elements) at the local level?
- what is the degree of influence of projected local changes on the global network parameters of the transport system?

The answer to the first question is obtained using a phased simulation at macro and micro levels.

In the first stage a static analysis at macro level is performed using the multimodal transport model that takes into account both traffic flows and public transport. As a result of this calculation, we procure an estimate of the distribution of transport demand by mode of transport and by the element of street and road network of the city. This also allows us to assess the level of service (LOS) of these elements (Fig. 1.15).

It is clear that if LOS exceeds 1, part of the demand for car trips remains unsatisfied. Since the initial calculation is multimodal, the effect of substitution of car trips with trips by public transport, i.e. the effect of modal split change, is a

Fig. 1.15 An example of traffic demand distribution over the city street and road network. *Red* corresponds to $1 \geq \text{LOS} > 0.85$, brown corresponds to $\text{LOS} > 1$ [*Source* Created by authors using Aimsun (TSS-Transport Simulation Systems 2016)]

priori. So, in reality, an LOS in excess of 1 means that peak loads are redistributed to later hours. In other words, the increasing demand is met with a certain delay.

Next, using the estimates of transport demand, obtained in the first stage, the modeling is carried out using the dynamic (microscopic) simulation model. As a result, we get a clear picture of traffic conditions at the local level: at a group of intersections, within a local transportation area, or on a side road (Figs. 1.16 and 1.17).

A standard set of calculations is described above to determine the average value of LOS, separate local traffic parameters and make certain practical conclusions on this basis.

For example, for the case study, presented earlier, these conclusions are:

- The shortage of capacity on major highways in the direction of Moscow's city center varies between 17 and 25 %;
- The congestion in the main direction of the main section of a highway determines the traffic situation on the adjacent local street network. In this case, the difference between estimated demand and supply is about 6 %.

Acquiring such estimates using standard transport modeling tools is now practically mandatory in Russian planning practice, unlike the way it was a few years ago. This advancement reflects some of the positive trends currently underway in our professional field.

The problem, however, is that these evaluations are not enough for the comparative analysis of the effectiveness of different scenarios at the network level. Specifically, they are insufficient to analyze the effect of a particular action on the operation of the transport system as a whole.

This defect is hard to see when for each arc (i) of the graph of the street and road network the convenient condition LOS $\ll$ 1 is present i.e.:

$$Flow_i = Demand_i \ll Capacity_i$$

In this case, traffic, by definition, is equal to demand, and the average speed (measured over time or space) can be estimated using a simple calculation (Fig. 1.18).

Unfortunately, in Russian cities, where the values of $в$ и s_v are extremely low (as has been noted in the Foreword), the situation is exactly the opposite. Thus, during peak hours, the major highways in Moscow (and in other major Russian cities) usually have an LOS > 1, and sometimes even an LOS $\gg$ 1.

Under these conditions (Fig. 1.19):

$$Demand_i > Capacity_i$$

and, for obvious reasons from traffic flow theory,

$$Flow_i \ll Capacity_i$$

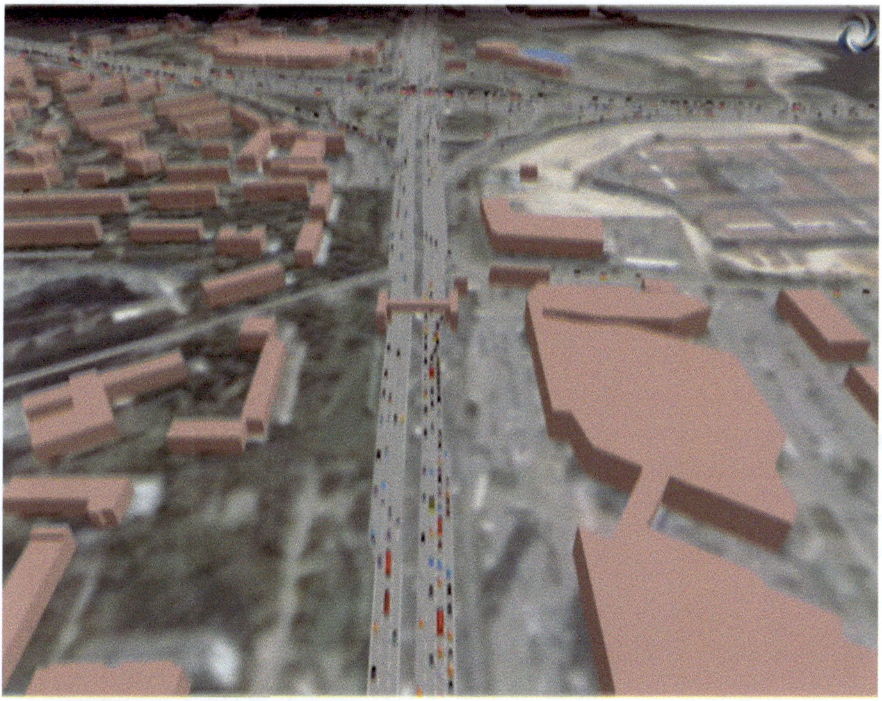

Fig. 1.16 Transport situation at the local level [*Source* Created by authors using Aimsun (TSS-Transport Simulation Systems 2016)]

Fig. 1.17 Transport situation at the local level [*Source* Created by authors using Aimsun (TSS-Transport Simulation Systems 2016)]

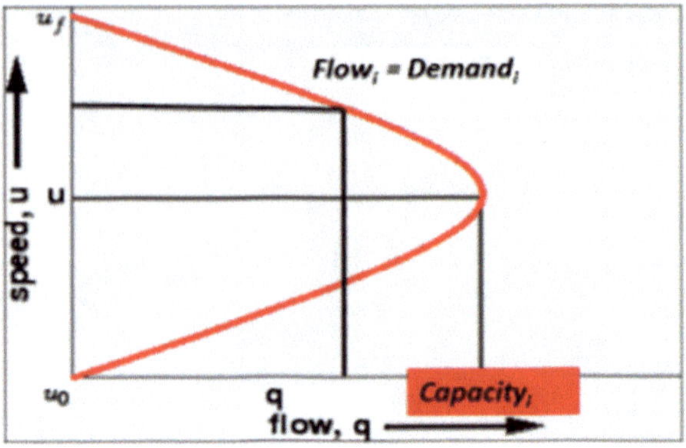

Fig. 1.18 The first scenario for the simulation of the street network [*Source* Reproduced by authors from Haight (1963)]

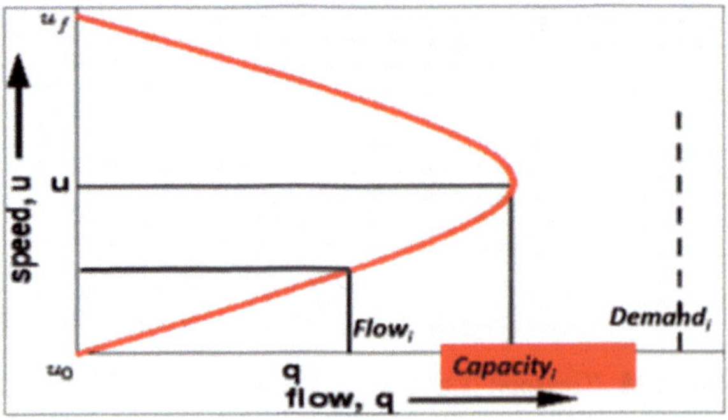

Fig. 1.19 The second scenario for the simulation of the street network [*Source* Reproduced by authors from Haight (1963)]

In typical Russian conditions of network overload there is an effect of delayed demand, and consequently, the formation of queues, as well as shock waves, generated by bottlenecks in the direction opposite to the flow of traffic.

Accordingly, we are faced with the non-trivial task of getting reliable estimates of certain factors, while taking into account all the aforementioned effects. These factors are:

- Traffic intensity in time dynamics,
- Average speed, flow density and delays at typical urban routes, passing through many successive arcs of the street and road network.

It is clear that it is impossible to carry out an objective analysis of the effectiveness of the proposed design solutions and to choose an optimal way to improve the transport system during a network overload without such data.

The solution to this problem has resulted in the creation of a special software module, which has become a key part of the multi-functional geographic information system we have developed for the analysis of the transport model results (Fig. 1.20).

The results of such calculations is presented to the user (analyst) for each direction and in each section of the street and road network, while taking into

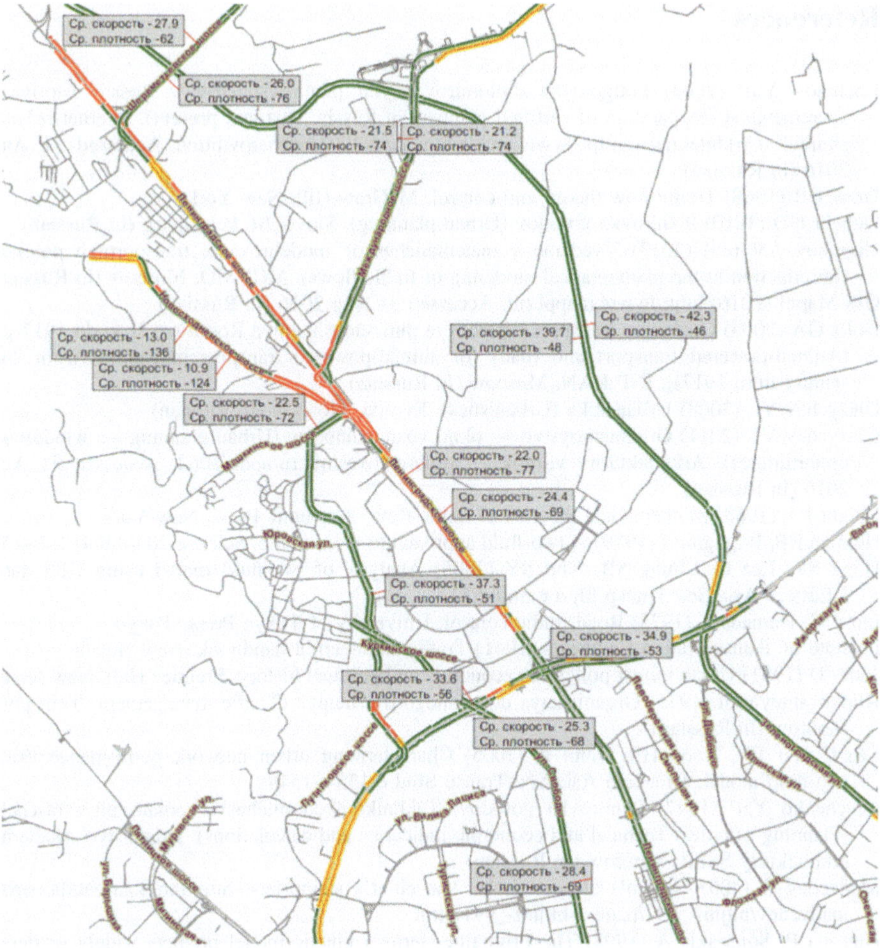

Fig. 1.20 Illustration of the results of the application of the special software module to calculate the intensity, density and speed of traffic flow in certain parts of the street and road network [*Source* Created by authors using the original module for the analysis of the results of static modeling using MapplGIS (GIS Mappl 2016)]

account the length of the overloaded chunk. Thus, users receive comprehensive data that enables them to undertake the objective assessment of the effectiveness of the proposed measures, comparative analysis of their implementation scenarios, and projected changes.

The use of the original module (of course, in conjunction with standard traffic simulation tools) allows a deeper understanding of the situation. It does so not only locally, but also for the city (or urban area) in general, and, accordingly, makes it possible to select some optimal solutions. In our practice, such decisions have usually had the aim of reducing the total VKT by improving connectivity of the street and road network and improving the conditions of public transport operation.

References

Bocharov YuP (2003) Podgotovka diplomirovannykh planirovshhikov v Rossii. Istoriya i sovremennost (Preparation of certified planners in Russia. Past and present). Internet-izdanie zhurnal «Arkhitektor». http://www.architektor.ru/ai/2003/bocharov.htm/. Accessed 31 Aug 2016 (In Russian)

Drew DR (1968) Traffic flow theory and control. McGraw-Hill, New York

Dubelir GD (1910) Planirovka gorodov (Urban planning). Slovo, St. Petersburg (In Russian)

Gasnikov AV (ed) (2013) Vvedenie v matematicheskoe modelirovanie transportnyh potokov (Introduction to the mathematical modeling of traffic flows). MCNMO, Moscow (In Russian)

GIS Mappl (2016) http://www.mappl.ru/. Accessed 31 Aug 2016 (In Russian)

Goltz GA (2005) Guzhevoy transport i guzhevye puti soobshheniya Rossii (so II v. do 1917 g.) [Animal-powered transport and roads for animal-powered transport in Russia (from 2nd sentury until 1917)]. INP RAN, Moscow (In Russian)

Glazychev VL (2008) Urbanistika (Urbanistics). Evropa, Moscow (In Russian)

Glazychev VL (2011) Gradoustroystvo — okno vozmozhnostey (Urban planning—a window of opportunities). Arkhitekturny vestnik 2. http://archvestnik.ru/node/3062. Accessed 31 Aug 2016 (In Russian)

Haight FA (1963) Mathematical theories of traffic flow. Academic Press, New York

Herman RP, Prigogine I (1979) A two-fluid approach to town traffic. Science 204(4389):148–151

Hong SY, Lee C, Chung SB, Kho SY (2005) Analysis of two-fluid model using GPS data. J Eastern Asia Soc Transp Stud 6:560–572

Inose H, Hamada T (1975) Road traffic control. University of Tokyo Press, Tokyo

Institute of Transportation Engineers (1941) Traffic engineering handbook. New Haven

Jones D (1981) Urban transit policy: an economic and political history. Prentice Hall, New Jersey

Klinkovshteyn GI (1975) Organizaciya dorozhnogo dvizheniya (Traffic management) Transport. Moscow (In Russian)

Lee C, Yu JW, Yoon HR, Sohn K (2005) Characterizing urban network performance using two-fluid model. J Eastern Asia Soc Transp Stud 6:1534–1544

Levchenko YaP (1947) Planirovka gorodov. Tekhniko-ekonomicheskie pokazateli i raschety (Planning of cities. Techical and economic indicators and calculations). Izdatelstvo Akademii arkhitektury SSSR, Moscow (In Russian)

Medvedev R (2005) Stalin's closest circle. The chief's supporters. http://bookz.ru/authors/roi-medvedev/blijnii-_991/page-5-blijnii-_991.html

Nelsona P, Sopasakis A (1998) The Prigogine-Herman kinetic model predicts widely scattered traffic flow data at high concentrations. Transp Res Part B: Methodological 32(8):589–604

Sheleyhovskiy GV (1946) Kompoziciya gorodskogo plana kak problema transporta (The urban plan composition as a transport problem). GIPROGOR, Moscow (In Russian)

Shmulyan B, Popkov Yu, Posokhin M, Gutnov A (1983) Sistemny analiz i problemy razvitiya gorodov (Systems analysis and problems of urban development). Nauka, Moscow (In Russian)

TransNet (2016) http://www.isa.ru/transnet/. Accessed 31 Aug 2016 (In Russian)

Transportation Research Board (1965) Highway capacity manual. Washington, DC

Transportation Research Board (2011). 75 Years of the fundamental diagram for traffic flow theory. In: Greenshields symposium. http://onlinepubs.trb.org/onlinepubs/circulars/ec149.pdf. Accessed 31 Aug 2016

TSS-Transport Simulation Systems (2016) Aimsun https://www.aimsun.com/. Accessed 31 Aug 2016

Vuchic VR (1999) Transportation for Livable Cities. CUPR/Transaction, New Brunswick

Williams JC (1997) Macroscopic flow models. In: Gartner N, Messer CJ, Rathi AK (eds) Traffic flow theory. Oak Ridge National Laboratory, Oak Ridge, pp 1–31

Wilson AG (1970) Entropy in urban and regional modelling. Pion, London

Zhukov AI, Savin DA, Chuev AV (2008) Inzhenernyj podxod k resheniyu gorodskix problem. Informacionno-analiticheskij portal (The engineering approach to solving urban problems)

Socpolitika.ru. http://www.socpolitika.ru/rus/social_policy_research/analytics/document6408.shtml. Accessed 31 Aug 2016 (In Russian)

Zilbertal AH (1932) Tramvaynoe hozyaystvo (Tram economy). Gostransizdat, Leningrad (In Russian)

Zilbertal AH (1937) Problemy gorodskogo passazhirskogo transporta (Urban passenger transport issues). Gostransizdat, Moscow (In Russian)

Chapter 2
Urban Spatial Structure as a Factor of Travel Behavior

Elena Koncheva and Nikolay Zalesskiy

Abstract Urban spatial structure is considered to be one of the determining factors of the transport demand volume and structure. Russian cities are traditionally characterized by high levels of public transport ridership, compared to the Western cities. Thus, it can be assumed that the spatial structure of Russian cities is a perfect illustration of the Transit Oriented Development (TOD). However, the spatial structure of the majority of the Russian cities, which has been developing during the rapid urbanization in the 20th century, currently preserves and reproduces the specific extensive models peculiar to the cities in the socialist countries. The authors analyze the spatial development patterns of 13 Russian cities in order to assess the current situation and the prospects for transit oriented development in the Russian Federation. A brief history of urban spatial development during the Soviet period is provided. Fundamental differences between TOD and Soviet Style Development (SSD) and their impact on transport demand are discussed.

2.1 Soviet Style Development Versus Transit Oriented Development

At the first glance, the spatial structure of Russian cities (as well as the spatial structure of the majority of the former USSR cities) seems to be a perfect illustration of the concept of transit-oriented development (TOD) as formulated by the American architect Peter Calthorpe in his book "*The New American metropolis*" (Calthorpe 1993), published two years after the end of the Soviet era. In fact, the parameters used in planning and development of the Soviet cities were thoroughly oriented at intensive public transportation usage. However, some fundamental differences between TOD and the Soviet Style of development (SSD) do exist.

E. Koncheva · N. Zalesskiy (✉)
Institute for Transport Economics and Transport Policy Studies, National Research University Higher School of Economics, Moscow, Russia
e-mail: nzalesskiy@hse.ru

M. Blinkin and E. Koncheva (eds.), *Transport Systems of Russian Cities*, Transportation Research, Economics and Policy,
DOI 10.1007/978-3-319-47800-5_2

Firstly, TOD became a historical antithesis to the tradition of car-oriented development paradigm (COD), longstanding in the cities in the US, Canada, Australia, and other parts of the Western world and was based on the extremely high motorization rate and, accordingly, on the transport self-sufficiency of households, particularly of those living on the periphery of cities and in the suburbs. The attention to urban spatial development as a factor of travel behavior in the countries of the Western world originated in response to the problem of urban sprawl in the second half of the 20th century which caused numerous social problems such as congestion, environmental degradation, and slow social and spatial integration. This issue has been closely associated with the phenomenon of car dependency (Schiller et al. 2010; Newman and Kenworthy 2006). Both problems became more critical in North America and Australia in comparison to Europe and Asia. The overall situation was a consequence of the post-war growth in the sales of cars and individual houses in the suburbs, that is, the phenomenon of the "*exodus of the middle class to the suburbs*" (Jones 1981). This fact explains the great contribution of the North American and Australian scientists to the research of this issue. At the same time, the adoption of TOD paradigm does not (even theoretically) imply the reduction of the motorization rate to the limits determined by the standards of the Soviet times. In the Western cities, the usage of public transportation can be encouraged, but it is impossible to limit the motor vehicle production and sales. The first studies on spatial planning and the influence of land use on travel behavior appeared in the 1970s (Handy 2005) when scientists noticed the interrelation between urban population density and public transport usage. The scholarly debate on the issue intensified in the early 1990s after the publication (Newman and Kenworthy 1989) which examined the interrelation between population density and fuel consumption in large cities. To date, TOD model has been tested in many cities around the world as a tool to reduce the popularity of private cars and make public transportation more attractive. The concept has been taken into account in urban spatial development management in North America, Australia, and Europe. The most developed countries and territories in Asia (Japan, South Korea, Singapore, and Hong Kong) have also reached great success in this field. The concept of TOD is being widely introduced in China, India, and Latin America. Nowadays, the development focused on public transportation is seen as the most sustainable form of urban spatial development (Cervero and Murakami 2008).

One of the major features of socialist cities as opposed to those in the West was the low motorization rate. The motorization rate in Russia remained extremely low throughout the Soviet era: by the end of it, in 1990, there were 80 vehicles per 1000 residents in the USSR, including no more than 50 personal cars. In the USSR, the low motorization rate was achieved with the simplest of means: national production of cars was severely restricted (the only plant intended for mass production of cars more or less advanced by the standards of the time (FIAT) was built in 1970 in Togliatti[1]); import of cars was banned; the price of a new car was equal to 3–5

[1]Before the construction of the plant the motorization rate was less than 10 cars per 1000 residents.

annual salaries of a qualified engineer. Thus, in the Soviet Union, car-oriented suburbs had never developed. Prosperous citizens would have two homes: a small apartment in an apartment building and a "dacha" (a modest country house) located on a 0.06 ha lot in the suburbs, usually with a small garden and without urban amenities. The less affluent citizens lived in communal apartments with a kitchen and a bathroom shared by several families. The housing stock in large cities was totally dominated by multifamily houses. In the 1950s and 1960s, 5-storey residential buildings were typical while since the 1970s and until the end of the Soviet era, buildings with 8, 12 or more stories prevailed. The share of the built-up area in a city reserved for urban street and road network was exceptionally low. It is necessary to mention that the term "*share of built up area in streets and roads*" was already used by Russian urbanists at the beginning of the 20th century. For example, Greg Dubelir, the Russian pioneer of urban studies stated in his monograph written in 1910:

> … if the value ε implies only the share in streets and roads and does not include parks, large open areas, etc., it can be assumed that the coefficient ε is equal to:
>
> - for tall buildings and wide streets with estates with small lots – 0.40,
> - for average conditions – 0.30,
> - for a rational plan with narrow streets and small houses with estates with relatively large lots – 0.20-0.25 (Dubelir 1910).

According to the recommendations of the USSR Academy of Architecture, the share of area in streets and roads had to vary from 15 % (in small towns) to 25 % of the total area of a residential zone (Levchenko 1947). The book by the acknowledged Soviet specialist in transport studies, A. Polyakov (1953) presents the data on the percentage of built up area in streets and roads for 111 cities in Russia as of 1946, according to which the average value of the indicator was 25.2 % of the total residential area, varying from 32.5 % in urban centers and 27.7 % in the intermediate zone to 23.0 % in the peripheral zone. Noting that all these values conform to the strictest requirements of the recommended design standards, A. Polyakov argued that the standards were insufficient rather than redundant. At the same time, the scientist was convinced that "*a significant saturation of cities with motor vehicles would entail an increase of this relative indicator*". Unfortunately, the optimistic forecast by A. Polyakov has only been fulfilled halfway through: the predicted "*significant saturation of cities with motor vehicles*" came true, but it only happened in the post-Soviet period. However, it was by no means accompanied by "*an increase of this relative indicator*". It does not exceed 8–12 % in any city in the Russian Federation. Thus, the cities in the USSR, unlike their Western counterparts, had not been shaped by private cars in the course of the 20th century; rather than that, they had been shaped by total absence of private cars.

Secondly, and perhaps most importantly, TOD, as well as COD (and all the other spatial development models applied in the Western cities), is based on the private ownership of land and unconditional freedom of choice available to the citizens. Neither the former nor the latter was present in the USSR. The TOD researchers focus mainly on factors such as housing density (density), land use diversity

(diversity), and designing pedestrian-friendly urban spaces (design). This set of factors is referred to in the academic literature as 3-D (Cervero and Kockelman 1997). Some publications also mention two additional factors: the distance to public transit stations (distance to transit) and travel time to the destination (destination accessibility), resulting in 5-D group of factors (Ewing et al. 2007; Hamin and Gurran 2009; Cervero and Murakami 2008). The importance of these factors transpires in the greater efficiency of TOD development in comparison with the traditional high density development near public transit stations (Cervero and Murakami 2009; Lin and Shin 2008; Loo et al. 2010). These characteristics enable the stimulation of public transport usage and reduction of private car usage by local residents and people visiting the area (Boarnet and Crane 1997). The distinctive features of the TOD such as high density, diversity of uses, and focus on pedestrians have a lot of influence on travel behavior and often encourage people to stop using a personal motor vehicle (Kenworthy and Laube 1999; Ewing and Cervero 2001): the generalized cost of travelling by public transport becomes much lower. These details, characteristic of TOD, seem quite alien to the realities of SSD. The Soviet housing estates were territories used exclusively for residential purposes, and mixed-use development was not practiced. Considering the low level of car ownership that was typical for Soviet cities, the residents had no choice whatsoever with regard to the transport mode. The Soviet people used the overcrowded public transport without thinking about the dubious convenience of such trips. Public spaces near subway and suburban railway stations were physically present, but the issue of their convenience and attractiveness was hardly relevant, as the service and trade sector was notoriously underdeveloped during the Soviet period. Within the limits of the development strategy implemented in the country there was no room for urban land markets. The absolute imperatives were the growth of industrial production. As to the housing construction, according to the observation by the same author, it was dominated by "*a kind of super-large villages near the factories*" (Glazychev 2008).

According to the classical models, the attractiveness of land intended for development dwindles moving away from the urban core, which leads to a decrease in density and development. In case of the cities that have developed in the countries with free market economy, this process is controlled mainly by the market forces. The result is a city where the maximal housing and population density occurs in the center while being most sparse at the outskirts. If the current density of economic activity does not correspond to the actual land value for any reason, effective mechanisms of renovation and redevelopment are launched, eventually forming a decreasing density profile. Under the free market conditions, developers aim at maximizing the value of the final product, but in the conditions of the administrative command system oriented at a proposal, the task of providing new housing is solved by simply adding square meters. Therefore, in the urban areas which have been actively developing for a long time in the conditions of planned economy the described decreasing density profile is not observed. According to the contemporary studies dedicated to the spatial structure of the post-socialist cities, particularly the works of Bertaud (Bertaud and Renaud 1995; Bertaud 2004), the cause for this dissimilarity is the rejection of private property and the absence of land markets in cities. In socialist countries, the types of land use were determined based on the perception of public importance of various objects.

The exotic Soviet urban development conditions have had a huge impact on the present day urban spatial structure, as rapid urbanization occurred namely during the Soviet period (the share of the urban population had grown from about 17 % in 1917 to about the current degree of 70 % in half a century). The development flaws of the Soviet cities such as the low values of the parameter ε were not critical while the transportation demand was almost entirely met by public transport (even though the latter was often of very low quality). Problems emerged with the rapid growth of motorization rate since the mid-1990s and in the 2000–2010s.

2.2 A Brief History of the Soviet Style Development

There are three important milestones in Russian history that triggered significant changes in urban spatial structures: (1) the abolition of the institute of private real estate ownership after the events of 1917, (2) The "*Khrushchev thaw*" and the beginning of the industrial housing construction era in 1957, and (3) the collapse of the Soviet Union and restoration of market economy and private property institutions in 1991.

The development of Russian cities before 1917 was progressive but, due to objective historical reasons, rather slow. The country's economy was actually an agrarian one and the level of urbanization was low. Typical FAR for cities in that period ranged from 1.3 to 1.8 in provincial cities (and even in some of the central neighborhoods in Moscow) to 3–4 in St. Petersburg; i.e. there was a fairly high-dense development similar to that in the European cities. The share of the area in streets and roads in the central part of a city was fully consistent with the contemporary European trends and amounted to 25–35 %.

The 1920–1930s saw intensive industrialization of the Soviet type, focused on the development of heavy and defense industries. This process led to the concentration of economic activity (and people) in cities. Production plants, including industrial giants, designed for tens of thousands of workers were usually built at the outskirts of cities. Later, many of them turned into vast brownfields.

Inadequate housing construction volumes were made up for either by constructing low quality temporary dwellings (e.g. barracks near big factories—the problem that still exists today) or (which was more common) by multiple densification of the existing housing stock. At that time, the dwelling area allotted for one person could be 3–5 m^2. Figure 2.1 shows a significant increase in population density in the central part of Moscow between 1907 and 1935, despite the fact that housing density had not changed much.

Since the mid-1930s, due to the development and enforcement of the Moscow General Plan and general plans of other major Soviet cities, fairly reasonable development and land use requirements were introduced along with standards for

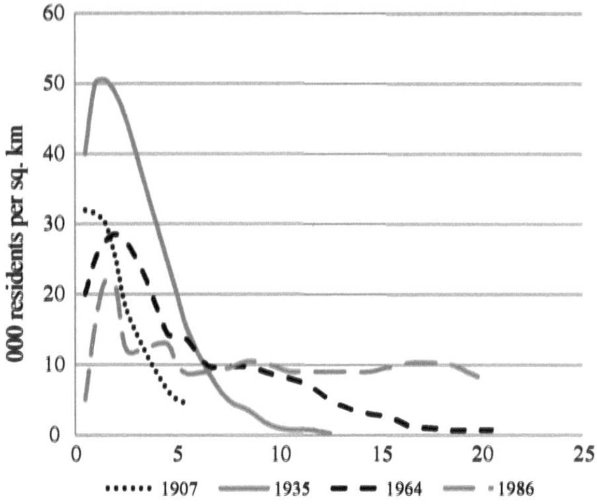

Fig. 2.1 Moscow density profiles in different years [*Source* Reproduced by authors from (Petrov 1988)]

insolation, green areas, and various urban infrastructure utilities. These new rules were obviously declarative (not to say, propagandistic) in nature. In reality, the high-quality housing construction in that period was concentrated almost exclusively in the elite segment, available solely to the higher strata of the Soviet bureaucracy as well as the cultural and academic elite (Khan-Magomedov 2006). The houses built during that period are located in what are the central areas of the respective cities today and are still perceived as high end real estate, which increases their price. Obviously, such kind of housing did nothing to solve the housing problem in general.

The era of cheap and rapid industrial housing construction started in 1957 with the construction of a pilot Cheryomushki neighborhood in Moscow. It was a real breakthrough, as millions of citizens were granted the previously unavailable comfort of living in a single-family apartment (Khan-Magomedov 2006). Per capita dwelling area in cities began to grow rapidly. Undoubtedly, mass housing construction required significant space, so the cities grew actively. Evolving cities overstepped the industrial zones, which led to mass construction at peripheral areas beyond the industrial areas.

Thereby a W-shaped urban structure height configuration, unique in the world, was created in the Soviet cities:

- multistoried (some of them historical) buildings in urban cores;
- “flat” industrial areas on yesterday’s periphery which evolved into intermediate urban zones by the 1950s;
- high-rise housing estates on the new periphery that were even more distant from urban centers.

The minimal personal freedoms granted by the government in the 1960s (“Khrushchev thaw”), which included, in particular, the relatively free choice of a working place combined with the preservation of the ban on private housing ownership, thereby preserving an extremely low residential mobility level. Citizens could now select their workplace, but it was still very difficult to move from one residential area to another. This circumstance had caused obvious problems and extremely high demand for public transportation in the Soviet cities.

At the same time, in the 1960–1970s, a package of regulations, quite advanced from a humanitarian point of view, was proposed, defining, inter alia, the required subdivision of urban areas into residential, green and industrial zones as well as the level of transportation infrastructure development.

In practice, these regulations resulted in the proliferation of neighborhoods that usually consisted of 5-storey apartment buildings accompanied by large green areas, with a typical FAR (Floor Area Ratio) value close to 1. The implementation of more advanced building techniques brought about the growth in the number of stories: while the BCA (Built Coverage Area) remained unchanged, the FAR value increased greatly and exceeded 10.

Obviously, these new high-rise housing estates were built on greenfield land, usually at a considerable distance from the city center and/or major concentrations of employment opportunities. In many cases, these neighborhoods were dominated by "*departmental housing stock*", i.e. houses designed exclusively for the employees of certain large industrial enterprises.

From a humanitarian point of view, these remote neighborhoods were a huge step forward compared to the barracks around the factories. At the same time, this type of development created demand for high capacity public transportation systems. In many cities, the initiative of developing such systems came from large industrial enterprises, the actual owners of the new residential areas who were required to ensure transportation of their employees from the place of residence to the production plant entrance.

In any case, local authorities and managers of the production plants (combined under the leadership of the Party bodies) were primarily concerned with the issue of commuting. The employment concentration at such enterprises was so high that it made quite effective the organization of regular bus routes (high-capacity buses) and even the construction of tram lines (with multi-car trams), designed exclusively to provide transportation links between a production plant and the housing estates associated with it. The most striking illustrations of this transportation policy were Samara and Nizhny Novgorod where the function of transporting workers between the production plants and the associated housing estates was carried out by the local subways. In other words, subways existed in these cities but did not play the role of universal city-wide transport systems. The task of subway extension to the historic downtown area, that is, the transformation of the underground transport system from a departmental to a city-wide was only addressed with the arrival of the new economic reality in the early 2000s.[2]

---

[2]It is worth noting that the opening of "*Gorkovskaya*" station in the central part of Nizhny Novgorod led to a sharp increase in ridership: from around 30 to 40 million passengers per year. For the reasons discussed below, this short-term growth could not reverse the trend of an overall decrease in ridership: the following year saw the restoration of the downward trend.

Another aspect, important for understanding the spatial structure of Russian cities, was the proposed motorization level, provision of parking spaces, and road network. The Soviet planning regulations (Gosstroy SSSR 1985) were based on the following directive hypothesis: "*… when calculating the capacity of the network of streets, roads and transport hubs, as well as the distribution of parking spaces, the level of motorization for the planning horizon should be assumed to be 150–180 cars per 1000 inhabitants*". These regulations were actually implemented: the citizens kept their personal cars either in the garages located in the specially allotted zones (for example, in the sanitary areas of railways and high-voltage power lines) or in microrayons,[3] beside high-rise apartment buildings. The actual level of motorization being 50 (maximum 80), this practice did not cause major problems, and there was no need for any parking restrictions. Today, with the level of motorization at 2–3 times higher than prescribed by the norms cited above, the psychological stereotypes of the citizens create acute problems when the authorities try to introduce appropriate parking regulations.

Planning regulations for microrayons presumed total exclusion of all transit-related traffic from a planning unit. For example, the planning regulations cited above required providing a local road network in multi-story residential areas based on a single 3.0 m wide lane. Given the extremely low level of motorization in 1970–1980s, such parameters were relatively harmless. Today, when 1000 residents of the same residential neighborhoods have 300–400 or more cars, such thoroughfares became a real obstacle not only for the residents departing their residential area but even for the passage of emergency vehicles and public services.

In the conditions when transportation demand was supposed to be met by public transport with almost no alternatives, the only parameter limiting the growth of planning units was the radius of access to public transport stations. Given this, the distance between the adjacent streets in the residential areas built at that time can exceed 1 km (Fig. 2.2). Such distances make it almost impossible to implement any kind of coordinated traffic control.

Moreover, in Soviet cities there was no segregation of urban network into urban streets and urban roads that has been present in the developed countries since the middle of the 20th century. The need for such segregation was appreciated by professional Soviet planners. A.A. Polyakov stated that "*when the saturation of the US cities with cars reached 200 cars per 1000 residents, construction of high class urban roads was launched in the largest cities. Those were urban highways, and usually they did not intersect with streets and roads at the same level*" (Polyakov 1967). He was confident that with the motorization in the USSR reaching the same level, the implementation of a hierarchical transport network subdivided into roads and streets will become inevitable. Unfortunately, this optimistic forecast has not come true. The preference was given to the so-called "*high street of continuous traffic flow*". This is a hybrid type of a street combining multi-lane carriageway and

[3]A microrayon is the main structural element of the Soviet residential areas that can be described as a collection of extremely large housing blocks without any typical block configuration.

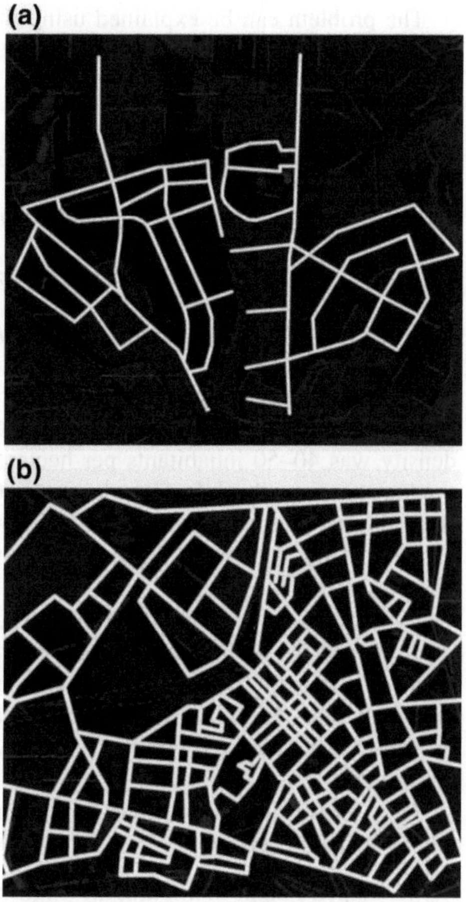

Fig. 2.2 Typical configuration of the road network in Moscow districts built (**a**) in late XIX–early 20th century (**b**) in 1970s. The illustrations are to scale. (*Source* Created by authors)

multi-level junctions with direct access to houses and intensive public transport traffic, which is not present in the planning requirements in the developed countries and in the World Road Association (PIARC) recommendations (Mikhailov and Golovnykh 2004).

As noted above, by the end of the Soviet period, the share of the built-up area dedicated to the transport network (ε) had not exceeded 8–12 % in any of the Russian cities. The decline of this index compared with the value of 25 % recorded in 1946 happened primarily due to the increase in housing density which took place without either heeding the recommendations by scientists or evaluating perspectives of the motorization rate growth. However, based on the Soviet motorization rate, the road network of so modest density was sufficient to meet any potential transportation demand. Phenomena such as heavy traffic or traffic jams were a rarity that would only occur due to a force majeure.

The problem can be explained using the following simple formula:

$$s_v = 10^7 * \frac{\varepsilon}{d * m}$$

$$s_v - \textit{area in square meters of streets \& roads per } 1 \textit{ vehicle}$$

$$\varepsilon - \textit{share of built up area in streets \& roads}$$

$$d - \textit{population density, people per } 1 \textit{ ha of built up area}$$

$$m - \textit{motorisation level, vehicles per } 1{,}000 \textit{ inhabitants}$$

Given m ≈ 80, at the end of the Soviet period, there were about 200–250 m^2 of streets and roads per 1 vehicle, even though ε was equal to 0.08 and population density was 40–50 inhabitants per hectare. The same situation is characteristic of modern cities in the US, Canada or Australia where motorization rate is several times higher but the same value of $s_v \approx$ 200–250 m^2 is being reached at the expense of fundamentally higher road density (ε ≈ 0.35) and lower average population density.

By now, when the motorization rate in the largest Russian cities crossed the mark of m ≈ 40 with the same road network density (ε ≈ 0.08–0.1) and growing population density (d ≈ 60–80), the average road network area per vehicle became extremely low. In Moscow (despite the active road construction during the last 20 years), the value s_v is close to 28 m^2, while in other large cities it is between 25 and 40 m^2.

Thus, the distinctive structure of Russian cities (SSD) was shaped by the 1990s, and is characterized by:

- W-shaped height configuration (high density city center and peripheral territories along with ineffective land use in the intermediate zone) (Bertaud and Renaud 1995);
- Extremely low share of built-up area in streets and roads.

2.3 Transformation of SSD During the Post-Soviet Period

The analysis of the spatial development in post-socialist cities was conducted based on the example of housing construction in the largest Russian cities, with the exception of Moscow and St. Petersburg. The source of information was the database of the State Corporation "*Fund for assistance to reforming housing and communal services*" (Reforma ZHKKH 2016). Moscow and St. Petersburg were excluded from the study because of their special status (these cities are also the subjects of the Russian Federation), as well as a significant difference in demographic, social and economic indicators and especially, in budgetary possibilities.

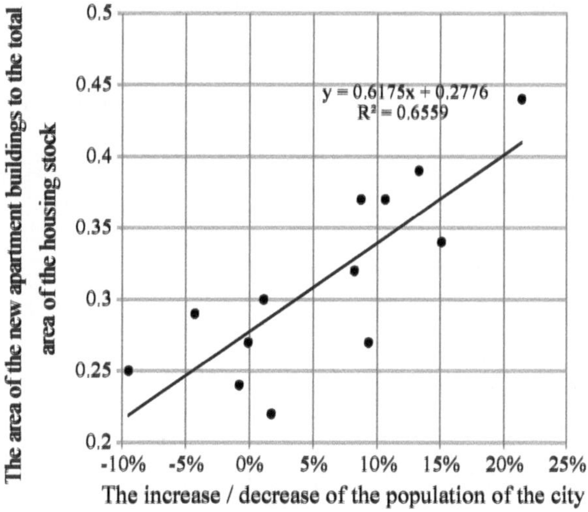

Fig. 2.3 The relationship of the change in the population and the volume of construction of new apartment houses, % by 1991 (*Source* Created by authors using data from Fund for assistance to reforming housing and communal services Reforma ZHKKH 2016)

Thus, 13 other cities with population over 1 million people were included in the study. The study analyzes information on residential buildings constructed in 1992–2015 in terms of their distance from the city center and public transport stations.

The total area of the houses built after 1991 comprises about one third of the total volume of the housing stock in the studied cities. This parameter is found to be in strong correlation with the dynamics of the city population growth during 1992–2015 (Fig. 2.3). The highest value of the indicator, 44 %, can be observed in Krasnoyarsk which experienced significant population growth in 1991–2015 (over 20 %) (Table 2.1).

Even in the cities where population decreased (Nizhny Novgorod, Samara) or remained essentially unchanged (Perm, Omsk, Volgograd) in 1992–2015, the share of the houses built after 1991 in the total area of the housing stock is 22–30 %. It can be argued that the construction of new apartment buildings in 1992–2015 reflects, on the one hand, the ongoing improvement of the standard of living within the financially prosperous stratum of the "settled" urban population, and on the other, the strong demand for high-quality new housing by the new, economically successful residents of Russia's largest cities.

Based on the available data on these cities it is possible to appreciate the degree to which housing construction in post-Soviet Russia has been able to change the spatial structure of its cities and influence the supply/demand of transportation services.

Creating the density profiles of residential development has shown that in all the cities on which data was provided the maximal density is registered in the central areas (Fig. 2.4). The density of residential buildings within the radius of about 1 km around the center is between 170 and 405 thousand square meters per 1 km^2.

Table 2.1 The characteristics of the housing stock changes in the largest Russian cities during the post-Soviet period

No	City	The share of the total area of the apartment buildings constructed after 1991 in the total area of housing stock (%)	Population, 2015 (thousands)	Population growth, 1991–2015 (%)
1	Novosibirsk	32	1548	8.2
2	Yekaterinburg	37	1446	10.6
3	Nizhniy Novgorod	25	1273	−9.5
4	Kazan	37	1191	8.7
5	Samara	29	1172	−4.3
6	Chelyabinsk	34	1169	15.1
7	Omsk	27	1166	−0.1
8	Rostov-on-Don	27	1110	9.3
9	Ufa	30	1107	1.1
10	Krasnoyarsk	44	1067	21.4
11	Perm	24	1042	−0.8
12	Volgograd	22	1018	1.7
13	Voronezh	39	1014	13.3

Source Created by authors using data from Fund for assistance to reforming housing and communal services and Russian Federation Federal State Statistics Service (Reforma ZHKKH 2016 and Russian Federation Federal State Statistics Service 2016)

The most distinguishable centers (335,000–405,000 m^2 per 1 km^2) are found in Samara, Novosibirsk and Yekaterinburg. Mean values of the indicator (303,000–324,000 m^2 per 1 km^2) are characteristic of Rostov-on-Don, Chelyabinsk, Voronezh, and Nizhniy Novgorod. The lowest density of residential buildings (168,000–275,000 m^2 per 1 km^2) is typical for the central areas of Kazan[4], Omsk, Ufa, Krasnoyarsk, Perm, and Volgograd.

All the cities except for Omsk are characterized by one or more (Ufa, Chelyabinsk) density peaks at various distances from the city center (Table 2.2). In most cities, these peaks can be registered using the density profile of residential development in general, but it is more convenient for this purpose to use the density profile of residential development only within the residential areas. The additional peaks of density (density of residential development within a residential area of 150,000 m^2 per 1 km^2 and more) at the distance of 5–7 km from the city center are observed in Ufa, Krasnoyarsk, Yekaterinburg, Voronezh, Rostov-on-Don, Kazan, and Chelyabinsk. Additional density peaks at the distance of 9–11 km from the city center are characteristic of Chelyabinsk, Ufa, Samara, and Nizhny Novgorod. In

[4]For Kazan, it is reasonable to use the density at 1–2 km from the city center, as the center of the city is clearly distinguishable, but the area within the radius of 1 km around the center has low density of residential buildings due to a number of reasons.

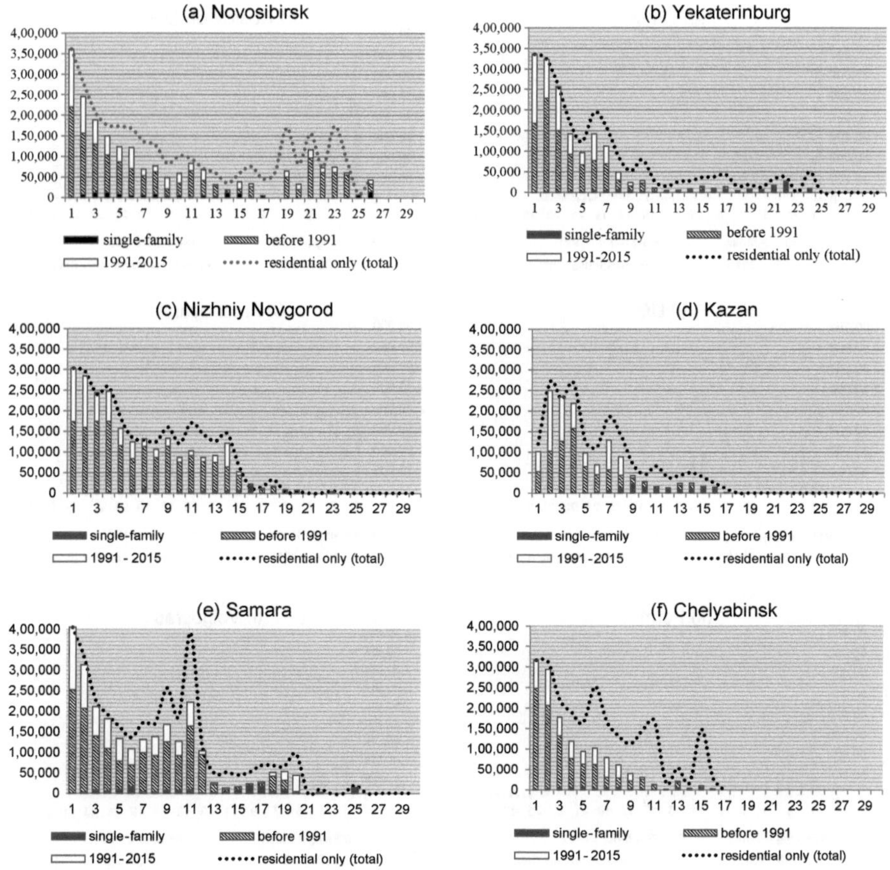

Fig. 2.4 **a–m** Residential housing density profiles of Russian cities [*Source* Created by authors using data from Fund for assistance to reforming housing and communal services (Reforma ZHKKH 2016)]

Perm and Chelyabinsk, additional peaks are located at the distance of 15–17 km from the city center. The presence of additional peaks at the largest distance from the city center (20–25 km) is typical of Volgograd and Novosibirsk. In such cities as Samara, Volgograd, and Ufa the density of residential buildings in residential areas within the peaks is comparable with the values of the corresponding figure in the central parts of the city.

The analysis of changes in density profiles has revealed that to varying degrees, the tendency of density growth in the central areas occurred in 1992–2015 in all the cities chosen for the study. The density of residential development within the radius of 1 km around the city center has increased by two and a half times in Kazan; by two times in Yekaterinburg; by more than one and a half times in Novosibirsk, Nizhny Novgorod, Ufa, Voronezh, and Samara; by 1.3–1.4 times in Omsk,

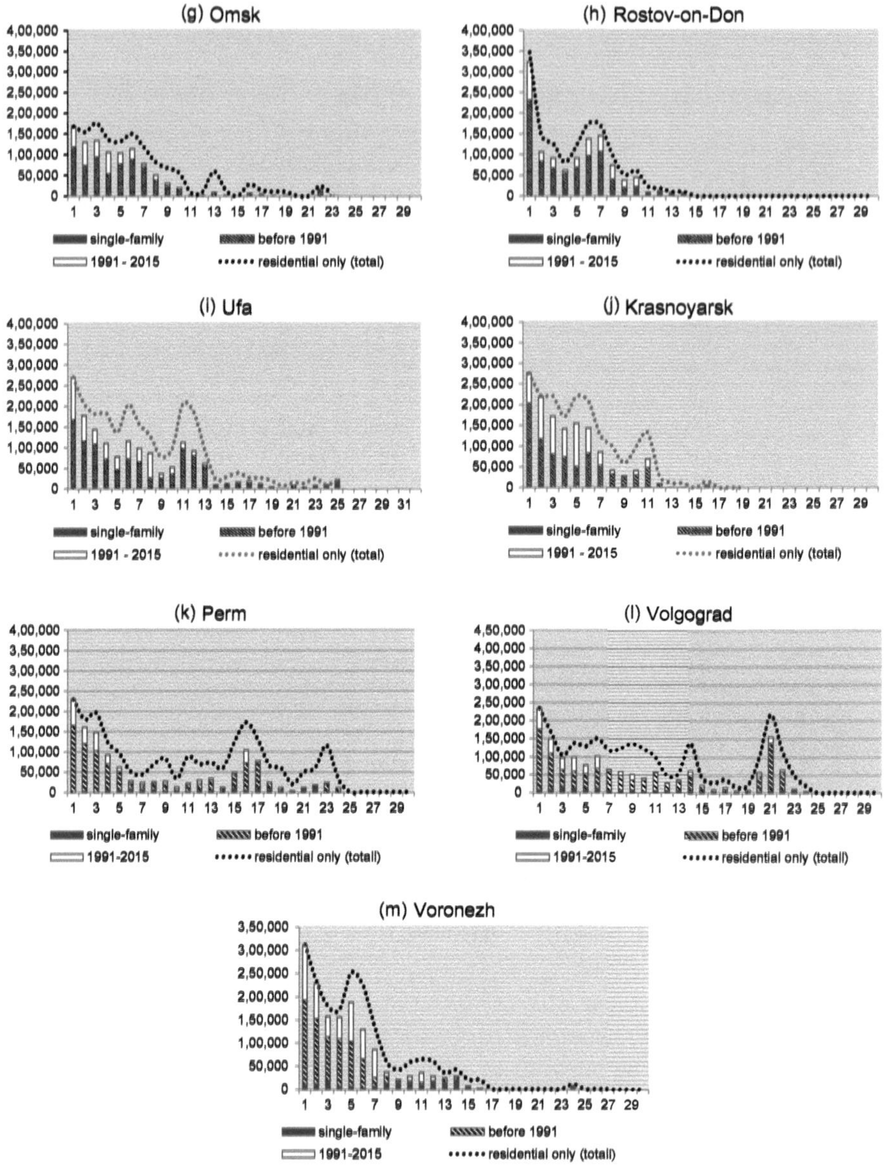

Fig. 2.4 (continued)

Rostov-on-Don, Perm, Krasnoyarsk, Chelyabinsk and Volgograd. Within the "*closest*" peaks in the cities, there is also a tendency of housing density increasing by 1.5–2 times. In some of the cities, the "*closest peaks*" appeared or became much more pronounced as a result of large-scale construction in the post-Soviet period

Table 2.2 Density peaks

City	Closest peak	Medium peak		Remote peak
Distance from the city center	5–7 km	9–11 km	15–17 km	20–25 km
Novosibirsk				+
Yekaterinburg	+			
Nizhniy Novgorod		+		
Kazan	+			
Samara		+		
Chelyabinsk	+	+	+	
Omsk				
Rostov-on-Don	+			
Ufa	+	+		
Krasnoyarsk	+			
Perm			+	
Volgograd				+
Voronezh	+			

Source Created by authors using data from Fund for assistance to reforming housing and communal services (Reforma ZHKKH 2016)

(for example, in Yekaterinburg and Voronezh). The "*medium*" and "*remote*" peaks have not seen such intensive growth of housing density.

Figure 2.5 shows the distribution of housing construction in the post-Soviet period in each city by zone at varying distances from the city center. As opposed to the general volume of housing construction, the share of the area of residential buildings in the studied cities within the radius of 3 km around the city center ranges from 15 % (Omsk, Krasnoyarsk) to 29 % (Perm). This kind of a zone is not leading in terms of the volume of housing construction in the post-Soviet period in any of the studied city. In some cities (Yekaterinburg, Nizhny Novgorod, Omsk,

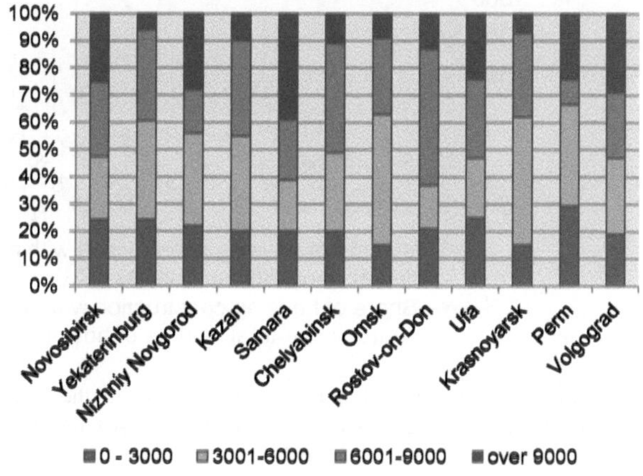

Fig. 2.5 Allocation of housing construction in the post-Soviet period by zone at the varying distances from the city center [*Source* Created by author using data from Fund for assistance to reforming housing and communal services (Reforma ZHKKH 2016)]

Krasnoyarsk, Perm), the largest volume of housing construction (34–48 %) is concentrated in the area of 3–6 km around the city center. In Novosibirsk, Kazan, Chelyabinsk, Rostov-on-Don and Ufa, the main volume of constructed housing (27–50 %) is situated in a zone of 6–9 km around the city center. Housing construction took place mainly at the distance of 9 km and farther from the city center in Samara (39 %) and Volgograd (29 %).

Thus, it is possible to discern two counterpoised trends. Both trends are illustrated by the example of the city of Yekaterinburg (Fig. 2.6).

The first trend shows that all the cities chosen for the study have seen an essential increase in housing density within their central areas. This trend prevailed in the early post-Soviet years, i.e. during the initial formation of the housing construction market.

The second trend is the continuation of the Soviet urban planning practice of the so-called integrated development of greenfield areas at the outskirts of cities (Kosareva et al. 2015). This trend prevailed during the housing construction boom induced by the high oil prices in the 2000s. The Russian model of the sprawl, contrary to its Western counterpart, is focused on dense multi-story development. In addition, development of new territories was only formally integrated: in fact, residential areas of the Soviet type were built with a minimal number of jobs nearby. Meeting the transportation demand in those areas (regardless of the level of motorization) is extremely difficult. Major housing construction projects in downtown areas are quite rare, but can still be found. Examples include the residential

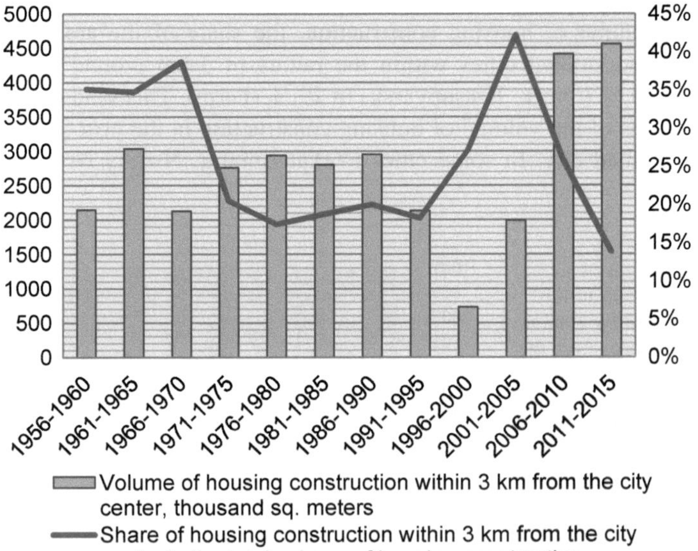

Fig. 2.6 The dynamics of housing construction in the central area of Yekaterinburg [*Source* Created by author using data from Fund for assistance to reforming housing and communal services (Reforma ZHKKH 2016)]

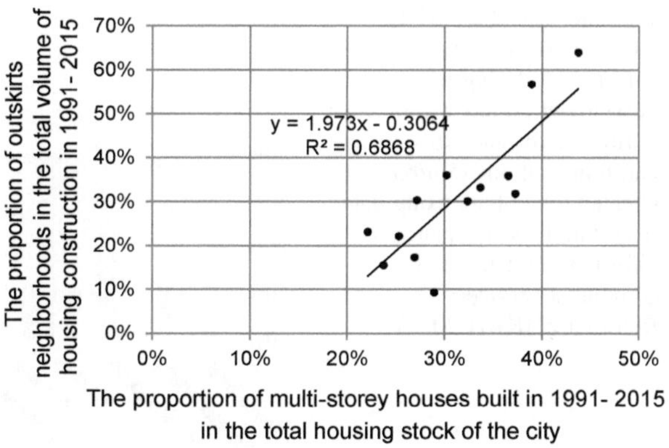

Fig. 2.7 The share of peripheral neighborhoods in the total volume of housing construction in the post-Soviet period [*Source* Created by authors using data from Fund for assistance to reforming housing and communal services (Reforma ZHKKH 2016)]

complex in Nizhny Novgorod (in the center of the city on the river Volga near the 2018 World Cup buildings) and new residential complexes in the district of New Savinovsky in Kazan (also in the city center and near the 2018 World Cup buildings, with the area of approximately 2000,000 m^2).

The share of the area of apartment houses built as part of the peripheral development of urban areas in the total amount of housing construction in various cities also confirms the notion that the large-scale construction in major Russian cities is associated with the sprawl. The biggest share of peripheral neighborhoods in the total volume of housing construction in the post-Soviet period is characteristic of the cities that have experienced the most significant growth of the housing stock (Fig. 2.7): Krasnoyarsk (64 %) and Voronezh (57 %). In some cities, the value of this indicator is significantly lower: Samara, 9 %, Perm, 15 %. In most cities, the value of the indicator is about 20–30 %.

The share of the total area of buildings constructed during the post-Soviet period within a walking distance (800 m) from subway stations in the cities where this transport mode is present amounts to 12–24 % (Fig. 2.8). The highest value (24 %) is registered in Novosibirsk where the subway network is moderately well-developed (2 lines, 16 km, 13 stations). In Nizhny Novgorod, which has a subway network comparable to that of Novosibirsk (2 lines, 19 km, 14 stations) the share of housing built in the post-Soviet period within the limits of normative accessibility is only 13 %. In Samara, Kazan, and Yekaterinburg there is only one subway line (10 stations in Samara and Kazan, 9 stations in Yekaterinburg). Among these cities, the leader in terms of the share of housing built in the post-Soviet period within a normative distance from subway stations is Yekaterinburg: 19 % versus 12 % in Kazan and Samara.

The proximity of the post-Soviet housing to tram stations varies greatly between the cities. In Samara, the share of the post-Soviet housing located within a walking distance from tram stops (400 m) is 60 %; in Krasnoyarsk, it is only 5 %. It should be noted that the factor of pedestrian access to subway stations has been and remains very important: in all these cities, subways operate with due frequency and capacity.

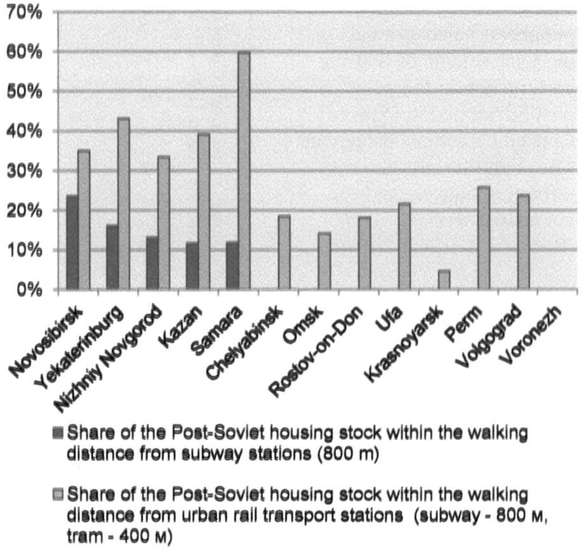

Fig. 2.8 The share of the total area of housing constructed during the post-Soviet period within a normative distance to subway and tram stations (*Source* Created by authors using data from Fund for assistance to reforming housing and communal services Reforma ZHKKH 2016)

At the same time, pedestrian access to tram lines is becoming an increasingly nominal factor: as shown in Chap. 3, tram lines in most Russian cities are currently in deep stagnation.

Another problem is providing transport services to the neighborhoods built on city outskirts. Usually the development of transport network is unable to catch up with housing construction in such areas: many of those, designed in the Soviet period, have been only recently provided with competitive public transport. Examples include the Botanicheskiy neighborhood in Yekaterinburg and the neighborhoods of Azino-1 and Azino-2 in Kazan.

The *Botanicheskiy* neighborhood is located in the southern part of Yekaterinburg at about 5 km from the city center. In fact, it is the last Soviet residential district in the city, as it was designed during the Soviet period: the development of the area became possible after the removal of the airport in 1985. The major part of the construction was carried out in the 1990s. The total area of residential buildings in the neighborhood is about 0.7 million square meters. The neighborhood is mainly built up with 10–16-story buildings. Jobs in the area are available only in the consumer services sector. Currently, the neighborhood is connected to the city center by the subway ("*Botanicheskaya*" station opened in 2011 but was designed back in the Soviet times). The part of the neighborhood that is most remote from the subway station is served by a tram line (no priority) connecting to the subway closer to the city center.

The neighborhoods of *Azino-1* and *Azino-2* are located to the east of the center of Kazan at a distance of about 8 km. The total area of residential buildings in these two neighborhoods is more than 2 million square meters. Similar to the *Botanicheskiy* neighborhood in Yekaterinburg, the development took place mainly in the 1990s. The area is located close to the Big Kazan Ring where light rail was launched after reconstruction in 2012. It provides access to the center of the city and to the southern subway station "*Prospect Pobedy*" (built in 2008).

In the more recently constructed peripheral neighborhoods there is currently no public transport which is able to compete with a private car in terms of convenience for the residents. At the same time, these neighborhoods are being built according to the Soviet planning regulations, for the most part. The residents willing to commute to the city center and back in their private cars will have to make use of the substandard road network built in the 1960–1980s. Competitive alternatives to private transport simply do not exist in most of the cases. The examples of such neighborhoods are *Krutye Kluchi* in Samara and *Akademicheskiy* in Yekaterinburg.

The neighborhood of *Krutye Kluchi* is being built at 21 km north-east of Samara city center. The construction began in the early 2010s; the project is expected to be completed by 2025. It is planned to build 5 million square meters of housing. Transport services in the area are currently provided by municipal buses and private bus operators with high frequency (with intervals less than 5 min) but without dedicated lanes.

The neighborhood of *Akademicheskiy* located at 7 km south-west of Yekaterinburg city center. The construction in the area began in the late 2000s, it is planned to build 9 million square meters of residential real estate and 4.2 million square meters of office and commercial real estate. The transport situation in the area today is similar to the situation in *Krutye Kluchi*. The construction of the planned tram line is using extra-budgetary sources, but the issue is not yet settled. In general, the involvement of developers and owners of commercial properties in the financing of public transport projects in Russia is extremely rare and limited to just a few cases.

Thus, in the spatial development of large Russian cities during the post-Soviet period, two key ambivalent trends can be indicated. On the one hand, there is a considerable density increase of residential development in downtown areas as well as in the areas properly served by public transport; on the other hand, extensive high-density development continues in suburban areas. These trends are evident in each individual city: it is possible to find positive examples of efficient land use in the urban centers characterized by high levels of transportation services and cases of large-scale greenfield development anywhere. The two trends in question are also characteristic of Russian capital cities, Moscow and St. Petersburg, which are not included in our study.

The first trend is a direct result of the transition to land and real estate market and is a very positive factor in terms of providing transportation services and reducing the need for travel. Many cities have public transport systems of quite a good quality in their centers, either inherited from the Soviet period and subsequently expanded or developed in the post-Soviet years. The development of the areas served by public transport can improve economic sustainability of public transport operators and help prevent the growth of car use. In the cities where, for some reasons, public transport systems have been underdeveloped for some time, the trend for density increase will significantly simplify the task of providing transportation services in the future.

The second trend, on the other hand, has a negative impact on transportation services. Only in rare cases cities manage to keep pace with housing construction and provide new areas with public transport of reasonable speed and capacity. New neighborhoods are served by bus operators at regular routes and schedules or by "jitneys". In both cases, the buses move in the general flow of vehicles, that is, have ROW-C.

The commuters—the residents of new neighborhoods—are usually private car owners, often in the first generation. It is very difficult to persuade them to use even high quality modes of public transport, and they are unlikely to opt for a bus route with ROW-C.

These circumstances lead to extensive use of private cars as means of commuting and, accordingly, to the further growth of private car ownership, including the addition of a second or third car in a household. Meanwhile, the new neighborhoods continue being unsuited for this kind of a situation. There is a chronic shortage of parking spaces as well as the obvious inconsistency between the adjacent road network capacity and the demand for it, triggered by widespread motorization.

Mass housing construction in remote areas is often combined with inefficient land use within the urban core which is properly served by public transport. An example is the *Ametevo* neighborhood in Kazan, located in close proximity to the subway station built in 2005. Here, despite the availability of transport resources sufficient for transit-oriented development, the low-density individual buildings constructed 40 or more years ago continue to prevail. The more typical examples of ineffective land use are disused industrial zones located in the intermediate areas of many Russian cities.

The described negative trends can be explained, in the first place, by the low quality of land use regulation, characteristic of the majority of Russian cities.

Currently, the governing territorial planning documents in a typical city include a "*general plan*" (the analogue of a master plan) as well as land use and development rules (zoning regulations). The quality of these documents and, moreover, the integrity of their enforcement have been fairly low throughout the post-Soviet period.

Moreover, huge parts of urban land remain the municipal property. As might be expected, the abolition of the planned economy and of the state monopoly on urban development did not immediately result in the creation of a fully-fledged land market. The administration received the right to grant land to private investors and developers and accept money and other sorts of compensation such as the construction of apartment buildings in return (Trutnev 2011). This kind of a situation increases a municipality's interest in the development of new territories. A similar problem can be observed in Chinese cities where land resources are also concentrated in the hands of municipalities. The cities of Central and Eastern Europe also experienced difficulties in formulating a clear policy of spatial development in the context of choosing between the development of peripheral areas in response to the demand for new construction and the need to make urban core denser (Bertaud 2004).

In Russian cities, things are being made worse by factors such as monocentricity and low residential mobility. With the increase of the share of the service sector in the economy, the employment is naturally concentrated in urban cores, which is evident in all Russian cities. On the other hand, the level of residential mobility remains very low, which does not allow citizens to optimize their transport needs by changing their place of residence. According to the estimates based on the level of prices and mortgage lending rates, in 2014, only 30 % of the Russian population had the opportunity to purchase an apartment or a house, and even that is the highest figure in a decade. In the first quarter of 2015, due to the decrease in income and increase in bank interest rates, the value fell to 24.7 % (Kosareva et al. 2015).

2.4 The Impact of Urban Spatial Structure on Travel Behavior (the Case of Moscow)

It is interesting to analyze the travel behavior in a post-socialist city, in areas with different levels of public transport availability. Our analysis is based on the data from surveys conducted in 2013 and 2015 by the Moscow Institute for Social and Cultural Programmes (Moscow Institute for Social and Cultural Programmes 2016). The study covers seven aspects of urban life, one of which is transport, and covers all the 146 districts of Moscow.

In 2015, the following questions were included in the group of questions concerning the transport sector:

1. Which of the following problems in your district are most critical to you?
   (a) poor thoroughfare situation (poor quality of roads, traffic jams)
   (b) public transport does not operate properly (few routes, overcrowding, low frequency)
   (c) not enough parking spaces
2. How do you usually get to work/place of study?
   (a) by private operator bus (jitney)
   (b) by bus
   (c) by taxi
   (d) by subway
   (e) by private car
   (f) by bicycle, scooter, roller skates
   (g) by tram
   (h) by trolleybus
   (i) on foot

Each respondent could choose more than one answer.

The results of the study present, in particular, the information on the distribution of respondents in each district based on the selected answers and, accordingly, on the modal split. This data can be used for aggregated analysis of the relationship between the transport behavior and a spatial structure and the urban features of an area.

The share of residents who use subway was chosen as a typical indicator of travel behavior. In order to describe the spatial structure of a region the following factors were chosen: the physical distance from the district to the city center (Red Square) and the weighted average distance from the housing stock to subway stations. Increased distance from the city center greatly impacts the attractiveness of subway in comparison to other public transport modes; for short-distance travel, land transport is often more convenient because of the more frequent stop locations, while the low (compared to the subway) speed of the land transport is not critical. The weighted average distance from the housing stock in a neighborhood to the subway stations shows the ease of access to subway stations for the residents of each particular area. In order to calculate this indicator, the data published by State Corporation "*Fund for assistance to reforming housing and communal services*" (Reforma ZHKKH 2016) was used. The study includes the districts of Moscow within the Ring Road or adjacent to it as well as the areas that are not adjacent to the Ring Road but are served by the subway (Yuzhnoye Butovo). The analysis results show that the increase of the average distance to a subway station by 100 m reduces the proportion of residents using subway by 0.7 %. The increase of the distance from the city center by 1 km prompts the increase of the number of residents using the subway by 0.8 %.

Given the high (more than 400 cars per 1000 inhabitants) motorization rate in Moscow and referring to the foreign research data (Ewing and Cervero 2001), one would expect a very large (up to 25–30 percentage points) difference between the number of car trips by the residents of the districts with the largest and, respectively, the smallest distance from the housing stock to the subway stations. This hypothesis is also supported by the fact that the Muscovites are very critical of the land public transport quality of service: according to the same survey, the residents of the districts served exclusively by land public transport are more likely to be worried by drawbacks of public transport such as low density of the route network, low frequency, and overcrowding. However, the analysis of empirical data on the modal split for the residents of the districts located between the Third Ring Road and the Moscow Ring Road has clearly illustrated that the hypothesis is not entirely correct.

Figure 2.9 shows the popularity of various transport modes among the inhabitants of the districts located between the Third Ring Road and the Moscow Ring Road with the maximum and minimum average distance from the housing stock to the subway stations.

Table 2.3 presents the average values of modal split for two typologically different sets of neighborhoods. The sum of all modal shares exceeds 100 % because of the accounting for multimodal trips.

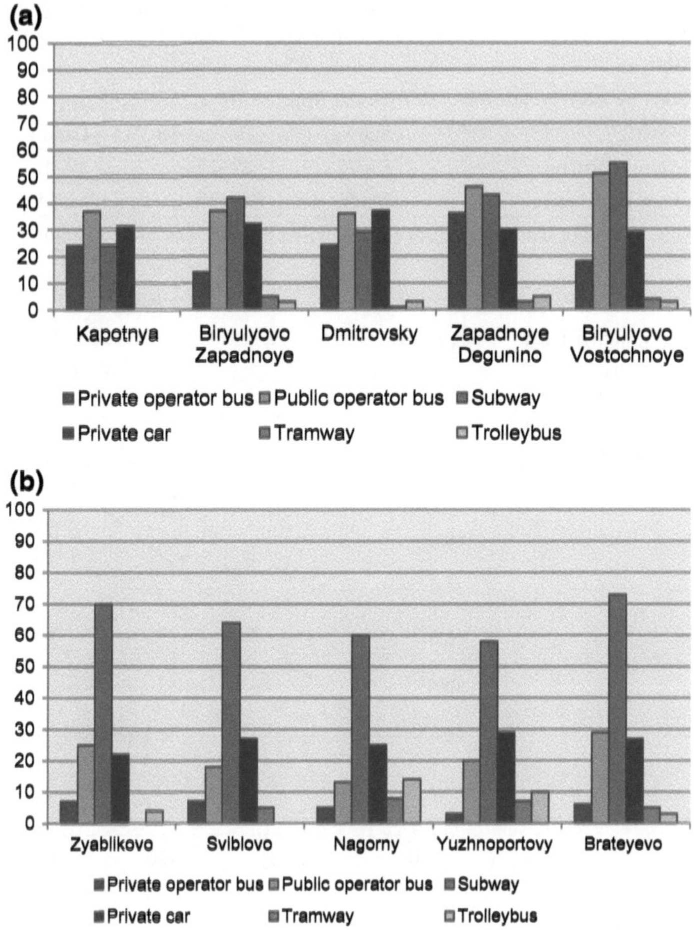

Fig. 2.9 The popularity of various transport modes among the inhabitants of the districts between the Third Ring Road and the Moscow Ring Road with (**a**) minimum average distance from the housing stock to the subway stations (**b**) maximum average distance from the housing stock to the subway stations (*Source* Created by authors using data from Moscow Institute for Social and Cultural Programmes 2016)

Thus, the residents of the neighborhoods not served by subway are more likely to use land public transport than private cars instead of subway (or in addition to it). This situation can be explained by the comparative analysis of modal split and motorization level of world cities.

Figure 2.10 presents the regression line based on a sample of 42 cities in the world and connecting between the motorization level and the share of car trips. Moscow's point on this graph is located substantially below the regression line, which means that the share of car trips in Moscow is significantly lower than in foreign cities with a similar level of motorization.

Table 2.3 Modal split in the neighborhoods located between the Third Ring Road and the Moscow Ring Road with minimum and maximum average distance from the housing stock to the subway stations

Neighborhoods by average housing stock distance from the subway stations	Modal Split		
	Subway (%)	Private car (%)	Land transport (%)
Minimum average distance	58–73	22–27	30–40
Maximum average distance	24–55	29–37	61–89

Source Created by author using data from (Moscow Institute for Social and Cultural Programmes 2016)

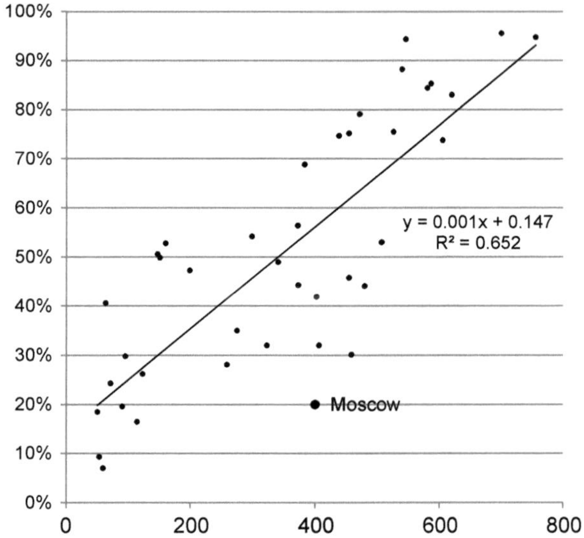

Fig. 2.10 The nominal level of car ownership and the share of car trips [*Source* Created by author using data from Institute for Mobility Research (2013)]

The cause of this phenomenon is likely to be explained by the extremely low area ratio of road network for one car, which is typical for Russian cities. In order to clarify the merits of the case, we have used the formula introduced above:

$$s_v = 10^7 * \frac{\varepsilon}{d * m}$$

This will help us to see that the real impact on the modal split is not produced by the nominal level of car ownership per se but rather by an adjusted value based on the parameter s_v.

For this purpose, we introduce a new indicator—"*real*" motorization level:

$$m_{real} = 10^7 * \frac{\varepsilon}{d * sr}$$

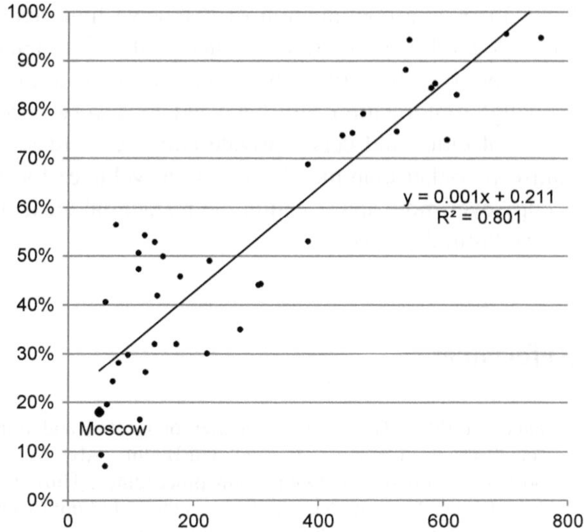

Fig. 2.11 The level of "real" car ownership and the share of car trips [*Source* Created by author using data from Institute for Mobility Research (2013)]

where *sr* is the road area, which maximizes the correlation of m_{real} with the share of car trips. In fact, m_{real} shows the number of cars (per 1000 inhabitants), which can be used on a daily basis, considering the actual values of population density and the road network development in the city.

The calculation shows that *sr* is about 200 m^2, and certainly, when switching from the nominal motorization level (*m*) to the real motorization level (m_{real}) the regression line becomes more valid ($R^2 = 0.801$ against $R^2 = 0.662$ in the previous case), and Moscow's point moves closer to the trend line (Fig. 2.11).

These results can explain the patterns of travel behavior in the neighborhoods not served by subway, presented above. Each car owner living in such district would probably use his or her private car for commuting. However, the road network deficit makes this choice either irrational in terms of generalized cost or physically impossible.

2.5 Conclusion

The main features of the spatial development of Russian cities by the beginning of the 1990s were as follows: (1) the W-shaped height and density configuration comprised of a high dense city center and peripheral territories and low-dense industrial areas in the intermediate zone; (2) extremely low share of built-up area used for transport network. During the post-Soviet period, there have been two key ambivalent trends in the spatial development of large Russian cities. On the one hand, there has been a considerable density increase of residential development in downtown areas and in the areas properly served by public transport; on the other hand, there is a continuation of extensive high-density development in suburban areas. A combination of factors such as the high-density development of peripheral areas, the low share of built-up area used for urban road network, and high (and growing) motorization rate have led to a very specific variety of the traditional "clash of cities and cars": private cars are increasingly becoming a part of urban landscape rather than merely serving as vehicles for everyday use. Cars continue to occupy ever more space within the residential areas, but are used for commuting to a very limited degree.

References

Bertaud A (2004). The spatial structures of Central and European cities: more European than socialist? *In* Tsenkova S, Nedoviü-Budiü (eds) Winds of societal change–Remaking post-communist cities. Symposium proceedings, University of Illinois, Urbana-Champaign, June 17–19, 2004, 45–64. Available on the web: http://www.alain-bertaud.com/AB_Files/AB_Central%20European%20Spatial%20Structure_Figures_2.pdf. Accessed 31 Aug 2016

Bertaud A, Renaud B (1995) Cities without land markets: location and land use in the socialist city, Washington, D.C. The World Bank. Available on the web: http://documents.worldbank.org/curated/en/991331468757791794/pdf/multi-page.pdf. Accessed 31 Aug 2016

Boarnet M, Crane R (1997) L.A. Story: a reality check for transit-based housing. J Am Plann Assoc 63(2):189–204

Calthorpe P (1993) The Next American metropolis: ecology, community and the American dream. Princeton Architectural Press, New York

Cervero R, Kockelman K (1997) Travel demand and the 3Ds: density, diversity, and design. Trans Res Part D: Trans Environ 2(3):199–219

Cervero R, Murakami J (2008) Rail + property development: A model of sustainable transit finance, University of California. Berkeley Center for Future Urban Transport: A Volvo Center of Excellence, Berkeley

Cervero R, Murakami J (2009) Rail and property development in Hong Kong: experiences and extensions. Urban Stud 46(10):2019–2043

Dubelir GD (1910) Planirovka gorodov (Urban planning). Slovo, St. Petersburg (in Russian)

Ewing R et al (2007) Growing cooler: The evidence on urban development and climate change. Urban Land Institute, Chicago. Available on the web: https://www.nrdc.org/sites/default/files/cit_07092401a.pdf. Accessed 31 Aug 2016

Ewing R, Cervero R (2001) Travel and the built environment: a synthesis. Transp Res Rec: J Transp Res Board 1780:87–114

Glazychev VL (2008) Urbanistika (Urbanistics). Evropa, Moscow (in Russian)

Hamin E, Gurran N (2009) Urban form and climate change: balancing adaptation and mitigation in the U.S. and Australia. Habitat Int 33:238–245

Handy S (2005) Smart growth and the transportation-land use connection: What does the research tell us? Int Reg Sci Rev 28(2):146–167

Institute for Mobility Research (2013) Megacity mobility culture: How cities move on in a diverse world Springer

Jones D (1981) Urban transit policy: an economic and political history Prentice Hall, New Jersey

Kenworthy J, Laube F (1999) Patterns of automobile dependence in cities: an international overview of key physical and economic dimensions with some implications for urban policy. Transp Res Part A 33(7–8):691–723

Khan-Magomedov SO (2006) Khrushhevskij utilitarizm: plyusy i minusy (Khrushchev' utilitarity: pro et contra). Academia. Arkhitektura i stroitelstvo 4 (in Russian)

Kosareva NB et al (2015) Osnovnye tendencii zhilishhnoj e'konomiki rossijskix gorodov (main trends housing economy of the Russian cities). Gorodskie issledovaniya i praktiki. 1:33–54 (in Russian)

Levchenko YaP (1947) Planirovka gorodov. Tekhniko-ekonomicheskie pokazateli i raschety (Planning of cities. Techical and economic indicators and calculations). Izdatelstvo Akademii arkhitektury SSSR, Moscow (in Russian)

Lin J-J, Shin T-Y (2008) Does Transit-Oriented Development Affect Metro Ridership? Evidence from Taipei, Taiwan. Transp Res Rec: J Transpo Res Board 2063:149–158

Loo B et al (2010) Rail-based transit-oriented development: lessons from New York City and Hong Kong. Landscape Urban Plann 97(3):202–212

Mikhailov AYu, Golovnykh IM (2004) Sovremennye tendencii proektirovaniya i rekonstrukcii ulichno-dorozhnykh setej gorodov (modern trends in the design and reconstruction of urban street and road networks). Nauka, Novosibirsk (in Russian)

Moscow Institute for Social and Cultural Programmes (2016) http://www.miscp.ru/en/. Accessed 31 Aug 2016

Newman P, Kenworthy J (1989) Gasoline consumption and cities: a comparison of US Cities with a Global Survey. J Am Plan Assoc 55(1):24–37

Newman P, Kenworthy J (2006) Urban design to reduce automobile dependence. Opolis: An International Journal of Suburban and Metropolitan Studies, 2(1):35–52

Petrov NV et al (1988) Prostranstvenno-vremennoj analiz sistemy rasseleniya moskovskogo stolichnogo regiona (Spatio-temporal analysis of the Moscow capital region settlement system). Institut geografii RAN, Moscow (in Russian)

Polyakov AA (1953) Gorodskoe dvizhenie i planirovka ulic (urban movement and street planning). Gosudarstvennoe izdatelstvo literatury po stroitel'stvu i arkhitekture, Moscow-Leningrad (in Russian)

Polyakov AA (1967) Transport krupnogo goroda (transport of the large city). Znanie, Moscow (in Russian)

Reforma ZHKKH (2016) https://www.reformagkh.ru/ (in Russian). Accessed 31 Aug 2016

Federal State Statistics Service (2016) http://www.gks.ru/ (in Russian). Accessed 31 Aug 2016

Schiller P et al (2010) An introduction to sustainable transportation: policy. The Cromwell Press Group, Planning and Implementation, London

Trutnev EK (2011) Logika razvertyvaniya sistemy pravovogo gradoupravleniya Moskvy: chem opasno i chem polezno proshloe dlya budushhego? (The logic of the urban legal system deployment in Moscow: what from the past is dangerous and what is useful for the future?). Proekt Rossija 62:149–156 (in Russian)

Chapter 3
Urban Public Transport Development: Trends and Reforms

Pavel Zyuzin and Alexander Ryzhkov

Abstract The aim of the article is to present the ways of urban public transport development in Russia within the context of the transition towards the market economy. The article consists of two blocks: the trends of urban public transport development in the Russian Federation and the problems of urban public transport management. The first block presents the survey of urban public transport development trends by transport mode with the attention to the mass transit. The quantitative and qualitative analysis of urban public transport in Russian cities since 1991 is provided. The positive and negative cases of urban public transport development are revealed. The most important features of urban public transport functioning connected to the urban planning and transport planning practice are discussed. The second part presents an overview of the main decisions made by public authorities in the field of urban public transport management. The key questions are: the general logic of administration, the role of private transport operators, funding and risks of the market participants. The general conclusions reveal the place of the Russian Federation within the context of world urban transport development trends.

3.1 Historical Background

In the 1950–1980s, the cities of USSR saw intensive development of all modes of public transport. This process was determined by the peculiarities of urban development in the context of the Soviet planned economy as well as by severe restriction of private car ownership, typical of totalitarian societies.

The restriction of household vehicle ownership was a systematically implemented policy during the entire Soviet period. Since the 1930s, the transport self-sufficiency of Russian population was almost down to zero: horses were owned

---

P. Zyuzin (✉) · A. Ryzhkov
Institute for Transport Economics and Transport Policy Studies,
National Research University Higher School of Economics, Moscow, Russia
e-mail: pzyuzin@hse.ru

M. Blinkin and E. Koncheva (eds.), *Transport Systems of Russian Cities*, Transportation Research, Economics and Policy,
DOI 10.1007/978-3-319-47800-5_3

Table 3.1 Private vehicle ownership in different countries (1928)

Indicators	Countries				
	USA	Great Britain	France	Germany	USSR
Number of vehicles (thousand)	22,137	1023.7	891	319	20.6
Cars per capita	200	29.4	21.7	5.1	0.14

Source Created by authors using data from Chudakov (1928)

by public bodies or kolkhozes (collective farms); cars had not yet arrived. In 1928, a researcher E. Chudakov wrote, "…in terms of vehicle fleet our [Soviet] Union occupies the last place among the more or less industrialized countries" (Chudakov 1928); he also provided a very articulate comparison data (Table 3.1).

By 1928, the total number of motorized vehicles in the USSR had not yet exceeded the figure reached by the Russian Empire in 1916. The number of cars per 1000 people was 22 in 1970, 52 in 1980, and 60 in 1990. In such conditions, regular bus routes were indispensable for maintaining daily activities in almost every community. Urban routes were organized in 1378 Russian communities. 42 cities on this list also developed trolleybus and tramway networks in addition to buses and taxis, and 5 cities built the subway systems.

The reasoning behind the initial introduction of the electrified public transport (trams and trolleybuses) in the USSR was distant from the present-day realities: artificially low electricity prices for public utilities[1]; chronic shortage of large-capacity buses, typical of planned economy; ensuring full-time employment for female population. For the urban administrators of that time, another important argument in favor of electric transport was the fact that unlike buses, trams and trolleybuses could not be ordered away from their regular service to transport people to gatherings, seasonal works, and other activities typical for the Soviet reality.

In 1990, 24.1 % of the world trolleybus networks and 24.2 % of the world tram networks were operated in Russia, with the total length of lines being 7300 km (Table 3.2) (Zyuzin 2012). In the same year, the number of buses working on urban routes in Russian cities had reached 80 units per 100,000 residents.

Despite the myth about the "*the Soviet public transport being the best in the world*", the actual supply of transport services never met the demand for public transportation at any respectable level of quality. The results of the surveys conducted in Moscow in 1985 show that 47 % of the rush hour passenger traffic occurred in the conditions of 0.15 m^2 of vehicle floor space per passenger, that is, 6.67 standee per 1 m^2 of the floor area. Another 15 % of the rush hour passenger traffic occurred in the conditions exceeding the technical capabilities of the bus: 0.125 m^2/passenger, that is, 8 standing passengers per 1 m^2 of floor space. Thus, over 60 % of the rush hour passengers of Moscow buses were transported in the

[1]In all Russian cities except Moscow, trolleybuses and trams belonged to the jurisdiction of the RSFSR Ministry for Housing and Communal Services.

Table 3.2 General information about public transport in Russian cities (1990)

	Bus	Tram	Trolleybus	Subway	Taxi	Total
Number of communities	1378	70	85	5	550	1378
Number of routes	9693	703	910	18	–	11,324
Length of routes (km)	43,373.6[a]	4369.6	3008.9	337.3	–	51,089.4
Number of vehicles (thousand)	60.3	14.8	13.8	5.2	54	148.1
Ridership (millions of trips)	22,869	6020	6000	3695	557	39,141

Source Created by author using data from Russian Federation Federal State Statistics Service (2016)
[a]536 largest cities only

conditions contravening the requirements adopted by the UITP (Blinkin et al. 1988). In European and North American cities, concentration of passengers in a bus was usually about 0.33 m^2 per passenger, or 0.25 m^2 per passenger in the worst cases during the same period.

Significant overcrowding was also observed on urban electric transport. Thus, in 1990, in 7 Russian cities, trams carried more than 3 million passengers per 1 km of rail tracks, in Izhevsk, 4.3, in Vladivostok, more than 4.9. On trolleybus routes, figures of 2.0–2.5 million passengers per 1 km of lines were typical for Russian cities; a record figure, 3.5 million passengers per 1 km of lines, was observed in Tula. To compare, the average figure for tram networks in EU countries is 0.7 million passengers per 1 km of rail tracks, and for EU-15, 0.6 million passengers per 1 km of rail tracks (ERRAC 2009).

During the Soviet period, passenger road transport companies as well as tram and trolleybus operators were public property. The former were supervised by the Ministry of Road Transport and the latter, by the Ministry of Housing and Communal Services of Russian Federative Republic of the USSR. The only exception was Moscow where all companies in charge of land public transport were supervised by the municipal authorities. Subways in all cities, including Moscow, were supervised by the former USSR Ministry of Railways.

The welfare of passenger transport companies was based on the planned supply of rolling stock and on "*state capital investments*" in infrastructure and production facilities, also planned and provided from the main funds of respective ministries. Thus, capital costs were almost fully covered by external sources.

Since the beginning of the 1970s, urban public transport in Russia operated with farebox recovery ratio below 1 in terms of operating costs. The actual value of this indicator in Russian cities varied between 0.6 and 0.8 (Blinkin et al. 1988). This situation was possible due to artificially low fares, "*frozen*" since the 1950s, the presence of numerous reduced fare categories eligible for free transportation, and the considerable, and ever-growing number of faredodgers, due to either misdemeanor or overcrowding.

When the farebox recovery ratio declined below 1, a procedure was adopted, according to which "*the planned losses*" of the companies serving urban bus routes were covered by:

- cross-subsidization, i.e. partial redistribution of income from profitable modes of passenger and freight road transport, made at the level of regional branches of the RSFSR Ministry of Road Transport;
- centralized funding, usually limited to covering the costs of transporting certain categories of concessionary passengers. This funding was provided within the respective annual plans; the companies obtained it through the same regional branches of the Ministry.

In the case of trolleybus and tram operators, the "*planned losses*" were covered by the funding coming directly from the Ministry of Housing and Communal Services of Russian Federative Republic of the USSR.

In 1991–1992, transport companies were transferred into the ownership of municipalities and/or the subjects of Russian Federation (Russian Federation Supreme Council 1991), while public transport lost its former source of funding, the centralized subsidies. At the same time, frankly populist legislative amendments were made to grant the right of free public transport usage to a considerable number of citizens. The number of concessionary categories was increased from 32 to 65 (Tarkhov and Derkach 1999). In total, 27.3 % of the population obtained the right to use public transport for free; in some cities, the figure rose to 35 % (Rodionov 2005). According to these amendments, the operators were to be compensated for the transportation of concessionary categories of passengers from the corresponding budgets: i.e. from the federal budget, for the concessions established by the federal laws, and from local budgets, for the concessions established by the local authorities.

Given the extremely difficult financial situation faced by the public transport operators at that time, the compensations promised by legislators had stimulated large-scale upward distortions. Many cities provided data for the official statistics that claimed transporting 6 million or more passengers per 1 km of rail tracks. The record was reached in 1995 in Tula, wherein the figure was 8.4 million passengers per 1 km of rail tracks, exceeding the same figure for St. Petersburg subway (7.9). However, all this data cooking soon lost all sense, as compensations for reduced fares, though established by law, were either paid inconsistently or not paid at all. Thus, the veteran operators who were managing the entire fleet of trams, trolleybuses and buses of high and extra-high capacity almost simultaneously lost centralized investments and operation subsidies as well as a significant part of their revenue. Chronic financial problems of these companies led to the decline in daily usage of vehicle stock available for service (Gwilliam 2000), and eventually, to a radical decline in ridership and none the less radical reduction of the total length of the public transport network routes.

There is every reason to say that in the 1990s, the smash-up of the public transport systems began in the majority of Russian cities. The trend of this process in the 1990–2010s resembles in many ways the similar trend observed in the USA in the 1950–1970s. In the US, the smash-up of the 1950–1970s was associated with the problems with subsidies and with the competition from privately owned jitneys, but mainly with the radical reformation of the settlement system and the development of suburbs whose residents switched to cars and stopped (fully or partially) using public transport (Jones 1981). A considerable share of residents moved from areas of transit-oriented development to areas of car-oriented development and, accordingly, chose car-oriented mobility as more adequate with respect to the new settlement patterns.

The Russian public transport smash-up of 1990–2010 was determined by the same factors with one important exception: the mass transfer to cars and the abandonment of public transport in Russian cities took place without any significant changes in the settlement system. Citizens who continued to live in the conditions of transit-oriented development chose car-oriented mobility because of the absence of any viable alternative.

These events initiated the process of replacing the services previously provided by municipal electric transport operators and municipal bus operators:

- firstly, by self-supplied transport, its share growing along with the rapid growth of motorization rate;
- secondly, by a specific Russian form of privately operated jitneys.

A key competitive advantage of jitneys lays in the fact that Russian "marshrutkas" (jitneys) were not covered by the populist legislation concerning free (concessionary) fares. Today, there are two parallel public transport systems operating in Russian cities: municipal operators providing concessionary fares and private operators working on the same routes but without providing concessionary fares. Since around 2002, the ridership balance began to shift in favor of private jitneys. Official surveys of jitney ridership have been conducted after 2010; the scope of operations on this market prior to 2010 can be evaluated using the circumstantial evidence. For example, in St. Petersburg, the number of jitney routes has increased by a factor greater than 30 (from 12 to 363) between 1989 and 2011, and the length of the streets served by jitneys has reached 962 km or about 34.5 % of the entire length of road network in the city (not including local networks) (Zyuzin 2012). According to the results of a survey conducted in 2014, jitneys accounted for 86 % of all passenger trips per day in Kaliningrad (Shulyaev 2014).

Jitneys work on the routes approved by the city administration but usually use the algorithm "depart when filled up completely" rather than pre-announced route schedules. For the most part, they operate low-capacity buses or minibuses, and their ticketing systems are not integrated with those of municipal transport companies. These private operators have to compete strenuously with their municipal counterparts as well as with each other.

As of 2015, in most Russian cities (except Moscow, St. Petersburg, Kazan, and some other big cities), urban electric transport is maintained primarily as a municipal service providing a virtually free transportation for the senior citizens, the persons with disabilities, and other disadvantaged groups. All the other categories of the population use jitneys or prefer their own cars.

In consideration of the above, there is every reason to claim that the transport regulatory framework in Russia is flawed and does little to provide reliable public transport management. The situation began to improve in recent years, especially in Moscow and other major cities.

3.2 The Current Trends

The motorization rate growth in Russian cities, which began in the early 1990s, accelerated even more around the turn of 2006 (Autostat 2016). This circumstance led to the decline in the public transport ridership in 2004–2008, which affected the municipal as well as private operators (Fig. 3.1).

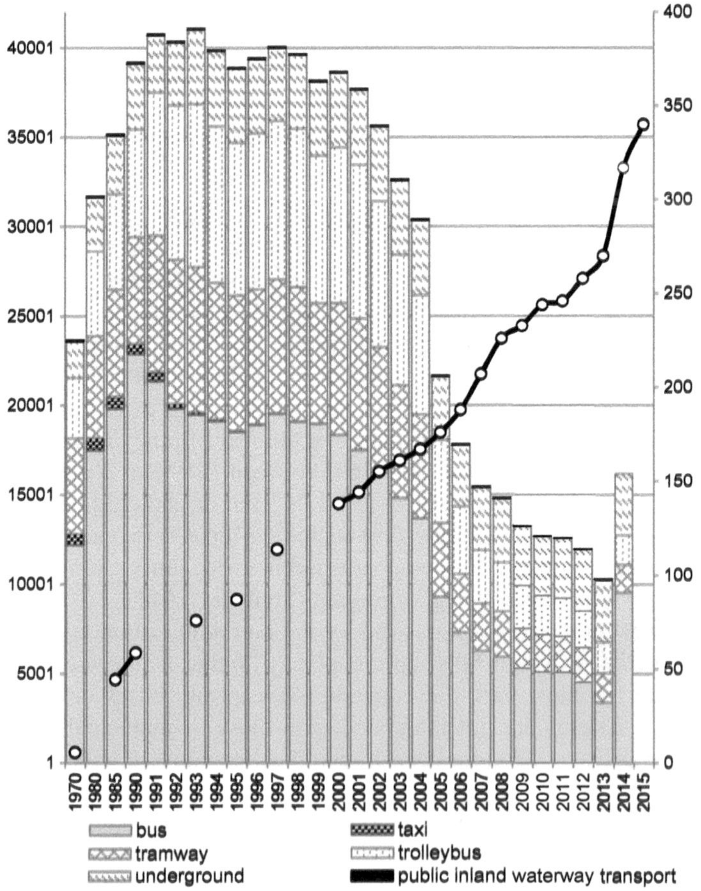

Fig. 3.1 Public transport ridership, modal split, and motorization rate, 1970-2015 [*Source* Created by authors using data from Russian Federation Federal State Statistics Service (2016), Autostat (2016)]

As of 2015, the motorization rate in Russia exceeded 340 cars per 1000 residents, but in some cities it reached the value of 500, while Vladivostok has the motorization rate of 550. According to the results of the survey conducted in this city in 2014, 36 of 224 hectares of the historic city center were occupied by parked cars (Gusev 2014). In Moscow, there are 3.8 million registered cars, but even 2.2 million cars hitting the road network generate impassable traffic jams all over the city (Polyakov 2016). In the current traffic conditions, the urban mobility model based on the degrading electric transport and the low-quality jitneys appears to stop working. Further viability of urban transport systems is becoming increasingly problematic without a strong surge of public transport attractiveness. However, achieving this objective is fraught with difficulties due to the lack of an established mechanism of cooperating with private operators on the public transport market, including their involvement in the electric transport operations. Also, modern transportation planning tools are only being used to a limited degree. For example, until recently, only rail transport modes (subways, railways and some ROW-A and ROW-B sections of the tram network according to the accepted international classification (Vuchic 2011) have had a priority status in the Russian transportation policies. However, adapting these networks to the changing conditions of motorization is associated with a number of problems characteristic of Russian cities.

3.2.1 The Publicly Owned Operators

Russia and other countries of the former Soviet Union have a typically high share of publicly owned companies in the public transport markets. All the subway, tram, and trolleybus services are provided by public operators. In the bus sector, the share of public companies is considerably lower. They provide services only on several routes while others are served by numerous private companies.

At the moment, publicly owned companies are regulated and managed by local authorities in a backroom manner. Local bodies do not formally regulate the operators. Usually they do not set short-term and medium-term goals for operators, do not monitor the results of their performance and, most importantly, do not formally establish financial obligations of operators as opposed to the local budgets. This latter issue is particularly critical because public operators receive support from local budgets to reimburse part of their expenditures. According to the available data, the farebox recovery rate of Russian public transport is 0.518 for tram and 0.478 for trolleybus operations (Higher School of Economics 2014). Subsidies are granted ex post to compensate for revenue reductions due to fare regulations and trips of concessionary passengers. At the end of each month (quarter, year), public companies calculate revenue reductions and apply to authorities for compensation. Deficit compensation does not allow authorities to plan the exact required amounts of public expenditures beforehand. For this reason, the planned sums of public expenditures often do not meet the estimations of operators. This results in numerous conflicts between operators and authorities with regard to the amounts of subsidies. So, from the operators' point of view, this granting system does not guarantee any stable financing. From the authorities' point of view, the problem lies in the quality of performance by the operators. In such conditions, deficit financing does not incentivize companies to improve the quality of services.

Both of these problems should be eliminated by increasing the financial responsibility of the authorities and the operators alike. Such reforms were made in developed countries, e.g., in the European Union, 20–25 years ago. Before the reforms, many developed countries faced similar problems. Public operators received deficit financing and were not incentivized to improve the quality of services (Gwilliam and van de Velde 1990). Problems were solved by a set of regulatory reforms oriented toward formalizing the relations between authorities and publicly owned companies. The researchers who summarized the EU experience (World Bank 2002; Inno-V et al. 2008) usually highlight the importance of the following reforms:

- Establishing adequately staffed and skilled public transportation services planning authority;
- Incorporating the passenger transport department of an authority into an in-house operator;
- Preventing political pressure on an in-house operator, aimed to achieve political and social, rather than economical, goals (reduction of fares, overstaffing etc.);
- Introducing competitive procurement of equipment and a range of support services by an in-house operator;
- Establishing a contractual relationship between the owner and/or transport authority and the in-house operator;
- Formulating an explicit policy and contractual aims within contractual texts;
- Determining ex ante lump-sum subsidy amounts to replace the former ex post deficit compensation;
- Gradually introducing incentivizing mechanisms linked to the realization of policy and contractual targets;
- Issuing an appropriate form of a competitive threat in some cases.

These reforms can help achieve clarity with regard to the level of competence of authorities and operators. As a result, it will allow improving public transportation services (vehicles, routes, timetables, fares, etc.) in return for available public funds. In the case of Russia, it can also prevent extensive and uncontrolled degradation of public transport infrastructure developed during the Soviet period.

3.2.2 Trolleybus Operators

Classic trolleybuses used in Russia serve more than 800 routes in 83 cities (2016). All the operators utilize the classical trolleybus power supply system based on the overhead contact network.[2] The total length of the trolleybus networks in the country grew by 364 km in 1990–2014.[3] This increase was caused by the inertia of the planned trolleybus networks expansion dating back to the late 1980s, as the majority of new lines had been introduced prior to 2000 on the basis of the late Soviet blueprints. Additional trolleybus depots were built (including depots of increased capacity for 200 vehicles) in Chelyabinsk, Ryazan, Petrozavodsk, Ufa, Cheboksary, Volgograd, and Penza at the turn of the 1990s[4] (Fig. 3.2). As a result, 11 cities (in addition to Moscow and St. Petersburg) have operated 3-4 trolleybus depots. During the period, the number of storage spaces has increased up to 430 in Novosibirsk and up to 475 in Chelyabinsk. By 2016, a particularly significant extension of trolleybus lines has been reached (and is maintained) in Cheboksary, Yoshkar-Ola, Sterlitamak, Murmansk, Vladimir, Bryansk, Kaluga, Ryazan, Ivanovo, Kirov, Saransk, Orenburg, Belgorod, Ufa, Tambov, and some other cities.

The expansion of trolleybus networks in Russian cities has been stimulated by the continuing shortage of bus routes. In some cases, the central streets were relieved from tramway lines by the 2000s, and the residual electric infrastructure was adapted for trolleybus lines. The largest amount of new trolleybus lines added according to this method has been introduced since 1991 in Ufa and Kazan, as well as in Lipetsk (Fig. 3.3).

However, the introduction of the new trolleybus routes in these cities had very limited, if any positive effects due to the removal of tramways and increased traffic congestion. For example, in 2014, trolleybuses transported no more than 190,000 passengers per 1 km of lines in Ufa, Kazan and Lipetsk (Fig. 6.3), which amounted to only a half of the respective figure in Novosibirsk where some lines have been decommissioned (about 9 km).

Among the regional Russian cities, the most ambitious urban transport development program has been implemented in 2008–2014 in Voronezh [Voronezh City Duma (2008, 2012)], and it included the restoration of the trolleybus network, but the effect of these developments was minor, similar to the cities mentioned above. Trolleybuses continued to operate without priority lanes with minimal efficiency: only 122,600 passengers were carried per 1 km of lines in 2014 (Fig. 3.4).

The problem of decreasing ridership and profitability is extremely acute for trolleybus operators in all Russian cities. The ridership values of 1 million passengers per 1 km of lines per year in 2014 are continuing to be recorded only in Chita. Izhevsk,

[2]Hybrid types of rail and non-rail public transport (translohr, guided buses) have not yet been introduced. In recent years, several cities started using trolleybuses with autonomous running in a test mode (Tula, Armavir, Maykop, Barnaul, Novosibirsk).

[3]In 2008, this transport mode reached a development maximum of 4809.6 km (adding 440 km to the 1990 level).

[4]Additional trolleybus parks at a high stage of readiness were finished in Kazan, Makhachkala, Vladimir, Tver, Novorossiysk, Lipetsk, Ulyanovsk (on the west side of Volga River), Yaroslavl and Grozny. Trolleybus lines have been intensively constructed in Kamyshin, Stary Oskol, Cherepovets and Novocherkassk. All these projects have not yet been completed.

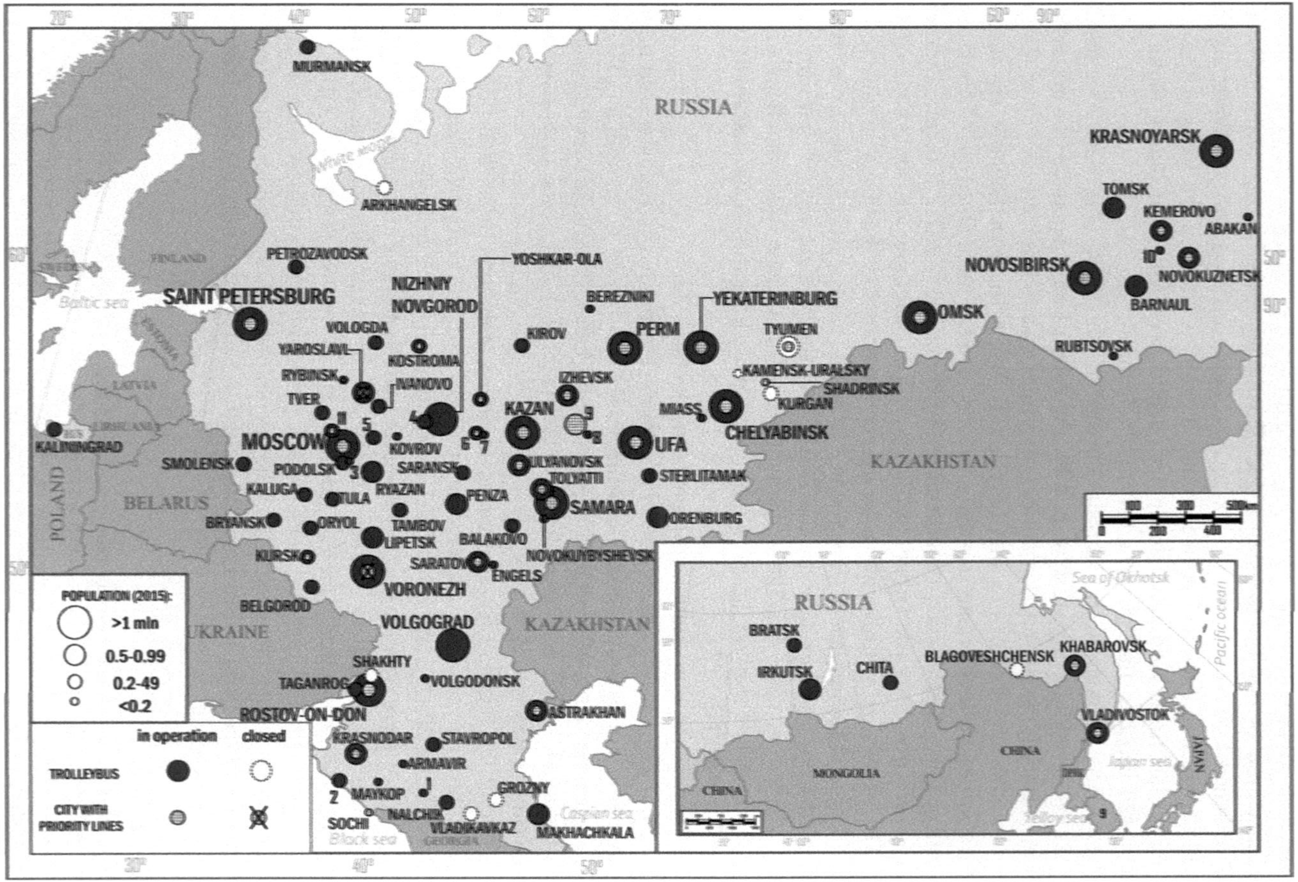

Fig. 3.2 Trolleybus networks in Russian cities (*Source* Created by authors)

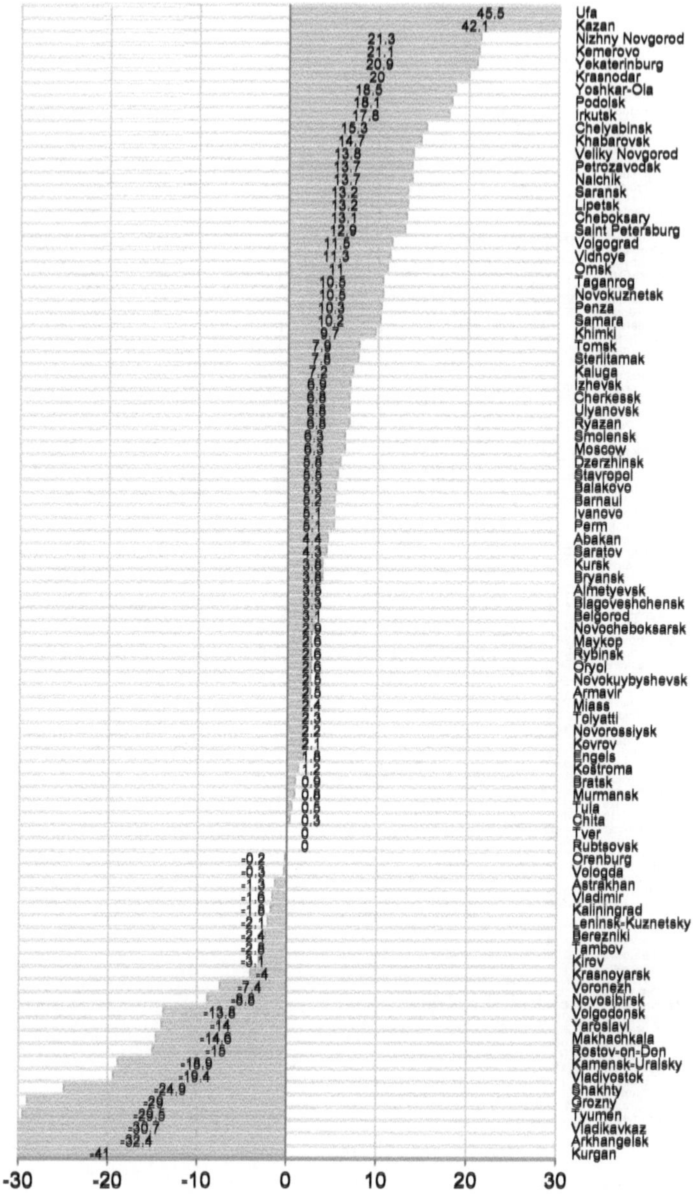

Fig. 3.3 Changes in trolleybus network length (km) between 1990 and 2014 (*Source* Created by authors)

Cheboksary, and Murmansk also manage to maintain a relatively high ridership (Fig. 3.5). In order to reduce the losses related to certain trolleybus routes, some cities decided to eliminate them rather than develop. Thus, the main, 20 km long trunk trolleybus route leading to the city center was closed in Vladivostok.

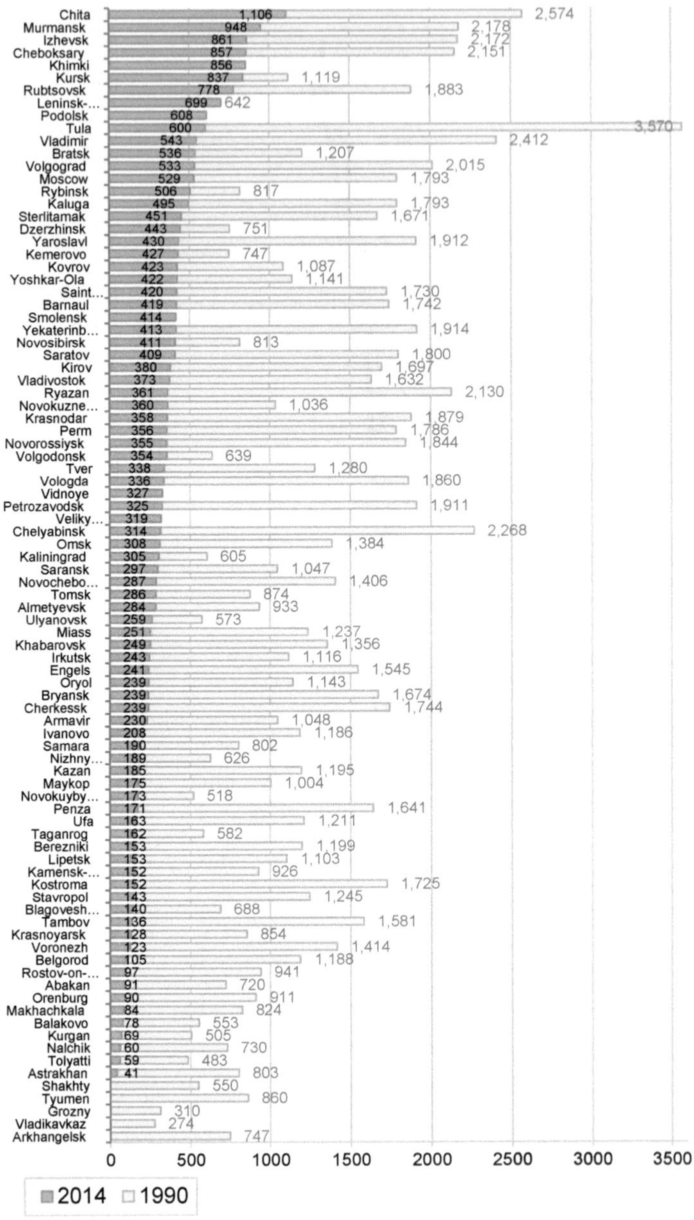

Fig. 3.4 Trolleybus ridership per 1 km of lines in 2014 and in 1990, thousand passengers (*Source* Created by authors)

The greatest drop in ridership, compared to the 1990 figures, has taken place in Astrakhan (by 94.9 %), Nalchik (by 91.7 %), Tambov (91.4 %) and Voronezh (91.3 %). Between 1990 and 2014, trolleybus ridership per 1 km of lines has fallen by 80 % in 35 Russian cities. The smallest decline has been recorded in Leninsk-Kuznetsky, Kursk, and Rybinsk (1, 25 and 38 %, respectively). The 2014 Kursk experience is especially notable, which included the introduction of a priority lane for trolleybuses in the city center in the opposite direction to main traffic. As a result, trolleybus ridership in the city decreased only by 0.7 % in 2011–2014, while the decrease observed in the neighboring cities without priority lanes was significant: 49 % in Bryansk, 40 % in Oryol, 33 % in Belgorod, and 22 % in Kaluga.

Generally, the implementation of priority lanes in Russia is constrained by the high cost of operating such lanes, particularly by the margin of the initial investment in automatic traffic control systems. Priority lanes for public transport have become widespread only in 4 cities[5] (2016): Moscow, Kazan, and, to a much lesser extent, Rostov-on-Don and St. Petersburg. Only 116.7 km of (two-lane) priority lanes have been introduced by 2016 in the cities with population over 200,000 residents (except for Moscow and Kazan) (Fig. 3.5), and their operation is more in the nature of an experiment, often with "risky political consequences".[6]

While Moscow and Kazan have created comprehensive networks of priority lanes, in other cities, such lanes operate in a test mode and only in certain areas. In fact, the entire Russian experience includes only several precedents. Thus, St. Petersburg, Novosibirsk, Krasnodar, and Tyumen have tested priority lanes for several years, and preparations are being made for introducing them on additional streets. In Chelyabinsk, Izhevsk, Kemerovo, Rybinsk, Cherepovets, Kostroma, and some other cities there are pilot projects on short sections of their respective road networks (usually only one lane in one direction). In some cities, priority lanes are used on the central streets in the direction opposite to main traffic for the convenience of operation (Kursk, Saratov, Rostov-on-Don, St. Petersburg, Shadrinsk, and Astrahan). In Rostov-on-Don and Tyumen, three streets (in each city) were selected for implementing the policy of uncontested public transport traffic in both directions of travel. In St. Petersburg, on Ligovskiy Prospect, there is the only pilot 2.2 km priority lane in Russia that combines bus and tram traffic in the center of the roadway. In Ulyanovsk, the same approach is being tested for the only priority lane in Russia running over a bridge, 700 m in length. Largely for these reasons, BRT systems (including those using trolleybuses) are not being built and operated in Russia.[7]

[5]Special guidelines for organizing priority lanes for buses and trolleybuses (ROW-C) were released in Russia in 1983, but such lanes remained exotic, and even in Moscow they were used on only 3 streets.

[6]In some cities, under the pressure of car owners priority lanes were eliminated shortly after their introduction (Yaroslavl, Voronezh) or are publicly discredited (Barnaul, Kirov, Khabarovsk).

[7]As of the 2015, such routes are only being discussed at the level of urban initiatives in Moscow and in Novosibirsk.

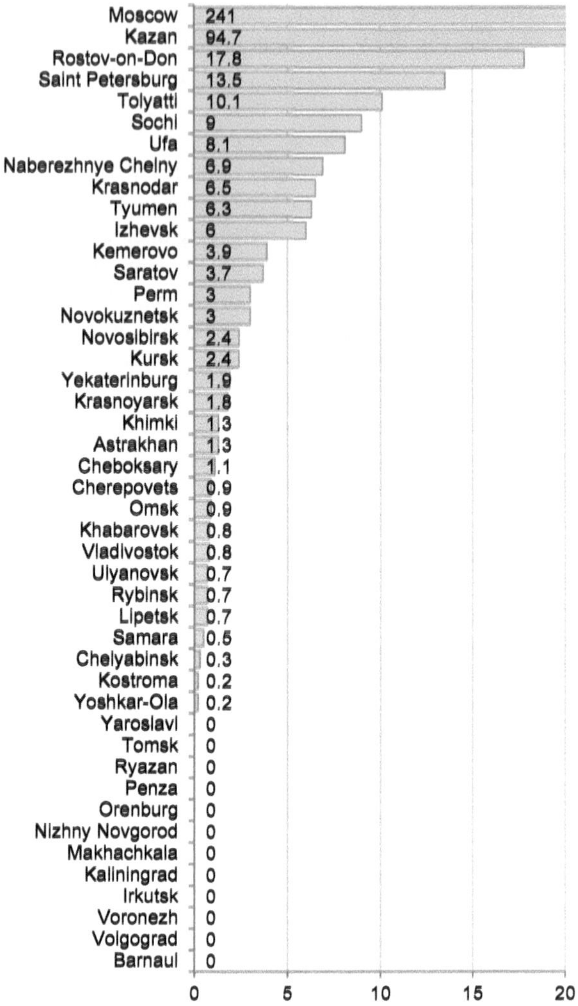

Fig. 3.5 Length of priority lanes (km) for non-rail public transport in major Russian cities, 2016, author's calculations (*Source* Created by authors)

3.2.3 *Urban Heavy Rail and Subways*

For a long time, railways in Russia served primarily industrial freight purposes and were used as urban heavy rail (UHR) very rarely and only as innovations promoted by the progressive local engineers. On the one hand, the situation with road traffic allowed for serving new urban areas by land public transport without significant problems. On the other hand, since the main function of the UHR was to ensure the evacuation of industrial enterprises and emergency mass transportation of the population to suburban areas, the provision of urban public transport services did not fit into this scheme.

As a result, the railway routes (with extremely rare exceptions) were not taken into consideration when new urban residential areas were planned (Ulyanovsk, Tolyatti, Naberezhnye Chelny), and in some cases, passenger stations were not built at all (Novocheboksarsk, Almetyevsk). Krasnoyarsk, Omsk, Novosibirsk, Perm, and Chelyabinsk, along with their suburban areas, do not have any "high" platforms serving local transport connections.

Thus, the entire Russian experience of using UHR includes only two practices. The first is using a special line transporting employees to an industrial facility within a city, but such examples are rare: Volgograd, Zarinsk, Bratsk, Zheleznogorsk (Krasnoyarsk Krai), Grozny (eliminated), Solnechny (now a part of Krasnoyarsk), and Sayansk. In other cases, suburban routes passing through UHR mainly terminate at a central railway station. In some cities, special transport hubs have been created (Nizhny Novgorod, Volgograd, Bratsk). Even in the major railway hubs, UHR with the characteristics closest to the German S-Bahn's are used very inefficiently. In Moscow, the share of heavy rail transport does not exceed 10 %, primarily due to the lack of transport hub development and the lack of connectivity between the rail lines (Zyuzin et al. 2016). In Vladivostok, in 2013, only 30,100 passengers per 1 km of double-track electrified railway line were carried within the urban area[8] (1,385,400 passengers per year). Moscow is close to completing the largest project of rail infrastructure development in Russia to be used for urban public transportation in 2016: preparations are being made to launch a ring line with planned ridership of 5.5 million passengers per 1 km of track per year. In other cities, even in those with a high potential for the UHR, its capacities are not being used for urban transportation.

For a long time, subways were considered an alternative to rail transport in major cities. All subways in Russia belong to the HRT category, i.e., to the networks of railway type. Their distinctive features are the considerable depth of the lines and, as a consequence, extremely high cost of new construction. In Moscow, the construction cost of 1 km of line ranges from 4 to 10 billion rubles[9] (129–333 million USD in 2011 rates), in Novosibirsk, 2.5 billion rubles (2012). Due to the combination of this and other reasons, the expansion of subways is carried out very slowly (Fig. 3.6), and most of them consist of one (Samara, Kazan, Yekaterinburg) or two lines (Novosibirsk, Nizhny Novgorod).

The guidelines for the development of subways in Russia were generally developed in the late 1980s. Since that time, in most cities, the previously initiated construction projects have been completed. The first stages of subway lines construction were completed by 1993; in a rich St. Petersburg, Frunzenskiy-Primorskaya Line took until 1998 to complete. Since 2000, one new station has been opened every 3–5 years in the cities other than Moscow and St. Petersburg, except in Kazan where subway construction was boosted in 1997–2005 in

[8]The section between the stations Vladivostok/Mys-Churkin and Artyom.

[9]Source: http://www.rbc.ru/rbcfreenews/20110915164754.shtml.

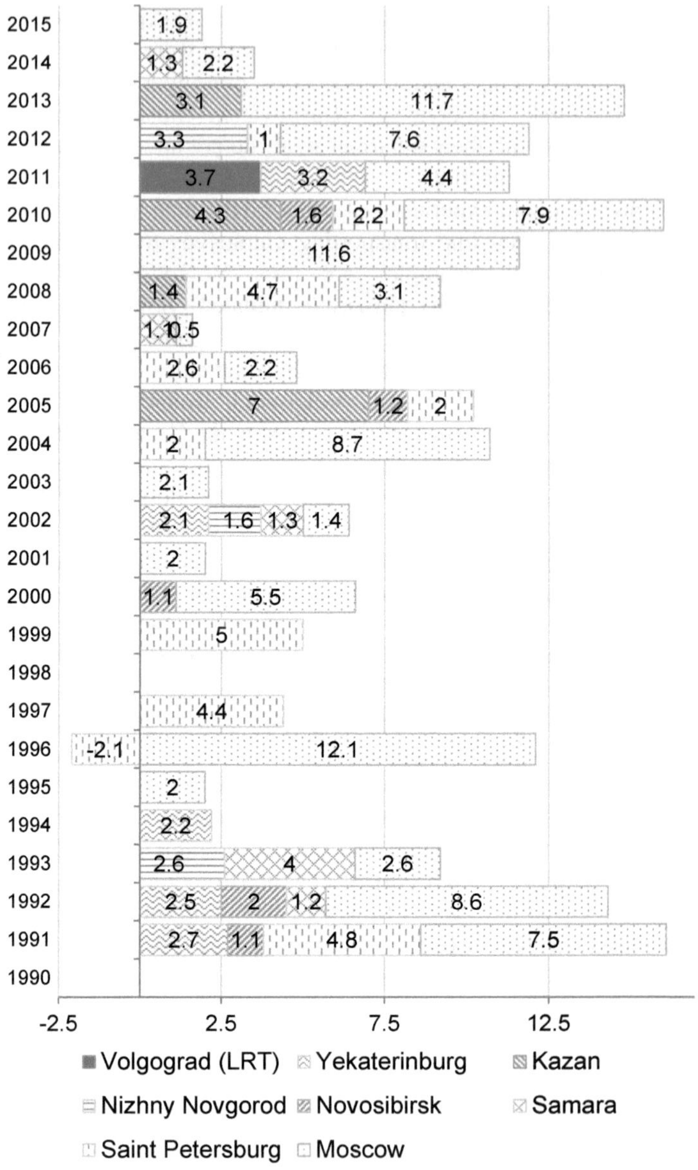

Fig. 3.6 Construction rates of subway networks in Russia (km per year during 1990–2014) (*Source* Created by authors)

anticipation of celebrating the millennial anniversary of the city, with 50 % of funds coming from the federal budget.

The second stage of completion of the construction projects initiated prior to 1990 began after 2007: the main lines were completed (Yekaterinburg, Kazan,

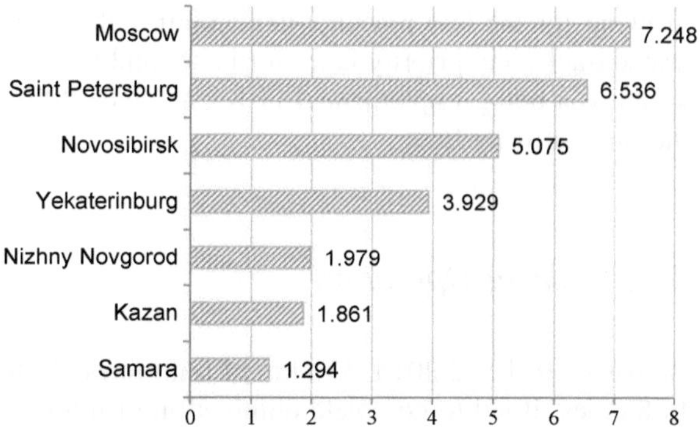

Fig. 3.7 Subway ridership in 2014 (thousands passengers per 1 km of lines) (*Source* Created by authors)

Novosibirsk); a powerful effect was observed in the cities where subway lines were built to connect the outskirts to the city center (Nizhny Novgorod and Volgograd[10]). Thus, in Nizhny Novgorod, subway ridership increased by 24.8 % after the construction of the station in the city center. In Samara, the subway line has not been extended to the central part of the city. As a result, the cost of transportation of 1 passenger was 41.25 rubles (2014), making Samara subway the most inefficient subway system in Russia. In 2007, about 260 million rubles were spent on covering the operating costs, while the revenues amounted to 80 million rubles only. There were years when the covering of losses from this project amounted to a quarter of all budget expenditures on urban transport in Samara, and its mode share does not exceed 9.6 % (2014) (International association "Metro" 2014). Largely because of this, modernization of the more popular tram infrastructure is not performed (Fursov 2015). The results of the studies show that in Russian conditions, the break-even point is achieved when subway ridership ranges from 6 to 8 million passengers per 1 km of lines (Yakushkin 1982). These ridership values have been achieved only in Moscow and St. Petersburg; Novosibirsk is close to achieving them (Fig. 3.7).

In other cities, ridership is much lower than the break-even point. It is not clear yet whether the break-even point can be achieved in Nizhny Novgorod, Samara, Kazan, and Yekaterinburg, since subways in some of these cities are focused, to a greater extent than in others, on servicing large industrial facilities. The mode share of the subway in the total number of transported passengers amounts to 9, 9.6, 12.5 and 24 % respectively (2014).

Thus, the previously used model of subway construction with a focus on HRT networks has exhausted itself in Russia outside the capital cities,[11] and demand is

[10]The city of Volgograd has an LRT line of ROW-A category within the city center that runs through tunnels built in accordance with subway requirements (transformation into a subway has been suggested).

[11]There is also a small pilot monorail line operating since 2005 in Moscow which suffers annual losses of several hundred million rubles. This form of transport has not been further developed, and its elimination has been announced.

growing for the less resource-demanding hybrid systems: LRT on physically separated tracks and priority lanes for buses and trolleybuses. A plan to introduce such systems is being implemented in Kazan since 2008, along with the completion of the only subway line.

3.2.4 Tram Operators

Between 1991 and 2015, 531 km of tram lines (17.8 %) were eliminated in Russia. In 8 cities, it led to complete elimination of tram transport,[12] which in some cases was done in the way of a forced optimization. Thus, there were situations when tram operators were owned by local industries and provided transportation services exclusively between factories and subordinated surrounding urban areas. After the tram transport became municipal property, it ceased operating; however, such cases are few (Ryazan, Karpinsk) (Fig. 3.8). In other cases, it resulted in the worst possible scenarios, including the termination of the tram operator either due to the high value of certain property assets, primarily land parcels (Voronezh, Arkhangelsk), or the necessity to relieve the road network in the conditions of growing motorization rate (Astrakhan, Ivanovo).

In order to achieve this, municipal operators were deliberately refused funding (Dzerzhinsk), which led to the termination of operation under the pretext of violating safety requirements (Shakhty, Karpinsk). Some cities started campaigning in the 2000s to eliminate tram lines and release road networks for private vehicles. The cities that have particularly excelled in this sense are Kaliningrad (76.4 % of the tram network has been eliminated), Tver (72.9 %), Vladivostok (69.6 %), Rostov-on-Don (43 %), Ufa (40 %), Lipetsk (37 %), Yaroslavl (33 %), and Smolensk (29 %). In all these cities, the first step was to remove the lines that were serving main streets in the city center. In some cities, the reduction of tram networks was carried out at the expense of low-demand sections in industrial zones without reducing the coverage of residential areas. Thus, in Angarsk, in spite of eliminating 31 % of the tram network in total, the lines were only discontinued in the industrial areas. The tram networks have been similarly optimized in Usolye-Sibirskoye, Perm, Osinniki, Salavat, and some other cities. In some cities, the part of the tram network that was removed did not exceed 5 % but the terminated lines had key roles in the topology of the local networks, which significantly compromised their integrity. Thus, in Kursk, in 1991, only 3.9 % of the network was eliminated (the single 1.6 km line was closed in the city center). This led to the disintegration of the cyclic framework (Tarkhov 2002) and reduction of the tram ridership by 67 % the following year. Closures resulting in topological flaws took place in Smolensk, St. Petersburg, Novosibirsk, and in 10 more cities; in 4 of those, the operation eventually ceased for good.

[12]The cease of operations in Kaliningrad was announced for April 2016.

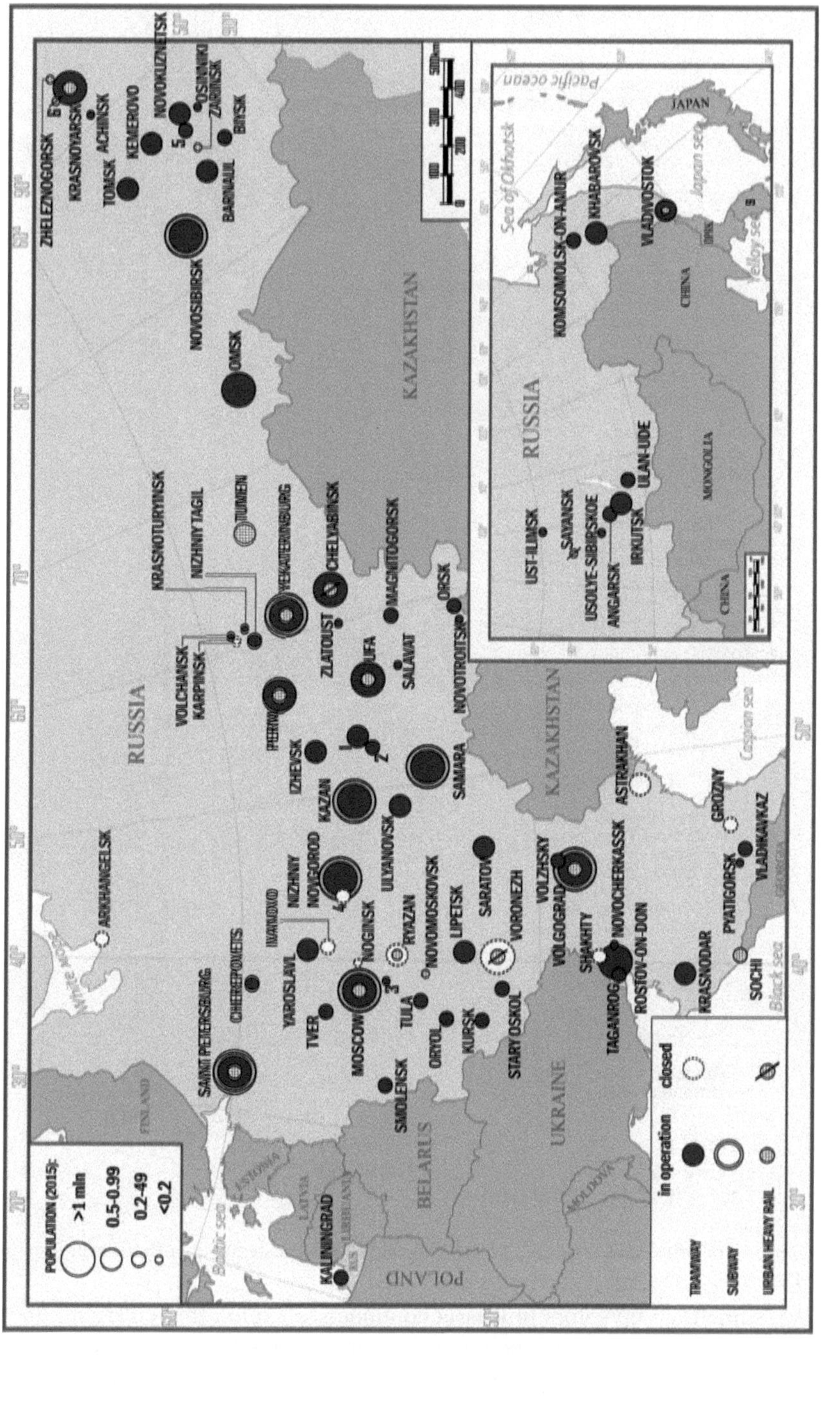

Fig. 3.8 Tram networks in Russian cities (*Source* Created by authors)

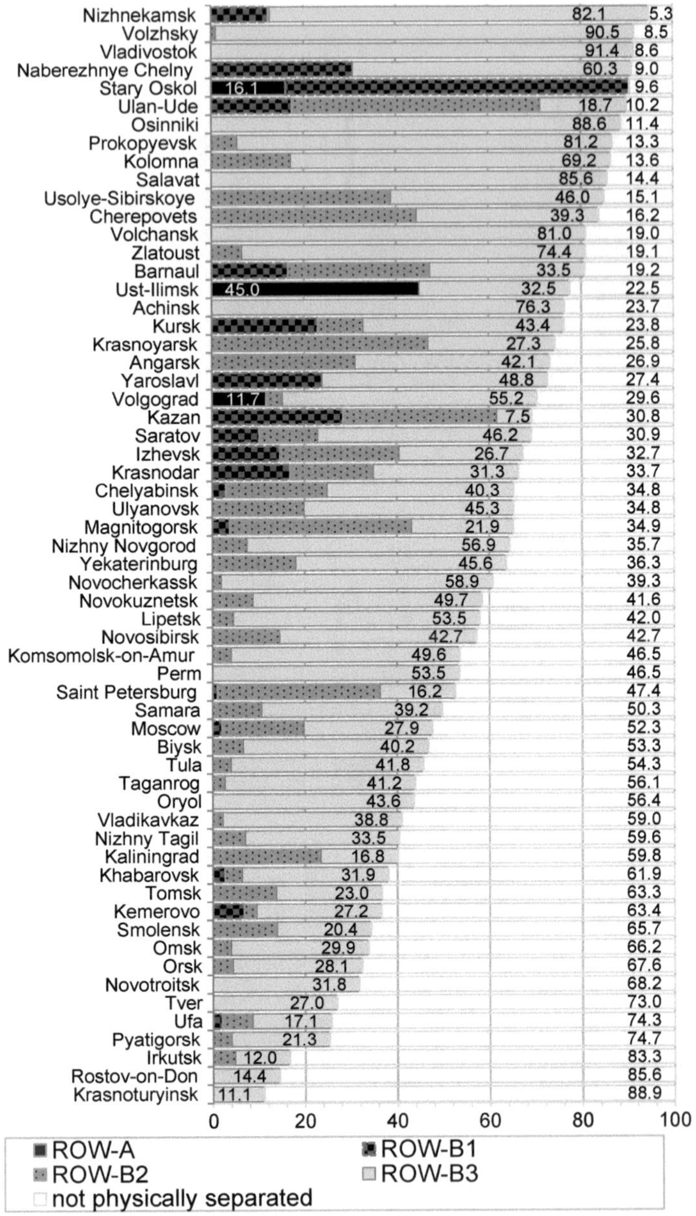

Fig. 3.9 Share of priority lanes for trams of ROW-A and ROW-B categories in Russian cities (%)

In 2015, some tram operators in Russia continued to exploit the remaining lines that were originally planned for specific reasons and sometimes even those that are no more required, but as a whole, the Russian tram transport has the highest share of priority and physically separated lanes in the country (Fig. 3.9).

With the increase of congestion, this advantage has contributed in some cases to the cancellation of plans for total elimination of tram operators (Lipetsk, Vladivostok, Omsk, St. Petersburg, Moscow, Tver, Kazan, Novocherkassk, and Rostov-on-Don). However, only a few cities have reconstructed tramway lines and increased the percentage of the right-of-way sections in the total length. The share of the right-of-way sections (ROW) (Vuchic 2011) of all types (A and B1-3) (Zyuzin 2016) is currently the largest (over 90 % of the total network length) in the tram networks of some of the "late Soviet" industrial hubs such as Nizhnekamsk, Volzhsky, Naberezhnye Chelny, or Stary Oskol; however, in some cases (Volzhsky) it still has no effect due to the insufficient service provided to residential areas. Despite the fact that three Russian cities (Volgograd, Stary Oskol and Ust-Ilimsk) have state-of-the-art light rail lines, in two of them, the potential of tram transport is not being used to the fullest. In Stary Oskol, 80 % of the modern tram lines run outside the residential areas, and in Ust-Ilimsk, the figure is 78.4 %. Thus, only 62,800 passengers are transported per 1 km of lines (2014) in Ust-Ilimsk, and 253,000 passengers, in Stary Oskol.[13] In some cities, tram lines have been reconstructed and the share of right-of-way sections (B1-3) has been increased: in particular, in Ulan-Ude, Kazan, Kolomna, Zlatoust, Barnaul, Kursk, Izhevsk, and Saratov. In some metropolitan areas such as Novokuznetsk, Prokopyevsk, and Volchansk, right-of-way tram lines serve routes connecting between remote neighborhoods located away from highways, thereby reducing their dependence on road transport. In some cities, only the right-of-way tram lines serving the most popular routes have been preserved, while most of the once extensive network was decommissioned. This was done in Yaroslavl, Moscow, and St. Petersburg. Despite the fact that by 2014, the tram ridership in Russia fell to 20.2 % of the 1991 value, in some cities it is maintained at a high level. For example, in Kemerovo, tram ridership comprises 83.2 % of the 1990 value (Fig. 3.10) but in other cities it is much lower.

Despite the significant decline in ridership in most cities in comparison with 1990, it can be argued that the high figures of the past can be explained by exceptional overcrowding of the tram lines, as discussed above. Thus, in 2014, despite the significant outflow, 7 tram operators still carry more than 1 million passengers per 1 km of lines yearly, and 20, more than 500,000 (Fig. 3.11).

In some cities, tram operators have certain infrastructural resources to increase ridership; however, there is a significant problem with disposing of the outdated rolling stock. In total, 50 % of tram cars were dismissed from service since 1990 (6594 cars). Only 7380 trams out of the remaining 8168 are still operated (Zyuzin and Shpakov 2016). Thus, in Tver, Vladikavkaz, Rostov-on-Don, and Lipetsk there is less than 1.6 car per 1 km of tram lines, while in 7 of the most successful tram companies in the country this figure exceeds 4.5: Yekaterinburg, Izhevsk, Samara, Volgograd, Oryol, and Moscow. In a number of cities, the rolling stock has declined significantly in relation to servicing possibilities despite the

[13]This is due to the tram transport in these cities designed to accommodate the needs of industrial enterprises.

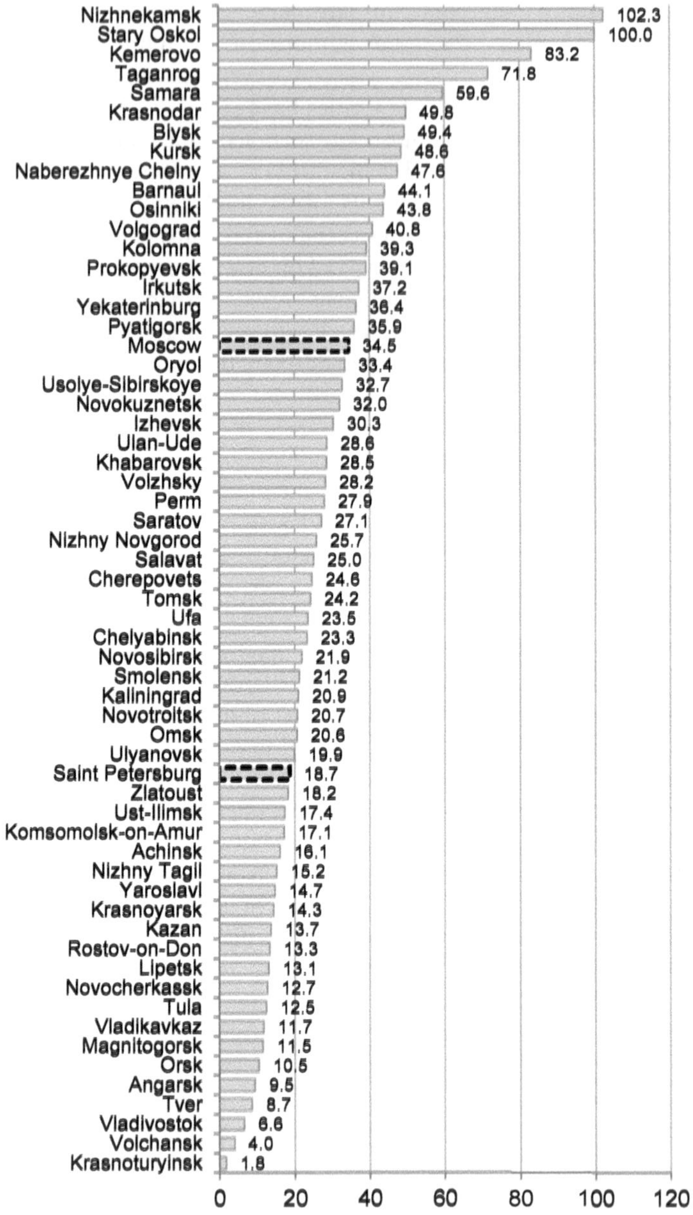

Fig. 3.10 Tram ridership in 2014 (in % compared to 1990) (*Source* Created by authors)

implementation of the program of renovation of the rolling stock co-financed from the federal budget in a ratio of 80 to 20. As a result, in some cities, there was a surplus of servicing capacity: in Rostov-on-Don, Kaliningrad, and Vladivostok, the

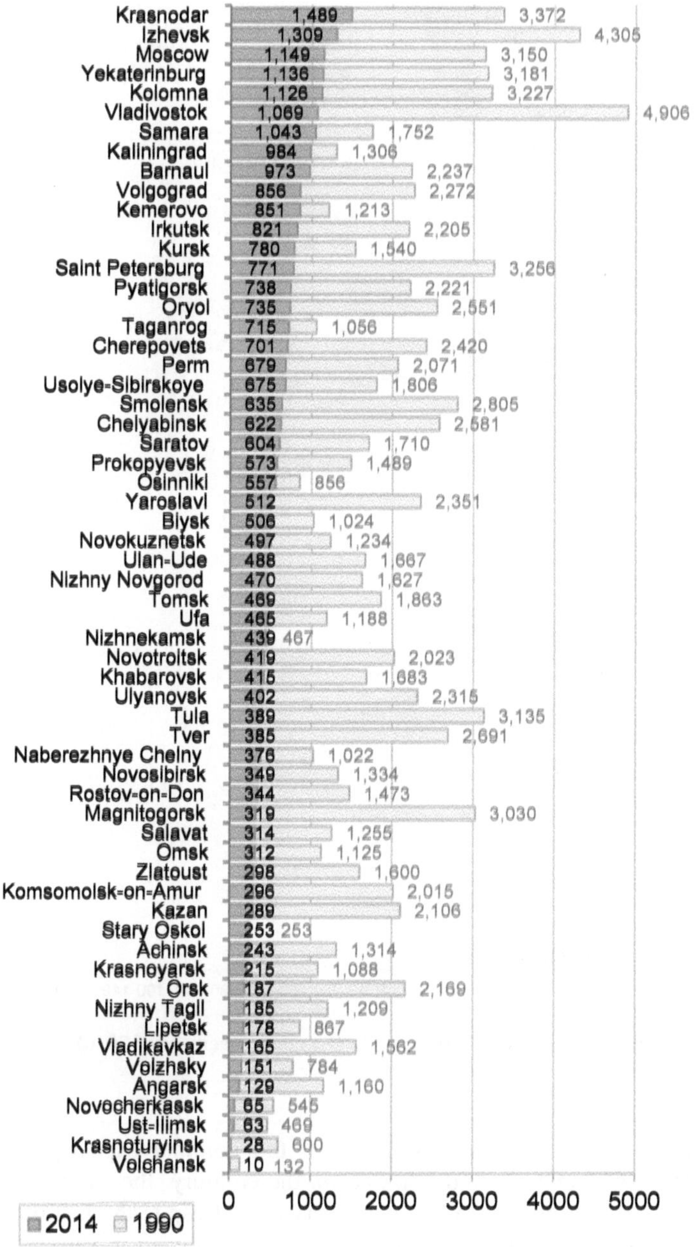

Fig. 3.11 Tram ridership per 1 km of lines per year in 2014 and in 1990 (million passengers) (*Source* Created by authors)

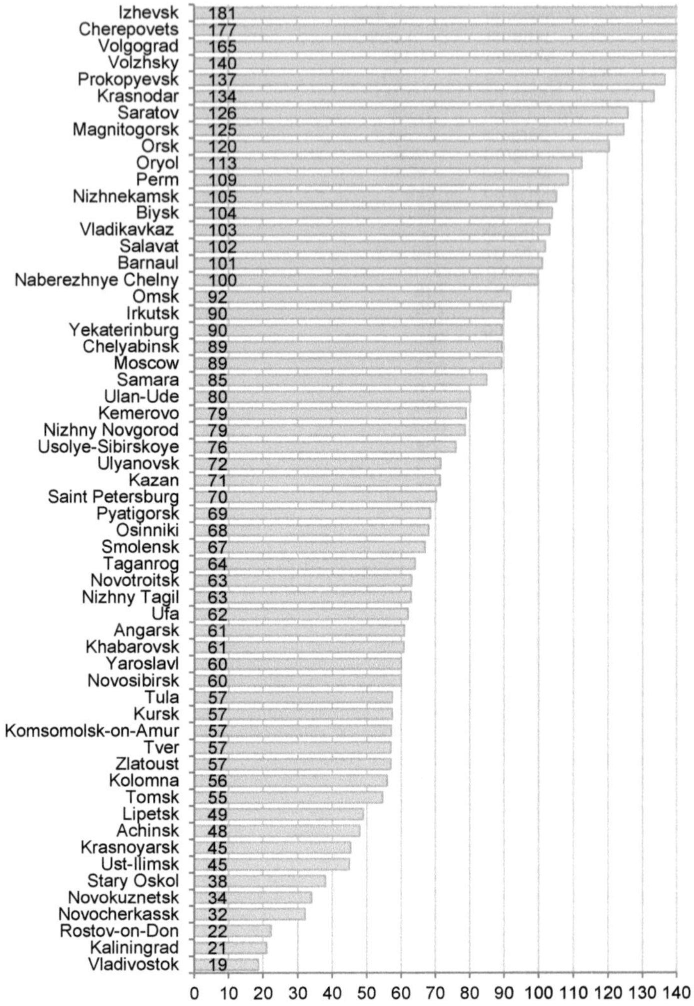

Fig. 3.12 Volume of rolling stock per depot (in % to the capacity as of 2016) (*Source* Created by authors)

existing fleet of cars is less than 25 % of the nominal capacity of the depot (Fig. 3.12). In Cherepovets and Izhevsk, on the contrary, the existing depots lack the capacity for high-quality rolling stock servicing. Despite the fact that 32 depots and 3219 storage spaces for tram cars have been eliminated in Russia since 1990, only a couple of cities have decommissioned relatively new depots built in the 1980s (Saratov and Orsk).

In large Russian cities, tram operators in charge with networks having high shares of ROW-A and ROW-B sections still have the potential for investment and reconstruction. Most of them are shown at the top of the list on Figs. 3.5 and 3.7. In large cities such as Rostov-on-Don, Irkutsk, Ufa, Omsk, Tver, and Tomsk, tram networks will require almost complete renovation, reconstruction, and expansion of the network to the areas with the highest demand for transportation.

3.3 Private Bus Sector: Challenges and Perspectives

Privately owned companies emerged on the Russian bus and minibus markets after the reforms of Russian economy in the early 1990s. Their development was also largely stimulated by the deterioration of the former Soviet Union transport companies that stopped receiving centralized subsidies. That meant that local authorities needed to find new and less expensive ways of providing public transportation services to people. Therefore private transportation companies experienced an enormous growth during the 1990s and 2000s. In 2000, private companies were involved in providing bus services in 307 out of 1290 cities of Russia (Rodionov 2000). In 2015, private companies provided bus services in almost every municipality. As a result, the 'traditional', that is, municipally owned bus companies lost their monopoly status.

Nowadays, the bus markets in big Russian cities are based on 2 types of players. Usually there is one large, publicly owned operator along with about 20–30 small private companies. A public company usually serves around 30–40 routes while an average private company serves 2 or 3 routes. Certainly, there are exceptions. For example, in Perm, all bus services are provided by numerous private companies while the in-house operator was dissolved between 2005 and 2009. A reverse situation is observed in Moscow and Kazan where the share of private companies is relatively low (Table 3.3).

During the 1990s and 2000s, the public transport regulatory system in Russia was fragmented. There was no comprehensive federal legislation on procuring public transportation services and on regulating private sector involvement in bus markets. For this reason, privately owned operators were involved in public transportation services under local legal acts. These acts were mostly directed at minimizing the amount of public transport funding coming from local funds. This practice resulted in considerable deterioration of quality of public transport services in Russia.

At the moment, the quality of bus services greatly depends on an operator. Public operators mostly use conventional buses and reduce fares for some groups of passengers. Private operators, in contrast, provide less convenient jitney-type services. These services are poorly planned and less safe for passengers. Drivers mostly use poorly maintained minibuses and do not accept concessionary fares. They do not follow timetables and clash with each other over passengers waiting at bus stops. Generally, privately owned bus operators have a dubious reputation from passengers' point of view. Nevertheless, in several cities, there are exceptions. For example, in Perm, private companies provide 'traditional' public transport services

Table 3.3 General information about bus markets of 15 largest cities of Russia

City	Population (01.01.2016) * 1000	Number of operators		Number of routes	
		Public	Private	Public	Private
Moscow	12,325	1	61	660	506
Saint Petersburg	5222	1	16	198	543
Novosibirsk	1584	3	65	23	103
Ekaterinburg	1444	1	29	29	60
Nizhny Novgorod	1272	1	39	59	60
Kazan	1217	2	8	40	24
Chelyabinsk	1197	1	27	32	50
Omsk	1180	4	96	47	133
Samara	1171	1	3	49	58
Rostov-on-Don	1118	1	13	31	75
Ufa	1115	1	11	41	89
Krasnoyarsk	1069	3	49	26	44
Perm	1036	–	46	–	71
Voronezh	1033	1	15	15	123
Volgograd	1018	1	80	40	99

Source Data from registries of bus routes (local authorities)

using conventional buses, fixed timetables and regulated fares. In Ekaterinburg, both private and public operators accept EKARTA smartcard. The most drastic changes occurred in Moscow where private companies were forced to provide services under gross-cost-route-based contracts with extensive service quality requirements.

The right to provide public transportation services in Russian cities is tendered by local authorities in a specific way. Operators are competing to get the franchise to operate on a particular route (or a bundle of 2 or 3 routes) and collect fares on it. In most cases, the duration of a contract is 5 years. In some cities, both private and public operators are involved in tendering process on equal terms. In others (e.g., Moscow), public operators do not participate in tenders as they have a privileged right to service their route bundle. Another important detail is that in most cities (except Moscow), contracts do not include any financial obligations on behalf of local authorities. However, most cities grant ex post deficit subsidies. The problem is that deficit financing does not allow authorities to plan for the exact required amounts of public expenditures beforehand. The planned sums of public expenditures often do not meet the estimations of operators. As a result, subsidies play a minor role in financing bus and minibus operators in comparison with farebox revenues. Subsidies, in fact, do not incentivize operators, especially those privately owned, to eliminate the competition "on the market" or to improve their services in any other way.

The winners of tenders are determined by a list of quality parameters. These parameters are usually related to the operator's vehicles, assets, and reputation. Thus, tenders do not simply determine the winner but also have a definitive influence on the quality of services that will be provided on the route during the term of the contract. Quality parameters are used in the tender documentation in two ways:

- As minimum (obligatory) quality requirements to bidders;
- As criteria for bids evaluation.

For example, a minimum requirement may refer to the availability of 10 minibuses, and the age of the vehicles may also serve as a criterion. The bidder that met all the minimum requirements and received the highest number of points based on the criteria established with the means of a special grading system wins the tender. The grading systems vary greatly between the cities (Fig. 3.13). Some cities have a "strict" grading system (the left side of the chart). In these cities (e.g., in Moscow), operators must satisfy a long list of mandatory requirements (e.g., vehicles type, color, floor level, emission standard) before their bid gets to be evaluated based on a very limited number of criteria. Other cities have a "flexible" grading system (the right side of the chart) with few mandatory requirements and multiple criteria for bid evaluation. The authorities organizing tenders in these cities (e.g. in Ufa, Chelyabinsk, Krasnoyarsk etc.), in fact, do not set any significant target requirements to the quality of public transport system.

One of the most important factors that make public transport attractive to people is providing standardized and high quality services. Therefore, the operators in many cities with developed public transport systems use only one or two models of buses (trolleybuses, trams) painted in unified colors, having the same equipment and subject to the same level of maintenance. If a public transport network is served by many private operators, the desired level of service can be achieved through

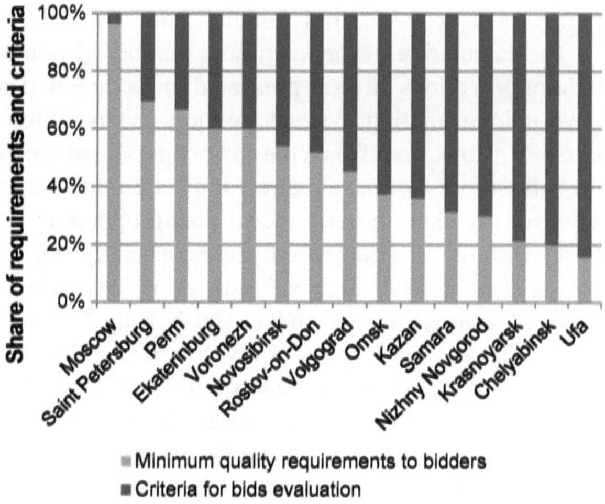

Fig. 3.13 Share of requirements and criteria (*Source* Created by authors using data from local authorities)

making operators comply with an extensive list of mandatory requirements and using a single criterion for determining a bid winner—the tender price.

In Russia, these principles are only being observed in Moscow within the framework of the so-called “new model of partnership with private operators” reform. With this reform, the Moscow authorities determined comprehensive mandatory requirements to operators (27 requirements referring to vehicle type, equipment etc.) and established the tender price as the only criterion determining the winner.

In other Russian cities, the situation is quite opposite. The authorities, for the most part, do not demand from operators to use modern and customer-oriented vehicles such as low-floor buses, buses with modern emission standards, buses equipped with air conditioning etc. The authorities prefer using “flexible” grading systems without strict requirements. The number of criteria prevails over mandatory requirements in many cities (Fig. 3.13). For example, the city administration of Ufa determined 3 requirements (quantity of buses, their capacity, and their maximum age) and 16 criteria. This policy results in assorted, poorly maintained, and poorly equipped buses on various city routes.

In most Russian cities, the winners of tenders are determined by a list of quality parameters that usually represent:

- Vehicle quality (age, equipment, emission standard, floor level);
- Route capacity (vehicle capacity and quantity);
- Operator’s assets, reputation, and other criteria.

The value of each criterion is determined by a number of points that may be earned for complying with certain requirements. For example, an operator will receive extra points under the criterion of vehicle age for employing brand-new buses. The grading system usually provides that maximum (i.e. 5) points are awarded if the vehicles employed by the operator were produced within 2 years before the start of the contract, 4 points for vehicles that are 2- to 4-years old, 3 points for 4- to 6-year old vehicles, and so on.

The calculation of the maximum number of points that a bidder may earn when bidding in various cities is presented in Table 3.4. The table shows that many cities have unique grading systems. Which one is better is a question for further discussion. However, from the passengers’ perspective, the optimal criteria are probably those that have a direct impact on the quality of services. This goal can be achieved by using criteria incentivizing operators to employ brand-new low-floor buses with modern equipment and ecofriendly engines. These criteria, however, do not have an equally high value in every city. They are intensively used in Perm (100 % of all points on tender) and Nizhny Novgorod (77 % of all points).

In other cases, on the contrary, the most valuable criteria refer to route capacity (Omsk, Volgograd) or the assets and reputation of an operator (Ufa, Novosibirsk, Kazan, Rostov-on-Don). For example, in many cities, operators receive points based on the number of years of experience on the market. This criterion neither improves the public transport service quality nor helps attracting new operators to

Table 3.4 Value of criteria for bids evaluation

	Tender price (%)	Vehicle age, equipment, emission standard and floor level (%)	Vehicle capacity and quantity (%)	Operators assets, reputation and other (%)
Moscow	100	0	0	0
Perm	0	100	0	0
Nizhny Novgorod	0	77	10	13
Voronezh	0	52	25	22
Chelyabinsk	0	52	13	35
Saint Petersburg	0	47	0	53
Ekaterinburg	0	46	54	0
Samara	0	40	25	35
Rostov-on-Don	0	35	0	65
Volgograd	0	27	52	21
Kazan	0	23	5	72
Novosibirsk	0	18	0	82
Omsk	0	16	73	11
Ufa	0	14	0	86
Krasnoyarsk	–	–	–	–

Source Data from local authorities

the markets. It gives privileged position to in-house operators that usually have extensive assets and work on their respective markets for decades.

Krasnoyarsk presents a special case. A local regulatory act (Krasnoyarsk Administration 2006) only establishes a list of criteria that a commission of local authorities must use for bid evaluation, but there is no grading system with a value of each criterion. This system is the most unreliable of all the discussed above.

Compared to other Russian cities, the Moscow experience looks innovative. The government of Moscow has reformed the public transport services procurement system. Private companies are financed by lump-sums defined in contracts while farebox revenues are transferred to the Moscow municipal budget (Moscow Government 2014). Thus, the Moscow Government fully integrates private operators into the public transport system. The quality of services provided by private operators became similar to that provided by the Moscow in-house operator MOSGORTRANS. Private operators use brand-new buses and minibuses painted in blue, based on the color scheme of the Moscow land passenger transport (Fig. 3.14). They accept smartcards and provide services for concessionary passengers. During the execution of the contracts, operators are also incentivized to follow timetables or else face penalties from the Department of Transportation.

The active stage of the reform took place in 2015 and 2016. In 2015, the Moscow government tendered 58 contracts to serve 211 routes. Buses and

Fig. 3.14 Minibuses of private operators working under "new model" in Moscow (*Source* Authors)

minibuses started to provide services in spring and summer of 2016. The remaining routes served by private companies according to the old rules will be gradually cancelled or included in the "new model" tendering list.

Private sector plays a significant role in Russian transportation systems. In many cities, private operators have surpassed the traditional, publicly owned companies by a number of quantitative indicators (number of routes, routes capacity, and coverage of urbanized area). At the same time, the emergence of the private sector posed new challenges to the authorities. After 25 years of Russian market reforms, these challenges remain unsolved. In most cities, private operators still provide jitney-type services. The only exception is Moscow with its successful reform. Hopefully, the Moscow experience will serve as the foundation for institutional reforms of public transportation in other Russian cities in the foreseeable future.

3.4 Conclusion

In recent decades, following the transformation of the economic and political system in Russia, its urban passenger transport has undergone significant changes and lost its status of a monopolist on the market of transportation services. In smaller towns, it has completely lost any significant role in urban transport systems. The main reason is the acquisition of larger civil freedoms in Russia, particularly the freedom of movement. This factor, stimulated by the availability of private cars, has shown itself in the growth of Russian motorization rate up to a level comparable with that of the developed countries. These fundamental changes have had some painful consequences for urban transport systems. The transition of a considerable part of the population to private vehicles due to the overcrowding of the public transport led to excessive decline in demand for the latter and the emergence of opposite extremes. The systems of land electric transport (trams and trolleybus) have been affected to the worst degree, being unable to respond flexibly to changes in the transportation demand. The problems of the land electric transport were exacerbated by inconsistent municipal and state policies: electric transport has been overloaded by social commitments and losing ground to the new bus operators breaking onto the market. As a result, many land electric transport routes and networks were closed and the overall quality of service declined as a result of aging infrastructure and rolling stock. In addition, the increase in individual mobility has had a very negative impact on traffic conditions for both private and public transport in large cities. This created a lot of new challenges for urban authorities. Public transport still cannot compete with private cars. Public transport is often not provided with priority lanes and separated tracks on key routes, and its level of comfort is absolutely incomparable to that of private cars. Eliminating these defects is impossible without addressing the planning of routes and infrastructure, issues of sustainable funding and regulation, questions about the role of the government, municipal and private operators. City authorities have not yet agreed on the role that the Russian public transport in Russia is supposed to play in the changing circumstances. Most likely, this role will be determined in terms of providing safe and comfortable service on the busiest routes, mainly the ones linking high-density residential areas with city centers. In order to achieve this goal, public transportation should undergo further transformation, not because of the pressure from external circumstances, but in accordance with purposeful transport policy of urban authorities.

References

Autostat (2016) Data on Cars, Bus and motorcycles for years 2008–2016. http://eng.autostat.ru/catalog_category/59/. Accessed 31 Aug 2016

Blinkin M, Gurevich G, Sarychev A (1988) Avtomatizirovannye sistemy transportnogo planirovaniya (Automated transport planning system). Itogi nauki i texniki (The results of science and technology), v. 13 Avtomobilnyj i gorodskoj transport (Road and urban transport)

Chudakov E (1928) Budushchee avtomobil'noy promyshlennosti v SSSR (The future of the automotive industry of the URSS). Za rulyom, Moscow

ERRAC, the European Rail Research Advisory Council (2009) Metro, light rail and tram systems in Europe. http://www.uitp.org/sites/default/files/cck-focus-papers-files/errac_metrolr_tramsystemsineurope.pdf. Accessed 31 Aug 2016

Fursov A (2015) Privlekatel'nost' stroitel'stva legkorel'sovogo transporta v krupnyx gorodax Rossii (The attractiveness of light rail construction in major cities of Russia). Moscow

Gusev A (2014) Nebol'shoe gorodskoe issledovanie transporta vo Vladivostoke (Small urban transport survey of Vladivostok), Vladivostok

Gwilliam KM (2000) Private participation in public transport in the FSU. In: TWU Series. The World Bank. http://siteresources.worldbank.org/INTURBANTRANSPORT/Resources/twu_40.pdf. Accessed 31 Aug 2016

Gwilliam KM, van de Velde DM (1990) The potential for regulatory change in european bus markets. J Trans Econ Policy 24(3):333–350

Higher School of Economics (2014) Transport data. Moscow. http://sophist.hse.ru/rstat/. Accessed 31 Aug 2016

Inno-V et al (2008) Contracting in urban public transport. In: Clean transport & urban transport. European Commission. http://ec.europa.eu/transport/themes/urban/studies/doc/2008_contracting_ urban_ public_transport.pdf. Accessed 31 Aug 2016

International association “Metro” (2014) Tehniko-ekspluatacionnye pokazateli metropolitenov (Technical and operational parameters of subways). http://www.asmetro.ru/metro/techno_ekonom/ (in Russian). Accessed 31 Aug 2016

Jones D (1981) Urban transit policy: an Economic and political history Prentice Hall, New Jersey

Krasnoyarsk Administration (2006) Resolution No. 828 of October 6, 2006 Ob utverzhdenii polozheniya o poryadke i usloviyakh konkursnogo otbora perevozchikov na osuschestvleniye perevozok passazhirov po gorodskim marshrutam regulyarnikh perevozok (On approval of regulation on procedure of competitive selection of public transportation operators). In: Krasnoyarsk Administration website. http://www.admkrsk.ru/citytoday/transport/Documents/%D0%9F%D0%9E%D0%A1%D0%A2%D0%90%D0%9D%D0%9E%D0%92%D0%9B%D0%95%D0%9D%D0%98%D0%95%20%D0%BE%D1%82%206%20.10.06%D0%B3%20%E2%84%96%20828.doc. Accessed 31 Aug 2016

Moscow Government (2014) Resolution No. 643 of October 31, 2014 Ob organizacii transportnogo obsluzhivaniya naseleniya v gorode Moskve (On Organization of Public Transportation Services in Moscow). In ConsultantPlus website. http://www.consultant.ru/cons/cgi/online.cgi?req=doc&base=MLAW&n=168652&from=158175-0&rnd=214990.5746480769956883&. Accessed 31 Aug 2016

Polyakov A (2016) Interview, MIIT transport conference (Interviewer Zyuzin P.). Moscow

Rodionov AY (2000) Pravovie osnovy organizacii transportnogo obsluzhivaniya gorodskogo naseleniya (The legal foundations of organization of public transportation services for citizens). The Institute for Urban Economics, Moscow

Rodionov AY (2005) Methodical recommendations on the organization of the transport service of the population of municipalities. Fund “Regional studies”, Moscow

Russian Federation Federal State Statistics Service (2016) http://www.gks.ru/ (in Russian). Accessed 1 Aug 2016

Russian Federation Supreme Council (1991) Resolution No. 3020-1 of December, 27 1991 O razgranichenii gosudarstvennoy sobstvennosti Rossiyskoy Federacii na federalnuyu sobstvennost, gosudarstvennuyu sobstvennost respublik v sostave Rossiyskoy Federacii, krayov, oblastey, avtonomnoy oblasty, avtonomnykh okrugov, gorodov Moskvy i Sankt-Peterbutga i municipalnuyu sobstvennost (On Separation of Public Property of Russian Federation on Federal, State and Municipal Property). In ConsultantPlus website. Available via DIALOG. http://www.consultant.ru/cons/cgi/online.cgi?req=doc&base=LAW&n=48179&fld=134&dst=1000000001,0&rnd=0.20456683737468007. Accessed 31 Aug 2016

Shulyaev V (2014) A comprehensive plan for the development of passenger transport in Kaliningrad until 2020. Research Institute of Transport and Road Infrastructure (NIPITRTI)

Tarkhov S (2002) Evolyucionnaya morfologiya transoprtnyh setey (Evolutionary morphology of transport networks). Universum, Smolensk-Moscow

Tarkhov S, Derkach A (1999) Gorodskoy passazhirskiy transport v Rossii i SNG (Urban electric transport in Russia and CIS). In: Basin E (ed) Russian Architecture and Construction encyclopedia. Bumazhnaya galereya, vol 6, pp 130–136, Moscow

Voronezh City Duma (2012) Long-term municipal program No. 687 of August, 10 2012. Development of urban passenger transport Voronezh City District for the period 2010–2014. http://www.regionz.ru/index.php?ds=463043. Accessed 31 Aug 2016

Voronezh City Duma. (2008) Resolution No. 438-II of December, 24 2008. Development of the Municipal District of urban passenger transport Voronezh for 2009–2013. http://www.gorduma-voronezh.ru/cgi-bin/press.pl/doc/n081224-00438-II. Accessed 31 Aug 2016

Vuchic V (2011) Transport v gorodax, udobnyx dlya zhizni (Transportation for Livable Cities). Territoriya budushchego, Moscow

World Bank (2002) Cities on the move: a World Bank urban transport strategy review. Washington, DC

Yakushkin I (1982) Passazhirskie perevozki metropolitenom (Passenger traffic on the underground). Transport, Moscow

Zyuzin P (2012) Prostranstvennaya transformaciya transportnyx setej postsocialisticheskix stran CVE i byvshego SSSR (The spatial transformation of urban transport networks in the Post-socialist countries of Central and Eastern Europe and former SU). Dissertation, Moscow state university

Zyuzin P (2016) Gorodskoj passazhirskij transport v Rossii v kontekste sovremennyx transportnyx paradigm (Public Transport Network in Russian cities in the context of modern transport paradigms). Paper presented at the international conference "Ethics, transport and sustainable development: the social role of transport science and the responsibility of scientists", Moscow State university of railways engineering (MIIT), Moscow, 2–3 Mar 2016

Zyuzin P, Shpakov I (2016) Sostoyanie otrasli gorodskogo e'lektrotransporta v RSFSR v konce 1980-x gg.: problemy i perspektivy (State of the urban electric transport industry in the USSR in the late 1980-ies: problems and prospects). In: Koroteeva N (ed) Kraevedcheskie zapiski, vol 2. Kursk, Investsfera, pp 61–68

Zyuzin P, Trofimenko K, Shafranov I (2016) Potencial pereraspredeleniya passazhiropotokov mezhdu peresadochnymi uzlami prigorodno-gorodskogo zheleznodorozhnogo soobshheniya v Moskve (Potential for the redistribution of passenger flows between urban and regional railway hubs in Moscow). Transport Information Bulletin: 71–78

Chapter 4
Integration of a Commuter Rail in the Transport System of the City

Larisa Morozova, Larisa Stepanenkova and Anatoliy Malashkin

Abstract In this chapter we will examine the prospects of transforming the railway transport of Moscow from the historically established commuter format into a more modern, suburban-urban format, focused on the transportation of passengers not only between Moscow and suburbs, but also between various city zones. We will also assess the possibility of creating new rail transport lines that could eliminate part of the peak load from the most packed elements of the city transport system.

4.1 Planning Features of Moscow and the Demand Specifics

Moscow for its transport and urban planning conditions can be attributed to the Asian city type, which is characterized by extremely high population density, and where the share of built up area in the streets and roads is not more than 10–12 %.

At the same time, Moscow has a number of important features of urban planning and settlement distinguishing it from other cities of the Asian type.

(a) An average value of the parameter ε for the city is about 9 %, the value of the parameter ε for the city central part is ≈28 %, which is comparable to European cities, while on the periphery of Moscow, between The Third Ring Road and the Moscow Automobile Ring Road—only 6 % (Fig. 4.1). Moreover, the density of the road network in the built up area beyond the Moscow Automobile Ring Road is even lower.

L. Morozova (✉)
Institute for Transport Economics and Transport Policy Studies,
National Research University Higher School of Economics, Moscow, Russia
e-mail: lmorozova@hse.ru

L. Stepanenkova · A. Malashkin
HSE Moscow Institute of Electronics and Mathematics (MIEM HSE),
National Research University Higher School of Economics, Moscow, Russia

M. Blinkin and E. Koncheva (eds.), *Transport Systems of Russian Cities*, Transportation Research, Economics and Policy,
DOI 10.1007/978-3-319-47800-5_4

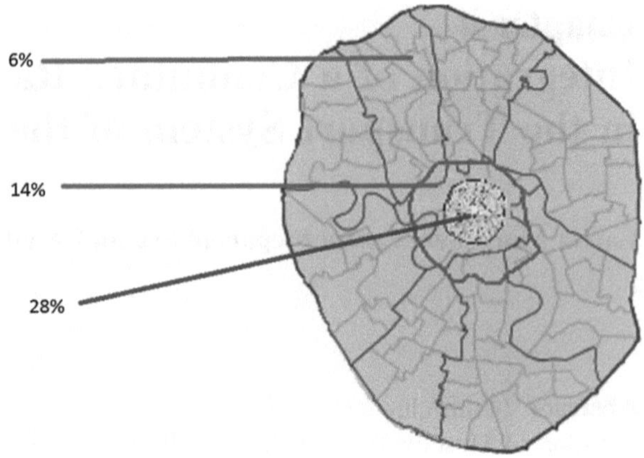

Fig. 4.1 The area of the road network in the city center, in the middle and on the periphery of Moscow [*Source* Created by authors using the data from Grigoryan et al. (2011)]

(b) A graph of Moscow's population density depending on the distance from the city center is fundamentally different from the same graph for Western cities. This phenomenon is explained by the fact that in 1930–1980 the processes of housing construction in Moscow, unlike these same processes in the western world, were not at all shaped by the forces of the market. As a result a significant part of Moscow's population has been concentrated in the outlying districts with an extremely high density.

From 1990–2010, Moscow's population increased by 28 % (Russian Federation Federal State Statistics Service 2016). The most rapid growth was in the outlying areas of the city and the inner suburbs.

Since the mid-1990s, nearby cities, and later remote areas of Moscow region began to change into commuter towns as a result of active construction of high-density multi-story buildings (20 or more floors). In the mid-2000s, Moscow region overcame Moscow to become the region with the most housing construction in all of Russia, a title it has since held for over 10 years. The core area of mass urbanization in Moscow's metropolitan area does not coincide with the administrative borders of the city, since it is so massive it includes even nearby satellite towns and new housing estates outside of Moscow.

(c) These very high rates of population density in the outlying areas are accompanied by an extremely high concentration of workplaces in the central part of the city: 40–55 % of all working positions in the capital are currently focused within a radius of 4.5 km from the Kremlin.

As a result of the imbalance between the locations of workplaces and housing areas, there is a daily massive conglomeration of traffic in the direction "periphery—center" (in the morning) and "center—periphery" (in the evening). A similar structure of daily work-related trips can be observed in every other major city around the world. However, even in this aspect Moscow is fundamentally different from similar cities abroad.

Instead of low-density family-oriented housing (single family, semi-detached, town house) typical of foreign suburbs, multi-story buildings, which are undeniably aspects of the format of transit oriented development, dominate the periphery of Moscow and all its suburban areas (see Chap. 2). At the same time, the people living in these buildings are not just car owners, but typical representatives of "car owners of the first generation."

We should keep in mind that the level of motorization from 1990 to 2010 became four times higher in Moscow and five times higher in Moscow region (Russian Federation Federal State Statistics Service 2016). In 2007–2008 Moscow region exceeded Moscow in this parameter; now the level of motorization in the region is even higher than in the capital. On the basis of emerging trends a level of 400 cars per 1000 residents in Moscow region can be reached in the upcoming 2–3 years. However, the level of motorization of residents of new built up areas is considerably higher than the average figures for the region. For example, there are more than 800 cars per 1000 residents in Odintsovo (as of 2015).

The explosive growth of motorization in Russia, particularly in the capital region, took place within the lifetime of a single generation. Accordingly, for the majority of current car owners it is not just a useful vehicle, but an object determining their lifestyle and social status. This psychological setting is particularly characteristic of the inhabitants of the new residential areas of Moscow and Moscow's suburbs, who are primarily middle class. Using public transport on an everyday basis is a very difficult choice for such people, especially from a psychological point of view, as it is equivalent to (as Moscow's youth likes to say), "going back to the USSR".

In foreign suburbs (especially in North America), there is always a developed road network, which includes a main frame composed of roads of higher technical categories (freeway, expressway).

In contrast, the density of the road network on the periphery of Moscow and all its suburban areas is very low. As for the main roads that join the new built up areas outside the Moscow Automobile Ring Road, their levels of transit traffic load is so high that there is no reserve capacity to maintain daily car trips "periphery—center".

Based on all this, here are some distinguishing features of the Moscow periphery and its suburban areas:

- Firstly, there is a supply shortage of transport resources due to an underdeveloped road network.
- Secondly, there is an excess demand for daily trips, generated by dense multi-story residential construction and a lack of jobs in the vicinity of these multi-story buildings. Therefore, there is also an extremely high daily commuting migration.
- Thirdly, inhabitants of the new residential development areas have a unique social and psychological mindset.

In official Russian transport strategy there is a distinct preference for "comprehensive balanced development of urban agglomerations with a priority of public transport infrastructure" (Ministry of Transport of the Russian Federation 2016). Unfortunately, this recommendation is usually just "an optimistic hope."

4.2 Moscow Railway System

Moscow railway system serves a region with a population of over 19 million people, which is 13 % of the Russian population. Tables 4.1 and 4.2 present a quantitative comparison of Moscow metropolitan area with the largest European agglomerations.

There is a well-developed urban transportation system in the metropolitan areas of Paris and Berlin. A similar system is hardly even outlined in Moscow metropolitan area.

Moscow railway system accounts for more than 50 % of commuter transportation in the country and a quarter of long-distance passenger transportation. The

Table 4.1 Moscow agglomeration in comparison to Paris and Berlin (population)

City	Population (millions)	Suburbia	Population (millions)	Population of the whole region (millions)
Moscow	12.2	Moscow region	7.2	19.4
Berlin	3.4	Brandenburg	2.5	14.1
Paris	2.2	Ile-de-France	11.9	5.9

Source Created by authors

Table 4.2 Moscow agglomeration in a comparison to Paris and Berlin (rail traffic)

A type of a rail transport	Moscow and Moscow region	Berlin and Brandenburg	Paris and Ile-de-France
High-speed interregional train	Sapsan	Intercity-Express Interregio-Express (IRE)	TGV/Ouigo
Traditional interregional train	Long-distance trains	Intercity Nachtzüge (City Night Line)	Intercity
High-speed regional	7000 Series trains, Aeroexpress	Regional-Express	Transilien
Traditional regional train	6000 Series trains	Regionalbahn	
Urban	6000 and 7000 Series trains (partly)	S-Bahn	RER

Source Created by authors

main carriers in Moscow's transport hub are the Central Suburban Passenger Company (TSPPK) and Moscow-Tver Passenger Company. The "Central Suburban Passenger Company" is responsible for 91.4 % of the suburban passenger traffic in Moscow's transport hub (Fig. 4.2).

The length of railways within the boundaries of "old" Moscow exceeds 250 km. The city has 9 railway terminals that receive electric trains from 11 directions. 5 of the 11 rail lines serving the city have dead-end railway terminals in the city center: Yaroslavskoe, Kazanskoe, Ryazanskoe, Paveletskoye and Kievskoe.

Railway communication has a significant underused potential: the railroad tracks are completely independent, which means they are exempt from the traffic condition in the city, the rolling stock has a maximum passenger capacity, and the platforms can accept 12-car trains (in contrast to 8-car subway trains). Moreover, most of the destinations have space for additional railways, which would increase the capacity.

Still, the share of rail transport is not more than 8–10 % of the total urban passenger transport (problems of urban heavy rail in other cities in Russia are discussed in Chap. 3). The most characteristic demand groups for commuter rail transportation are:

- Moscow region residents' commuting labor trips to Moscow; up to 30 % of the passengers arriving in Moscow by trains change from the railway station to the first Metro station;
- Moscow residents' commuting labor trips, departure or destination points of which are in the peripheral areas of the city that do not have an easy access to metro stations;
- Moscow residents' trips, people who do not have their own car, from home to a summer house.

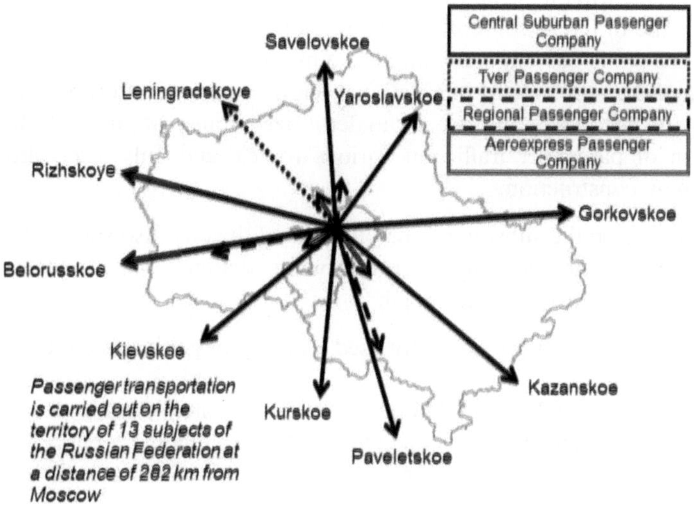

Fig. 4.2 Regional railway traffic of Moscow region (*Source* Created by authors)

A combined route selecting in the first of these segments is determined by the following circumstances:

I. Changing to the subway leads to a noticeable reduction in total trip time; a particular difference that is especially noticeable to passengers is connected with the time it takes for trains to arrive at terminal railway stations. The significance of this difference is determined by the technology employed by railway stations.
II. The Moscow subway has a single fare that does not depend on the distance of travel, commuter trains, on the other hand, charge a zone rate. Under these conditions a combined trip from Moscow region to the first interchange station by railway train and then by metro is significantly cheaper than a direct trip by train to the railway station. Many experts note that this flawed tariff policy stimulates early transfer from passenger commuter trains to the subway.

Currently, there is a single fare zone for intercity commuter rail transportation: 32 rubles per trip regardless of distance within the area shown below (Fig. 4.3). Different zones have different fares, but at the same time the fare is independent from the distance within that same zone. The passenger pays 20.5 rubles for each intersection between zones; therefore, the more zones are crossed, the more expensive the trip is. This means that for a passenger travelling from outside Moscow, there is a serious incentive to exit at the first railway station, which has a connection to a subway station. The distance between the zones is 10 km, except for the first zone (about 5 km) and zone zero, which is where the railway terminals are located. Zone zero is another reason for the declining popularity of commuter train routes to terminals.

Let us compare the two transportation modes: the trip by commuter rail from Reutov to the Kursky railway terminal takes 20 min and a single ticket costs 61 rubles, the same trip on the subway (Novokosino-Kurskaya) takes 24 min and costs 50 rubles. The difference in cost is also one of the reasons why subway trains are overcrowded while passenger commuter trains are half-empty in the same direction. This policy is the opposite of international ones: countries in Europe and elsewhere are trying to encourage the use of underutilized transport methods for uniform distribution of passenger traffic on various routes, and only afterwards do they invest in new construction.

III. Changing to the subway on the outskirts of the city also saves time: outlying stations are not deep belowground while stations located close to the city center are at a far greater depth.

Thus, rail transportation is only really used as a way of getting to a more convenient type of transport.

In April 2016 the Russian Public Opinion Research Center (Russian Public Opinion Research Center 2016) conducted a study (including a public opinion poll found below) on the quality of services provided by suburban rail transport and criteria of its rating (Tables 4.3 and 4.4).

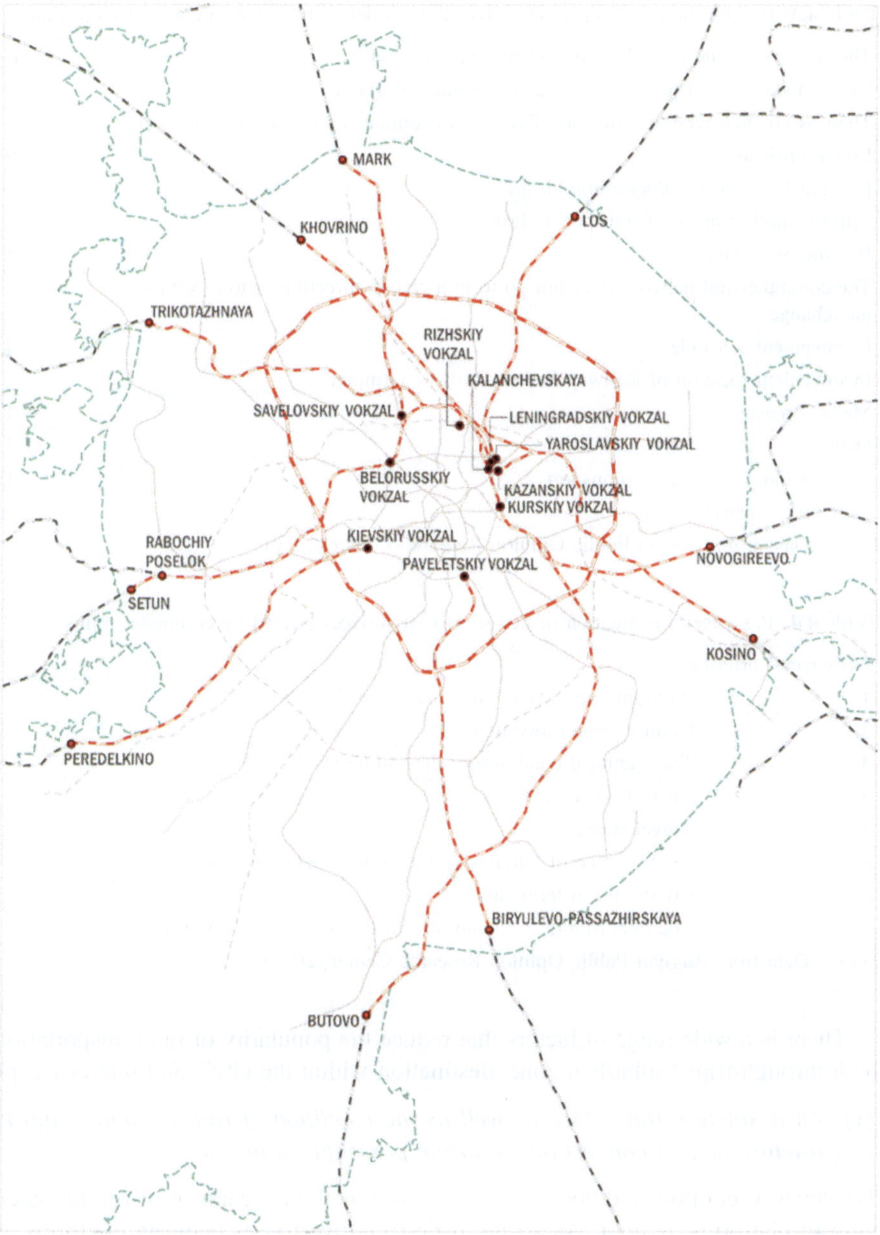

Fig. 4.3 Moscow's Single Fare Zone

Table 4.3 Poll on the reasons people prefer other modes of transport to commuter rail transit

The reasons for the refusal of the use of commuter rail	%
Other modes of transport are more convenient/comfortable	31
There is no such type of transport (There is no commuter rail transit in the area)	30
I own a private car	9
I do not have the need/possibility/desire	7
Trip by another mode of transport is faster	6
It is too expensive	5
The commuter rail network does not go to in a certain direction or to a certain interchange	5
Inconvenient schedule	4
Inconvenient location of Railway Station/Railway terminal	2
Medical reasons	1
Other	1
I do not like this mode of transport	1
Difficult to answer	3

Source Data from Russian Public Opinion Research Center (2016)

Table 4.4 Passengers' evaluation of the priority of services offered by commuter trains

Passengers' priorities	
1.	Providing security (at stations and inside trains)
2.	Comfort when traveling
3.	The technical condition of the rolling stock
4.	Cost of services
5.	Travel speed
6.	Convenience of purchasing travel documents (tickets)
7.	Conditions at terminals
8.	The opportunity to obtain information on navigation in trains

Source Data from Russian Public Opinion Research Center (2016)

There is a wide range of factors that reduce the popularity of rail transportation, both through-trips "suburban zone–destination within the city", and intercity trips.

(1) *An obsolete rolling stock as well as the condition of railway station infrastructure and its connections to urban passenger transport.*

The level of comfort and convenience, as well as the appearance of rolling stock, railway platforms, is poor. As a rule, pedestrian overpasses between platforms, as well as transitions from train stations to the Metro stations and road public transport stops are in even worse condition. In most cases, almost nothing has changed since the Soviet era, which dramatically contrasts with the appearance of city streets, cars and subway stations, which have all been updated in recent years.

The reason for this is that the development of the Moscow Metro and municipal road public transport, as well as street improvement (including the areas surrounding Metro stations, road public transport stops, etc.) are financed by the city of Moscow, which has had the necessary resources for relevant programs in recent years. At the same time, railway platforms and adjoining facilities are Federal property, and are therefore updated and maintained by state-owned company Russian Railways, which does not consider these costs as a priority.

According to many experts amendments to the existing legal framework are necessary in the field of contractual relations between the owner of the infrastructure and the carrier. This applies to the maintenance and operation of the individual elements of the complex infrastructure. In some cases the carrier should take the responsibilities of infrastructure owner. The carrier must be able to implement projects for the reconstruction of passenger platforms, railway stations, in particular, by lengthening platforms, if the planned passenger flows require 14-car train, and most of the platforms are able to accept only 12-car trains. And these costs should be included in the cost of services provided by the carrier in accordance with the contract.

This is especially important now, when the concept of "Responsible Carrier" is implemented. The government of Moscow and the Moscow Region have entered into long-term contract with company-carrier for a period of 15 years. The carrier will take over the purchase of new rolling stock, renewal and modernization of passenger infrastructure at the stations and stopping points, improvement and refurbishment of suburban platforms: lungs construction, opening of pavilions and canopies, provision of benches, urns and the establishment of new fences). Prior to that, only functions of the complex operation of passenger arrangement, cleaning, maintenance and repairing of electric trains were transferred to the carrier.

But in general, controversial issues between the subject of the Russian Federation, company-carrier and Russian Railways arise mainly due to the fact that mandatory minimum standards of passenger service are not specified at the federal level. These standards should include requirements for the maintenance of infrastructure used in commuter rail services, the technical and commercial requirements to passenger rolling stock: categorization of technical and commercial characteristics of the rolling stock, including the requirements on equipment; regulation of the maximum occupancy and minimum provision of rolling stock; frequency of repair; the minimum guaranteed standards of service to passengers in trains.

(2) *Dead-endtracing*

Railroads are not deeply enough embedded in the center of the city. Examples of the "pass-through" tracing rail lines through the city are rare. Reasonable time and comfort transfer between directions are rare as well.

This problem is of fundamental nature: in the General Plan of Moscow of 1971 the question was raised about the conversion of radial lines into diametrical through the construction of the tunnels, and the use of railway infrastructure of industrial use. Unfortunately, this decision was not made in the next decades.

(3) *Trains' schedule poorly integrated into the standards of urban life*

For a city dweller, accustomed to underground schedule with a frequency of 40 pairs of trains per hour, the intervals between the trains seem unacceptably large. Moreover, on the railway lines, there are breaks for maintenance with a length of up to 3 hours. These circumstances, of course, do not stimulate the popularity of rail transportation in the city.

(4) *Defects of the tariff-ticketing menu*

In recent years an urban passenger gets used to a flexible multi-modal tariff and fare system covering Metro and all kinds of road public transport.

Tariff and ticket fares for the suburban Railways also became very flexible and well adapted to the different segments of passenger demand. The problem, however, is that these two options are not integrated. This problem deals primarily with issues of ownership and intergovernmental relations: Metro and the largest road public transport company are the property of the city, private bus carriers are connected with the city by contracts, but rail transport is operated by companies with mixed capital, affiliated with the state company "Russian Railways".

(5) *A low level of security*

The level of personal safety of passengers when traveling by suburban trains surrender to the standards familiar to Moscow Metro users. Security threat (or, at least, travel discomfort) is created, as a rule, by anti-social elements and aggressively tuned free-riders. They can easily get on a train because not all the stations are equipped with turnstiles. The frequency of conflict situations increased greatly as the number of police officers patrolling the trains was reduced, which took place in recent years. It was planned to pass ensuring the safety of trains on to private security companies working under contract with the carrier. Unfortunately, the change was not equivalent: private security guard does not have the authority to disembark an irritating passenger from a train.

As a final conclusion we can say the following: on the basis of the experience of foreign countries the following items will contribute to the growth of loyalty degree, and demand for passenger rail transport:

- improving the quality of service at a fixed price level
- simple and clear tariff menu for a passenger
- flexible system of discounts
- attracting passengers from the group of car drivers by an option "parking + travel"
- a "single" ticket for several types of transport (single tariff system of urban and suburban railway transport taking into account technological integration)
- interval reduction of a train departure
- linking the commuter rail lines and other types of high-speed transport and off-road transport routes, and creating the modern high-tech transport hubs
- increasing of the trains, speed in the city.

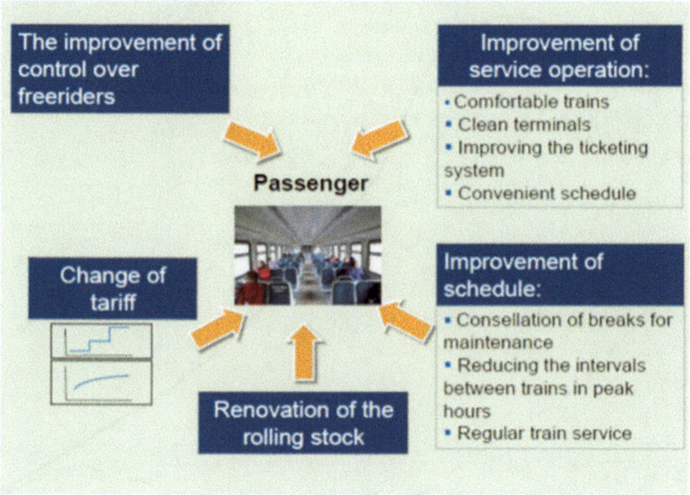

Fig. 4.4 Measures on increase of passengers' loyalty of railway transport

These necessary measures on increasing passengers loyalty of railway transport are schematically shown in Fig. 4.4.

4.3 The Capacity and Budget Priorities

It is advisable to analyze the ability of the alternative types of transport to carry out the task of traffic maintenance on the routes "periphery—the center" and "the center—periphery". We will analyze it by the typical example of the Moscow transport corridor from the Metro station "Vykhino" and a train platform "Vykhino" to the city center.[1]

Traffic in the direction "periphery—the center" is served here by the following highways: Tagansko-Krasnopresnenskaya Metroline, the direction of the Moscow Kazan Railway and Ryazan Avenue, which is a multi-band urban street with traffic light control (Fig. 4.5 and Table 4.5).

The calculation of carrying capacity of the road transportation is based on three lanes, including reversible lane that are available at Ryazan Avenue. The capacity of one lane regulated by traffic signalization is accepted at a rate of 900 vehicles per hour; a car filling ratio, based on the common values of Moscow, is 1.3 passengers. The calculation does not include transportation by ground urban passenger transport; this assumption is made on the grounds that a passenger is not inclined to use

[1]The area adjacent to the corridor corresponds to the area within the boundaries of the Garden Ring.

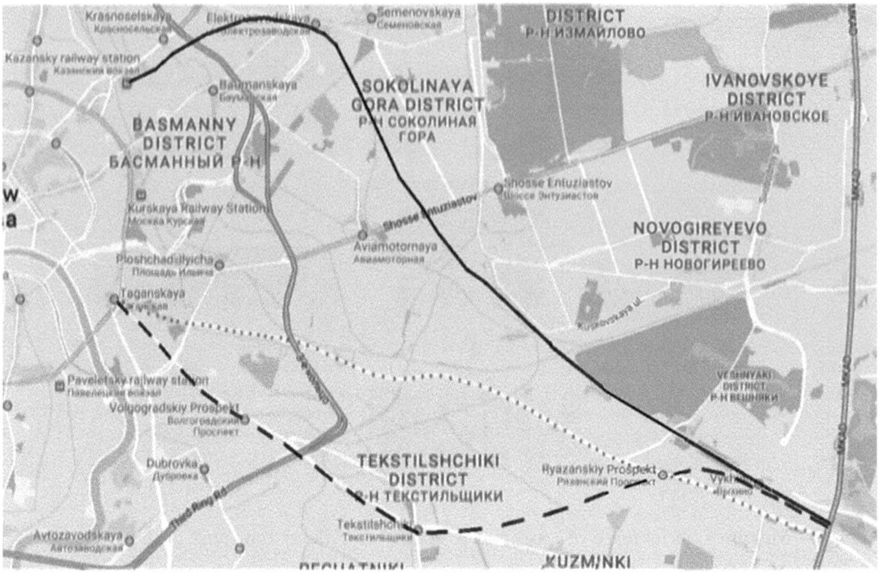

Fig. 4.5 Transportation highways serviced the flows "periphery—the center" in the south-east (*dotted line* Ryazan Avenue, *solid line* Kazan direction of The Moscow Railroad, *dashed line* Tagansko-Krasnopresnenskaya Metroline) (*Source* Created by author)

Table 4.5 Features of the transport highways servicing the flows "periphery—the center" in the south-east (the status quo)

	Time to "center" in the morning rush hour on the Vykhino station	The carrying capacity (thousands of people/hour)	The share of the total carrying capacity of destination[a] (%)
Ryazan Avenue, Passenger cars	18–22 min[b]	3.5	3.9
Kazan direction of The Moscow Railroad	17/25 min[c]	33.7	37.3
Tagansko-Krasnopresnenskaya Metro line	25 min	53.1	58.8
The total carrying capacity (thousands of people/hour)	90.3	100 %	

[a]The sum of all types of transport
[b]Depending on LOS
[c]Electric train with accelerated motion (in the accepted terminology "Sputnik")/normal train

public road passenger transport if there is off-road passenger transport, anyway, public road passenger transport without a "bus lane" and BRT format for a long-distance travelling.

The calculation of capacity of the suburban railway comes from the carriage filling of 216 people (the number of seated passengers, plus a number of standing passengers from the calculation "Area Assumed for One Standing Passenger" = 0.2 m^2/standing passenger, from 12-car trains, and traffic interval of 4.5 min.

The calculation of capacity of Metro line comes out of the car filling of 158 people (the number of seated passengers, plus a number of standing passengers at the rate 0.2 m^2/standing passenger, from 8-car trains, and 42 departures per hour.

Thus, the distribution of a total capacity of the transport corridor in the morning rush hour is that the share of Metro line is 57.6 % (more than 50 thousand passengers), the share of railway is 38.4 % (about 34 thousand passengers) and only 4 % (3.5 thousand people) is the share of car trips on Ryazan Avenue.

In the period of 2013–2017 in the reconstruction of Ryazan Avenue and its transport interchange with the Moscow Automobile Ring Road it should be invested,[2] respectively, €135[3] ml (2013–2015) and €85 ml (2013–2017). Through these investments, the capacity of Ryazan Avenue in one direction will increase up to 4500 cars (Moscow Complex of Urban Planning Policy and Construction 2014).

In the first scenario the entire resource of a reconstructed highway is used by cars, road public transport works in the general flow and, accordingly, is not used for long-distance trips. In the second scenario specified in the city program Moscow Complex of Urban Planning Policy and Construction 2016), one of the lanes is used for the organization of the BRT. From the point of view of the total transport volumes there are two possible scenarios (Table 4.6).

Thus, the reconstruction of Ryazan Avenue will increase the total transport capacity of the corridor by 2.6 % under the scenario #1, or by 4.8 % under the scenario #2. Even the best of scenarios will increase the carrying capacity of 4370 passengers per hour, which is equivalent to the capacity of two commuter trains.

It should be noted that in those same years, the Government of Moscow has invested heavily in the extension of the Metro line mentioned above. The question, however, that the line extension, which has significantly improved the Metro accessibility for the residents of the far periphery of the city, could not increase the line's carrying capacity which is already exhausted by now.

Projects aimed to extend the existing subway lines have significant negative impact on the functioning of the transport system of the Moscow agglomeration. Adding new passenger flows to underground line which is already experiencing sizable congestion during peak hours will considerably worsen the quality of life for residents of the city intermediate zone.

As practice shows, the opening of new metro stations in the undeveloped areas will inevitably lead to a massive housing construction within a radius of connections to the new stations. At the same time supply of new passengers will be carried by private car and bus transport, which significantly worsens the situation on the road network. From the organizational and technical point of view, the longer the

[2]Programme of transport complex development of Moscow region for the period up to 2020 with the definition of the activities associated with the expansion of the territory of Moscow.

[3]The average exchange rate of €1 = 75 rubles.

Table 4.6 Features of the highways serving flows "periphery—the center" in the south-east direction (after the reconstruction of the road infrastructure)

	The carrying capacity (thousands of people/hour)		The share of the total carrying capacity of destination, %	
	Scenario #1: passenger cars	Scenario #2: passenger cars + BRT	Scenario #1: passenger cars (%)	Scenario #2: passenger cars + BRT (%)
Ryazan Avenue	5.85/2.34	7.88/4.37	6.3	8.3
Kazan direction of The Moscow Railroad	33.7	33.7	36.4	35.6
Tagansko-Krasnopresnenskaya Metro line	53.1	53.1	57.3	56.1
The total carrying capacity (thousands of people per hour)	92.63/2.34	94.66/4.37		

line is, the more difficult and expensive it is to organize its operation, maintenance and repair. The extension of lines beyond a certain limit may cause an increase in operating costs exponentially (up to the necessity of building new depots, etc.).

The case analyzed above illustrates well the contradictions between the objective transport needs of Moscow region and typical solutions in the field of transport policy. The only real way to serve the powerful (and growing) traffic flows of the directions "periphery—the center" and "the center—periphery" is a commuter rail line, primarily through the exploitation of its underused parts within the city borders.

Under these conditions, the city continues to "fight with traffic jams" by investing considerable financial and land resources in the reconstruction of its arterials (avenues or highways), which can provide an extremely modest increase of carrying capacity of transport corridor as a whole.

4.4 Rail and a Subway: Synergy or a Competition?

A typical speed for the road public transportation of rush hours in Moscow is about 12–18 km/h, the same figure for private cars does not exceed 20 km/h. Only HTR systems work within the speed range of about 40 km/h: a railway and a subway.

To the present time the subway capacity is depleted on the majority of the lines (Figs. 4.6 and 4.7). Peak loads of the critical hubs, exceeding their calculated capacity, are largely connected with a transfer of the passengers from the rail to the subway.

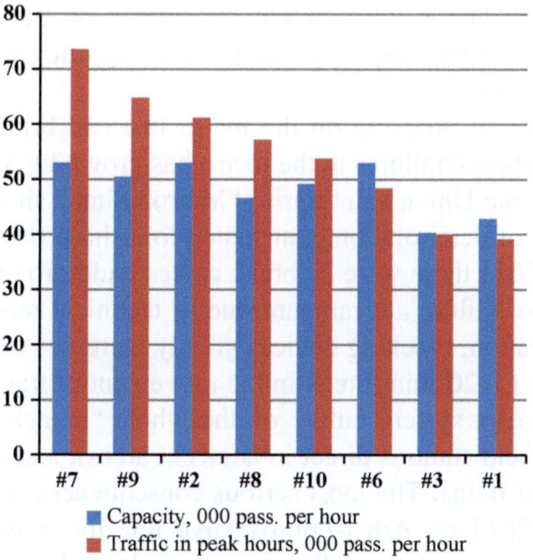

Fig. 4.6 Carrying capacity and the actual peak traffic on the metro lines which have interchange connections with suburban railways. (The carrying capacity is calculated on the basis of peak frequency of 8-car trains motion, and cars' filling at a rate of 0.2 m^2/standing passenger). *Note* #7 Tagansko-Krasnopresnenskaya, #9 Serpukhovsko-Timiryazevskaya, #2 Zamoskvoretskaya, #8 Kalininskaya, #10 Lyublinskaya, #6 Kaluzhsko-Rizhskaya, #3 Arbatsko-Pokrovskaya, #1 Sokolnicheskaya (*Source* Created by author)

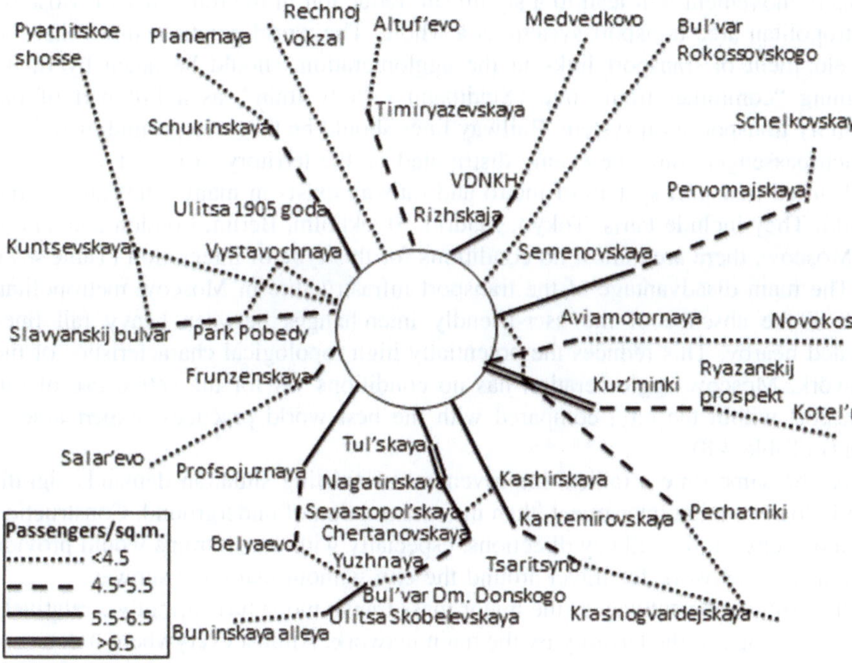

Fig. 4.7 Occupancy of subway lines (*Source* Created by author)

Excess occupancy above the nominal (calculated on the basis of the rate of 0.2 m^2/standing passenger) capacity leads to a significant reduction in travel comfort for all passengers.

A further increase in ridership on the metro line can bring to a very sad consequences. The number of failures in the metro has grown by 3.5 times according to statistics from the State Unitary Enterprise "Metropolitan" since 2004 for 10 years.

The high degree of deterioration contributed to a sharp rise in the power supply system failures: in 2011 there were 24 break cables underground, in 2012–40 cases. The consequences of failure are rampant due to technical reasons: malfunction of signalling, centralization, blocking devices greatly increases the range of motion on the line from 2.5 to 15–20 min (break in the movement of trains at the station from 5 to 10 min, involves a system failure on the whole "branch" (accumulation passengers at stations) and stations direct in adjacent areas), and in the worst case can lead to a collision of trains. The most serious consequences are caused by the poor state of tracks: in 2014 on Arbat-Pokrovskaya line the movement of trains was suspended for 3 days due to the collapse as a result of the splitting the switch.

The greatest number of failures of the rolling stock is marked on Tagansko-Krasnopresnenskaya (the most loaded line). Since 2011, it has doubled in the accident on the Circle Line. This figure is threatening, as experts predict that ring line will be overloaded twice in the next few years.

From the above we can conclude that a further increase in the number of passengers carried by the subway without working out alternative schemes of passengers movement will lead to a significant reduction in the reliability of Moscow metropolitan area transport system as a whole. The world practice shows that the development of transport links in the agglomeration, should be taken by transforming "commuter train" into "Stadtbahn"—"city train", as a full part of the intercity transportation system. Railway lines should be the "overground metro" by which passengers must be evenly distributed on the territory of the city.

Unified transport system of metro and railway exists in many cities around the world. They include Paris, Tokyo, Madrid, Stockholm, Berlin, London and so on. In Moscow, there are almost no conditions for this type of integration (Table 4.7).

The main disadvantage of the transport infrastructure in Moscow metropolitan area is the absence of the user-friendly interchanges between heavy rail lines located nearby. This reduces the potentially high topological characteristics of the network. Moscow agglomeration has no conditions yet for the active use of rail transport within the city, compared with the best world practices (expert assessment) (Table 4.8).

At the same time a radical improvement of existing situation demands significantly lower capital investment, than the construction of underground. Construction of hubs between the railway directions, especially with transit trains would provide a connected network for travel around the city without using the subway.

In addition, the subway as the major mass transit mode has currently a relatively small coverage of the territory by the main network. Almost everywhere it does not extend beyond the circle line, the area within the city which is about 3 % of Moscow area only within the Moscow Ring Road. The core network covers the

Table 4.7 The main approaches to the formation of an integrated transport rail system

The main approaches to the formation of an integrated transport rail system	World practice	Moscow
Unified contractor of transport work	+	±
Unified payment of transport work	±	−
Unified tariff system	±	−
Single requirements to comfort and security	+	−
Single requirements to frequency	+	−
Unified approach to operation and development	±	−
Unified approach to financing	±	−
Unified approach to navigation and information system	+	−

+ fully executed
± partially executed
− no executed

Table 4.8 Comparing Moscow with best world practices in the organization of intercity railway traffic

Index	Best world practices (%)	Moscow (%)
The share of t station-interchanges between subway and commuter train in the total amount of the city stations	15–20	8
The share of passengers who used both systems (subway and commuter train) to travel within the city, of the total amount	15–20	7
The share of passengers using heavy rail in the city of the total amount of city passengers	15–25	8
The share of commuter train passengers of total subway traffic within the city	40–60	20

peripheral area in the area of the connection Kaluzhsko-Rizhskaya and Serpukhovsko-Timiryazevskaya, Zamoskvoretskaya and Lublinskaya lines. It shows about the difficulties of peripheral connections by metro.

In terms of the reliability the full integration of all types of heavy rail passenger transport in Moscow increases the number of the network cycles outside the Ring subway lines and significantly increases the rate of network connectivity, which is particularly vitally important in the peripheral areas of the city.

To increase the attractiveness and availability of rail transport, expert community discussed projects of the construction and arrangement of railway platforms. The experts prepared proposals for the construction (transfer) of platforms for organizing the new interchanges between the railway and the subway, between railway lines, new hubs across multiple lines of railway and subway at the same time. In addition, proposals for the construction of the platforms in the center of the city are considered to increase the density of the overall network. In total there are about 25 platforms.

4.5 Moscow Central Ring

A unique event is the opening of passenger traffic on the Moscow Central Ring (earlier: Moscow Ring Railway) in September 2016.

Moscow Central Ring (hereinafter MTsK) was built at the beginning of the last century. The regular traffic on it was opened in July 20, 1908. In those days it was a big and ambitious project of the railway traffic organization around Moscow, which has integrated radial directions that existed at that time. Despite the fact that the MTsK at the time was held outside the city limits (the boundaries of Moscow on MTsK were recognized only on May 23, 1917), it was already envisaged that this line would not have one-level intersections with other railway lines and major highways. Totally 8 bridges across the river and 32 overpass across the road and rail line were built.

Initially, the passenger traffic on the Moscow Ring Railway existed. Some built stations are even preserved until today. But in the 1920s, the ring was set on uninhabited places, passenger traffic was small and in some point it was closed completely. At the same time, freight transport on the Moscow Ring Railway developed rapidly. At first it was just transit, then supply of goods to many industrial enterprises, which were built near the line, began to be carried out on the ring.

Due to the Moscow's expansion the ring itself is currently situated almost in the city center. Ideas to adapt it to the passenger traffic permanently existed, but during the Soviet era due to the extremely intensive freight traffic on the ring it was unable to organize also passenger transportation without substantial reorganization.

Moscow Railways, placed in the 19th century on the outskirts or outside the borders of the city, are now at the center and the intermediate zone (Fig. 4.8).

In the 1990s cargo movement began to decline due to the closure of businesses and the transfer of production to other regions. The idea of passenger traffic returned several times, and in 2011 at last the project of passenger traffic on the Moscow Ring Railway finally got its development. In the new project, currently being implemented, the absolute priority is given to passenger traffic. The length of the ring is 54 km, completely on the ring will be laid two tracks, to be used only for passenger traffic. Over 30 km in several areas there is also a third track, which will be used primarily for freight traffic.

Moscow authorities see the Moscow Central Ring as another, second circle metro line; its functions are seen as unloading itself the Circle Line and interchanges from the radial lines on the ring and unloading limiting (the busiest) spans of radial lines immediately before the change to the Ring subway line. Historically trail of the Moscow Central Ring slightly shifted to the north with respect to the Circle Line subway and in the future will naturally be complemented by the Third interchange circuit underground, the ring, which will be shifted to the south. Both lines will form a kind of "eight" with the intersection in two points—Horoshevo and Ryazanskaya.

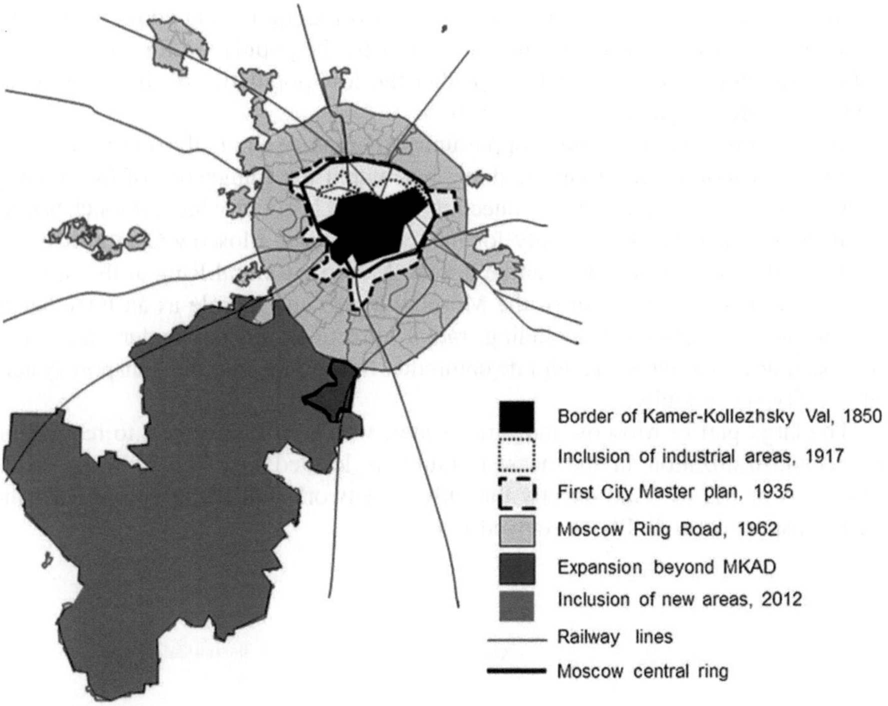

Fig. 4.8 Expansion of Moscow (*Source* Created by authors)

The stations (31) of the Moscow Central Ring will be built with interchanges (in various combinations) on the metro line, on the radial direction of the railway and land transport. 17 hubs will have the interchanges to metro, including three ones being under the construction stations (District, Shelepiha, Nizhegorodskaya). Interchange to commuter trains will be on nine of ten radial directions, while the station Testovskaya of Smolenskoe direction will have change from two stations: City and Shelepiha. Park and ride parking facilities will be on 13 stations. The challenge in this project for subway is to implement the full integration of Moscow Central Ring and the metro.

This means that it will work as a single ticketing system: change stations between MTsK and the Moscow Metro will be free. At Moscow Central Ring passengers will be able to make two free changes (metro-MTsK-Metro), but only between the nearest "interchange" stations. There will be also restrictions on travel times to transfer—90 min after the first entrance. According to calculations it should be enough for any route options.

Navigation in the subway and on the MTsK will not differ from each other, Moscow Central Ring will be present just like another subway line with the same style of navigation.

At the same time it is declared that the Moscow Central Ring could be used by commuters. But if you calculate the time spent by the population for changing one railway to another, it is possible to predict the low popularity of the ring for the majority of train's passengers (Fig. 4.9).

In the future, there is a unique opportunity to organize a radially-ring-route traffic on MTsK. Theoretically it can be done, but within the framework of the running project the electrification of the connecting branches is not provided. It touches only about four of twenty, that is, only for movement on the Moscow Central Ring.

The following conclusions can be drawn: Moscow Central Ring at the moment is seen as interchange circuit of the Moscow Metro and its role as an interchange circuit is underestimated including radial directions of commuter rail. Such approach does not allow to integrate commuter rail service into the transport system of the city successfully.

The large part of Moscow industrial zones, which will be subject to renovation or re-industrialization in the nearest future, is located near the railway—radial directions or MTsK. And exactly the railway network will be able to provide the best transport accessibility to revived territories.

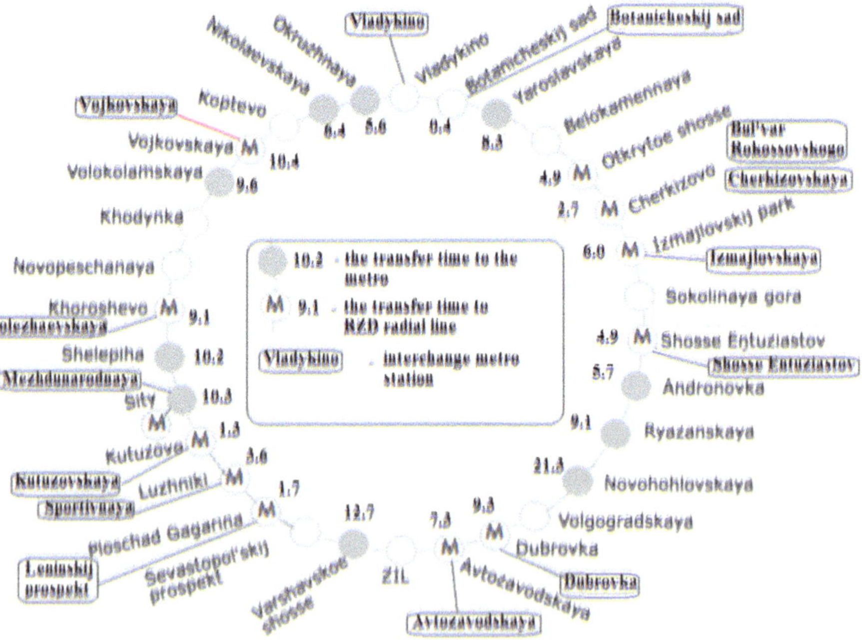

Fig. 4.9 Changes to Moscow central ring (*Source* Created by authors)

4.6 Scenarios for the Development of Passenger Rail Services in the Moscow Transport Hub

The crucial situation with the organization of passenger transport in Moscow agglomeration demands the integrated approach to development of the Moscow railway system, taking into account international experience. There are several visions for the further development of passenger rail services in the Moscow transport hub.

4.6.1 Public Stock Company "Russian Railways"

In general development scheme of Moscow railway system there are several directions:

- Construction of additional main tracks on existing radial lines.
- Development of the existing railway terminals.
- Formation of "terminal stations" in some radial directions (Fig. 4.10).

The transport experts have no doubt that transport problems of Moscow agglomeration can not be solved only by the construction of additional tracks. All

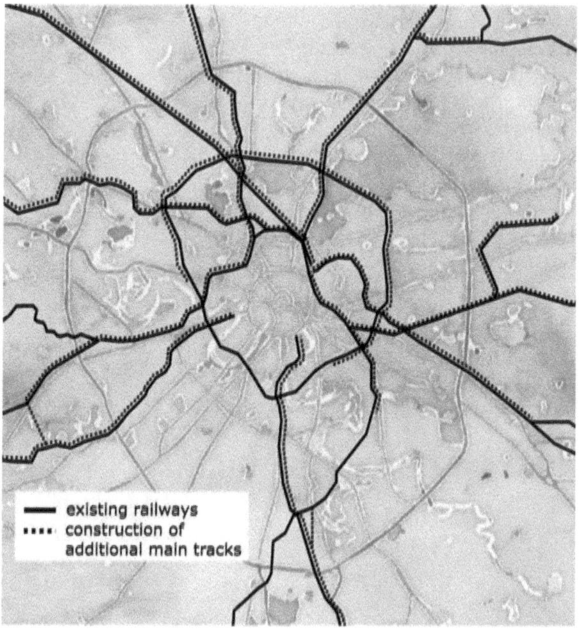

Fig. 4.10 Scenarios for the development of passenger rail services in the Moscow transport hub (Public stock Company "Russian Railways") [*Source* Created by authors using images by Stamen Design (2016)]

heavy railway lines in their current state have interchanges to the subway network, the capacity of which has already been exhausted. The capacity of the additional passenger flows can only worsen the transport situation in the city.

4.6.2 State Unitary Enterprise "Research and Project Institute of Moscow City Master Plan"

The strategy of this institute is based on following steps:

- Creating a new urban off-street transport system with the use of economically justified types of rolling stock, by combining the existing ones, including the blind and low-density lines;
- Combining the payment systems of all types of urban transport;
- Formation of "terminal stations" of all radial directions for a full rotation (a 12-car train will be changed by a 4-car train at the stations).
- The reorganization of a distant passenger transport;
- The reorganization of urban areas occupied by the railway infrastructure for the development of new business areas (Fig. 4.11).

This concept has more opponents than supporters. Arguments against: a forced additional transfer for passengers travelling to the city center. There is high

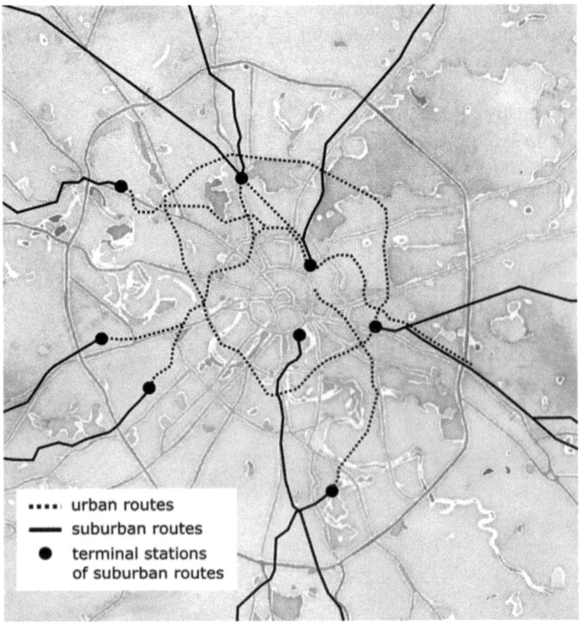

Fig. 4.11 Scenarios for the development of passenger rail services in the Moscow transport hub (State Unitary Enterprise "Research and Project Institute of Moscow City Master Plan") [*Source* Created by authors using images by Stamen Design (2016)]

probability that people will prefer to use the subway and will not go further by rail. That is, the city will obtain the opposite result instead of reducing the load on the subway.

4.6.3 The Expert Center "PROBOK.NET" (Moscow State University of Railway Engineering)

The experts suggest following activities:

- The development of diametrical routes of commuter trains.
- The rejection of the formation of terminal stations.
- Integration of the urban ticketing system with the commuter ticketing system to provide payment on commuter trains in the city of Moscow (in the range of a single tariff).

Most of the rail lines cannot be used as an independent mode of transport for the traveling within the city due to its "dead-end" and low connectivity. Not all the rail lines, with the ability of transit traffic, use this potential. Therefore, many experts believe that it is necessary to develop diametrical routes of suburban trains in the Moscow railway system.

Researchers from Moscow State University of Railway Engineering have suggested the organizational of diametrical movement in 3 stages:

1. *Intensification of traffic of existing diameters*
2. *Formation of new diameters*
3. *Integration of radial directions by tonnel lines.*

(1) Currently, the traffic of transit trains through the city is represented by three routes. The Belarussian-Savyelovskiy diameter (Fig. 4.12, route 4) connects western and northern parts of the city by direct links. Kursko-Rizhsky diameter (Fig. 4.12, route 2) connects northwest, south-eastern and southern parts of the city by direct links. Belarussian-Kursky (Fig. 4.12, route 3) diameter connects western, south-eastern and southern parts of the city by direct links.

The number of suburban trains may be significantly increased on Belarus-Savyelovskiy and Kursko-Rizhsky diameter. However, Moscow Metro has a competitive advantage due to the convenience of scheduling and a land use model in Moscow. The commuter trains are in a losing position. Passengers do not win the time using existing "diametric" routes that connect Rizhskoe and Kurskoe, Belarusskoe and Savelovskoe rail directions, so there is little demand for the intercity travel.

(2) The core of the competitive ability of railway lines is to allow moving at maximum speed. In particular, the reason for the unpopularity of "diametric" routes via Moscow is a big distance while a passenger can overcome a similar

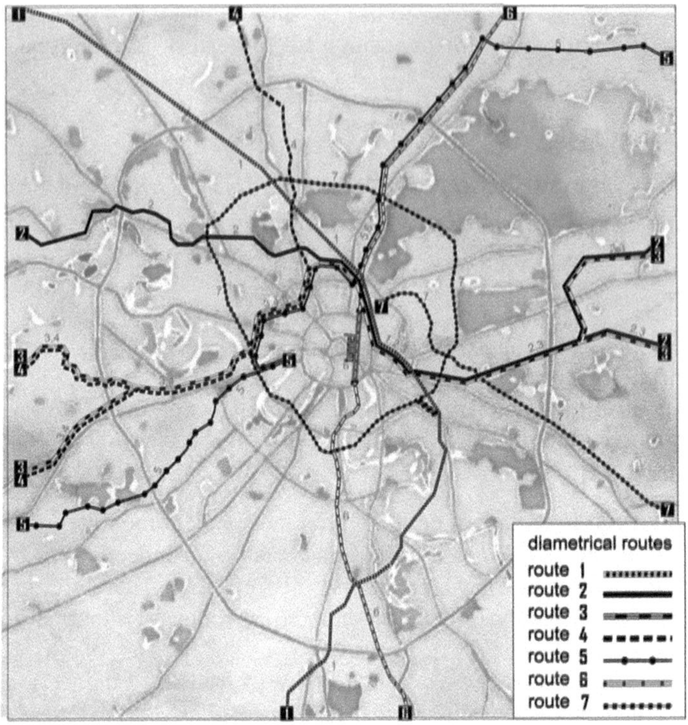

Fig. 4.12 Scenarios for the development of passenger rail services in the Moscow transport hub (Moscow State University of Railway Engineering) [*Source* Created by authors using images by Stamen Design (2016)]

way by Metro faster. The most popular routes would be the routes that take less time than Metro. It could be in such cases as: the absence of relatively short connection by Metro between the area of origin and the area of demand and when the tracing of the railway lines occurs in areas poorly served by Metro.

For example, the route that unites the Leningrad and Kazan directions. This route connects the outlying areas of the city on the way closest to shortest one, also serving areas, which are located further from the subway stations, especially in the northern sector.

Currently, the traffic of transit trains from Leningrad to Kazan direction is not performed. At the same time, these two directions are linked through a connecting branch Mitkovskaya. Now all the routes of the trains to Kazan and Leningradsky railway stations are a dead-end, which leads to an overload of the Metro station Komsomolskaya and makes commuter trains unattractive as an intercity transport.

Formation of Leningrad, Kazan diameter (Fig. 4.12, connection of routes 7-1) will allow:

- Significantly reducing travel time from the north-western districts of Moscow to the eastern and south-eastern parts of the city.
- Providing passengers of Leningrad direction an alternative route in the city with changes at Electrozavodskaya metro station and Aviamotornaya metro station.
- Providing passengers of Kazan direction an alternative route in the city with changes at Rizhskaya metro station and Petrovsko-Razumovskaya metro station.

Gorkovskoe direction has the ability to enter the transit Kalanchevskaya line with a subsequent transition to Rizhskoe (through Alekseevskaya line) or Belarusian direction, but this opportunity is not used (Fig. 4.13). There will be the ability to skip transit passenger trains after small reconstruction of a railway switches in Kursky railway terminal.

Currently, there is no transit traffic between Kievskoe direction and Savyelovskoe direction. These two areas are linked through The Moscow Ring Railway and other connecting lines. Now all trains of Kiev railway station is a dead-end that has a number of shortcomings: overloading of Kievskaya Metro station, low attractiveness of the trains of this direction as an inner-city transport. Creation of the Kievsko-Savyolovsky diameter will allow:

- Significantly reducing travel time by moving from the northern districts of Moscow in the western and south-western areas of the city.
- Providing passengers of Kievsky direction an alternative route in the city with a direct change at Kutuzovskaya, Mezhdunarodnaya, Begovaya and Belarusskaya Metro stations.

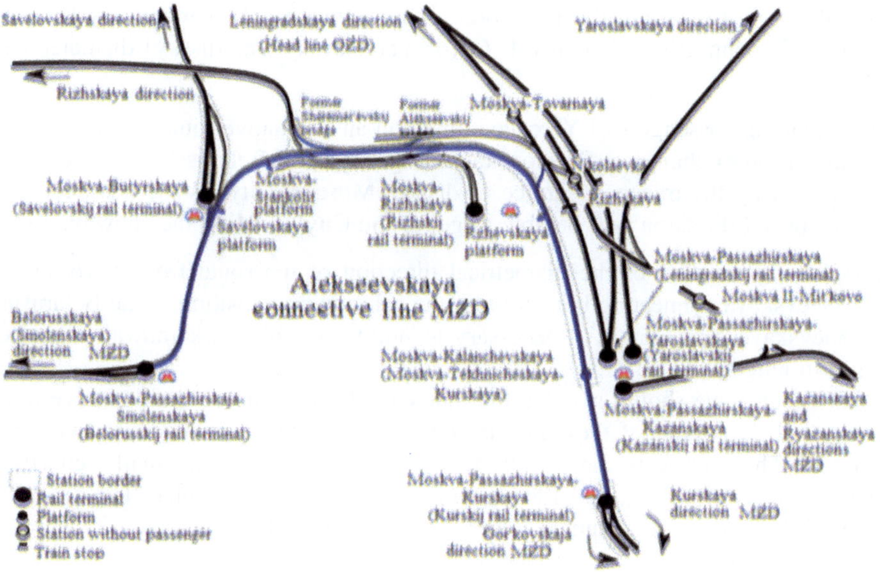

Fig. 4.13 Alekseevskaya connective line

Table 4.9 Comparison of travel times using the Metro and the perspective railway diameter

Route	Through the diameter (min)	By Metro[a] (min)
Lianosovo-Ochakovo	46	47
Lianosovo-Solntsevo	52	53

[a]Time comparison with Metro is indicated relatively to Solntsevo Metro line (under construction at the moment)

Comparison of travel times using Metro and the perspective railway diameter shows that the selected railway route takes about the same amount of time in comparison with the route using the subway (Table 4.9). At the same time using subway one needs to make at least two changes, and in some cases 3 changes. Taking into account the overloading of all Metro hubs during peak hours, it can be concluded that the direct railway route will be more attractive to the passengers.

The formation of the railway diameter will enable some passengers (mainly from the region) to get to the destination avoiding Metro.

(3) Creation of Yaroslavsky-Paveletsky diameter (Fig. 4.12, route 6), Yaroslavsky-Kievsky (or Yaroslavsky-Belorussky) diameter.

Yaroslavskoe is the leading direction of the Moscow railway system on the number of passengers transported daily by commuter trains. Every day it serves more than 300 thousand people. At this time Yaroslavl railway does not have any direct subway change except the railway station (change to metro Komsomolskaya). The planned increase of rail freight capacity radiuses (opening of the 4th main road of Leningradskoe direction and the planned construction of the 5th main track in the direction of Yaroslavskoe) will increase the inflow of passengers on the overloaded complex of Komsomolskaya station. The creation of Yaroslavsky-Kievsky (Fig. 4.12, connection of routes 4, 6) (or Yaroslavsky-Belorussky) diameter will allow:

- Providing passengers of Yaroslavl direction an alternative route in the city with direct metro change at Savelovskaya, Begovaya and Belarusskaya stations.
- Improving the transport service of district Moscow-City, as the passengers of Yaroslavl direction will be able to get to the City avoiding the subway.

For the organization of the diametrical direction of the route from Yaroslavl to Alekseevskaya connecting branch (with the subsequent transition to the Belarusian or Kievskaya direction) it is necessary to build a double-track railway tunnel of 2800 m length.

In general, we should say that the diametrical routes are unlikely to save time. They will be popular if they are disjoint and will allow passengers to avoid the subway. These projects can be carried out if they require some small reconstructions (railway switches, etc.) for organization of diametrical routes. The construction of the underground tunnel raises the question of the economic efficiency of such a project.

Acceleration of movement of trains within the city can increase the attractiveness of the intercity routes. Some engineering institutes develop various proposals. In particular:

- Elimination long parking at railway stations
- Widespread use of the electrodynamic braking to reduce the length of the electric braking distance
- Reduction the length of the block sites from existing 500–700 m to 300–350 m
- The use of rolling stock, where at least 3 pairs of doors per car to speed up the boarding process
- Equipment of all passenger platforms located in the curved sections with video cameras which will reduce the parking time at halting points, since the existing technology an assistant engineer has to check the boarding process in the last cars of the train personally
- Reduction of processing time on a dead-end suburban train passenger station (from 24 to 6 min), etc.

The basic principles of passenger service organization in the presented models of passenger rail in the Moscow transport junction (hereinafter—Model) are diametrically opposed (the organization of "zone pricing stations", the organization of "urban" rail connection as a separate mode of transport). Currently there is no confirmation of the proposed models in the form of forecasts of mobility needs of the population of Moscow agglomeration.

4.7 Transport Policy in Agglomerations

The Moscow authorities have long standing problem of optimal allocation of financial resources to the development of the Moscow agglomeration transport infrastructure. Due to the fact that the various transport modes have different advantages and disadvantages, it ought to be carefully chosen the mode of transport which should be developed in order to achieve a specific goal.

In Germany rail transport is in a priority. At the governmental level, it was recognized that the capacity of the national transport infrastructure (in the first place—highways), is insufficient to meet future transportation needs. It is assumed that rapid development of railway lines could eliminate the deficit.

Despite the substantially higher level of motorization in Germany, the mobility of the population in the suburban rail transport is 30 trips per person annually, which exceeds the Russian index in 5 times.

In recent years, passenger traffic in suburban rail transport in Germany is growing steadily. In Russia, in the period from 2000 to 2014 the number of passengers in the suburban traffic was reduced by 25 % and passenger turnover—by 36 %. At the same time in the Moscow transport hub, which carried 65 % of all commuter traffic, there is an increase in passenger traffic (from 2005 to 2014), the number of passengers increased by 14 %.

In general, in all countries the carriage of passengers by suburban rail is recognized as an unprofitable activity requiring financial support, since a passenger, as a rule, is not able to pay the economically reasonable ticket price. Such support can be provided either by profits from other activities (for example, freight transport), or by providing public subsidies. In Russia, the transportation of passengers by commuter rail is allocated and implemented by suburban passenger companies. Cross-subsidization of passenger transportation by truck is missing.

In the Russian Federation responsibility for organizing the commuter passenger transport, including rail, is assigned to the subjects of the federation. This suburban rail passenger complex also gets assistance from the federal budget.

One of the major drivers of suburban rail transport development in Germany is carried out government policy. It is the long-term financial support of the commuter passenger transport functioning from the federal budget. The size of state support and the rate of indexation are fixed by law for a period of 15 years.

In Germany, as one of the countries with the most developed commuter rail connections, there is a federal law that describes the model and the system of organization of suburban passenger transportation, including rail, the rights and commitments of the parties. In Russia, there are certain provisions relating to the procedure for organizing the commuter passenger transport, regulation of transportation tariffs, compensation of decreased income to the carrier, provisions describing the rights and duties of the concerned parties. But they are all contained in various legal acts. The document that describes a complex system of organization of commuter passenger rail transport is not currently available. The draft of federal law "on the organization of regular passenger rail service in the Russian Federation" was developed in February 2010 and is still under discussion.

Compensation from the federal budget for the cost of the infrastructure services provided to suburban passenger companies, as well as establishing rates of value added tax (VAT) services for the carriage of in suburban at 0 %, almost allowed the suburban railway complex to reach break-even in 2015. However, the decision on compensation for the cost of infrastructure services was adopted annually by one year, the rate of 0 % on the VAT was established for a period of 2 years. Only the predictability of government policy and long-term use of the state support measures to commuter passenger complex will develop and transform the railway transport of Moscow from the historically established commuter format into a more modern, suburban-urban format.

References

Grigoryan Yu, Pavlova A, Uglovskaya E (eds) (2011) Old Moscow. The Inventory 2011. Project Russia, vol 62, no 4, Annex. http://www.prorus.ru/news/the_inventory2011-pr62.pdf. Accessed 31 Aug 2016

Ministry of Transport of the Russian Federation (2016) Transportnaya Strategiya (Transport Strategy). http://www.mintrans.ru/activity/detail.php?SECTION_ID=2203. Accessed 31 Aug 2016

Moscow Complex of Urban Planning Policy and Construction (2014) Propusknaya sposobnost razvyazki na peresechenii MKAD i Ryazanki uvelichitsya na 40 % (The capacity of the interchange at the intersection of Moscow Ring Road and Ryazanskiy Prospekt to increase by 40 %) http://stroi.mos.ru/news/propusknaya-sposobnost-razvyazki-na-peresechenii-mkad-i-ryazanki-uvelichitsya-na-40 (In Russian). Accessed 31 Aug 2016

Moscow Complex of Urban Planning Policy and Construction (2016) Ryazanskiy prospect http://stroi.mos.ru/road/trunk/22 (In Russian). Accessed 31 Aug 2016

Russian Federation Federal State Statistics Service (2016) http://www.gks.ru/ (In Russian). Accessed 31 Aug 2016

Russian Public Opinion Research Center (2016) Rossiyskiye zheleznye dorogi: kachestvo uslug v zerkale obshhestvennogo mneniya (Russian Railways: Public Opinion on Service Quality). https://wciom.ru/fileadmin/file/reports_conferences/2016/2016-04-20-rzhd.pdf (In Russian). Accessed 31 Aug 2016

Stamen Design (2016) http://maps.stamen.com/. Accessed 31 Aug 2016

Chapter 5
Travel Demand Management in the Largest Cities

Tatiana Kulakova and Ekaterina Reshetova

Abstract The largest Russian cities are starting to implement travel demand management instruments to reduce car use. Today the non-pricing instruments are more popular in Russian cities: low emission zones, car sharing, pedestrian areas, public transport development, bicycles infrastructure development, etc. Paid parking is the main pricing instrument of travel demand management in Russian cities. Apart from the general description of pricing methods of the transport demand management which are currently being introduced in Russian cities it is necessary to develop the understanding of the measures effectiveness. Thus the experience of the paid parking implementation (about 20 cities, including Moscow, St. Petersburg, and Yekaterinburg) requires the serious analytical approach. Both official reports on effectiveness by the government and independent surveys are of interest. Not only should the operational parameters be examined, but also the public opinion should be analyzed. Taking into consideration the deficit of the information on the projects of paid parking implementation, the article would be interesting both for Russian and foreign specialists. The survey also deals with organizational, legal and institutional issues.

5.1 Transportation Demand Management: Definition

Features of social and economic development of Russian cities and regions at the present stage, lifestyle changes, and availability of private cars lead to a dramatic increase in vehicle traffic causing acute problems of lack of effective solutions in the area of private cars demand management in the segment of daily work-related trips from the periphery to the center.

T. Kulakova (✉) · E. Reshetova
Institute for Transport Economics and Transport Policy Studies,
National Research University Higher School of Economics, Moscow, Russia
e-mail: tkulakova@hse.ru

M. Blinkin and E. Koncheva (eds.), *Transport Systems of Russian Cities*, Transportation Research, Economics and Policy,
DOI 10.1007/978-3-319-47800-5_5

The indicated problems are exacerbated by the absence of common policy and the lack of effective tools for technical, tariff, tax and investment regulation in this area.

The review of international studies indicates that in order to reduce transport costs and eliminate congestion by controlling the structure of demand for the movement of vehicles in urban areas and changing the balance between personal and public transport many effective ways have been developed and implemented in the world over the recent decades.

Urban and transport planners began to use demand management programs in the 1970s. There is no uniform definition of "transportation demand management" or "mobility management" (TDM or MM) in the literature. Thus, the study performed by the Victoria Transport Institute suggests that TDM combines strategies for efficient use of transportation resources (Online TDM Encyclopedia 2014). According to some authors, the main goal of TDM is the reduction of the number of car trips (Single-Occupancy Vehicles (SOVs)) and of the number of people using the car for single journeys (Winters 2000).

According to Vuchic (1999), two sets of measures are to be applied in order to build a balanced urban transport system:

- promoting the use of public transport;
- resisting travelling by car.

At the same time, measures to encourage the use of public transport are to create conditions for the development of pedestrian movement.

Therefore, all TDM tools can be classified according to the criteria of the development of alternatives to individual vehicles to help redirect people to the use of another type of movement (Shift mode): public transport (Transit-oriented), bike (Bicycle-oriented), walks (Walking-oriented) and group travel (Ridesharing-oriented). In turn, all tools can be divided into two types: motorized (motorized strategy) and non-motorized (non-motorized strategies) (Litman 2010). The second type includes only travel by bicycle and on foot, and the rest belong to the first type.

There are other classifications of TDM strategies: political attitudes, regulatory requirements, financial, economic, organizational, legal and engineering ones. In this study, all TDM tools will be divided into three types: regulatory, pricing, and planning ones.

The main focus of the relevant literature available are the effects caused by the introduction of transport demand management programs (Table 5.1).

Anderson (2014) found out that there would be a 47 % increase in the average time spent in traffic jams (average delay) if public transport stopped functioning. This result is very different from the earlier studies estimating the impact of public transport on the traffic congestion as minimal. Thus, Parry and Small (2009) used modeling to find a 5 % effect (of 0.025 min per mile), and in the theoretical study by Nelson et al. (2007) the result was 2 min per mile. Empirical studies by Winston and Ashley (2006) and Duranton and Turner (2011) also argued that the

Table 5.1 The effects of transport demand management programs

#	Effect	Studies
1	Economic development: creating jobs, generating tax revenues to budgets of all levels, generating business income, reducing costs for road maintenance	Litman (2013), Garrett-Peltier (2011), Weisbrod and Reno (2009), Cambridge Systematics (1999, 2002) and Smart Growth America (2013)
2	Financial stability: individual cost reduction of car use and incidental expenses	Litman (2013), Weisbrod and Reno (2009) and APTA (2007)
3	Reducing fuel consumption	Davis and Hale (2007) and Shapiro et al. (2002)
4	Reducing traffic congestion	Litman (2013), APTA (2007) and Anderson (2014)
5	Health improvement	Litman (2010, 2012, 2014)
6	Providing the parking	Litman (2013) and Walker Parking Consultants (2011)

development of public transport was not a significant factor in reducing congestion. Winston and Ashley (2006) found that the net effect on the development of public transport was even negative. Duranton and Turner (2011) tested the fundamental law of road congestion according to which the vehicle mileage increases in exact accordance with the length of new highways.

As a result, the shifting car drivers to public transport leads to the saving from 0.24 to 2.76 US dollars per mile (depending on assumptions) in 2011 prices according to Litman (2013) estimates.

However, there are several factors that limit or promote the effective implementation of TDM instruments: funding opportunities, political acceptability, leadership and partnership (Gärling and Schuitema 2007; Jaffe 2013; Litman 2013; Taylor 2007; Zali et al. 2013).

It was these factors that had a significant deterrent effect on the use of TDM measures in the Russian cities.

5.2 Planning Measures: Traffic Calming

5.2.1 Speed Limit

In 1683 Emperor Peter the Great issued a decree banning fast driving in Moscow. In 1730 Empress Anna Ivanovna introduced penalties for speeding—fans of fast driving were to be fined, flogged and exiled to Siberia.

The "Guidelines for the Moscow Police" of 1883 prescribed policemen to monitor the speed limit.

In 1900, the procedure for movement of passenger and commercial vehicles was approved in the city of St. Petersburg. Maximum permitted speed for Moscow was set at 20 versts per hour for passenger cars (1 verst equals 1066.8 m), 12 versts for vehicles weighing more than 350 poods, and at night 10 versts per hour, in St. Petersburg—15 versts, in Kiev—10 versts per hour in busy streets and 20 for all others.

The first Soviet document related to road safety was "Instruction on the use of motor vehicles and motorcycles and on the traffic in the city of Moscow and its suburbs" approved by the Moscow City Council on July 31, 1918. Following that document, in 1920 Lenin signed the Decree "On automobile traffic in the city of Moscow and its suburbs (rules)" setting the speed limit for passenger cars at 25 versts per hour, for trucks—15 versts per hour, for moving at night—10 versts per hour.

In 1939, speed limits were imposed of 100 km/h for cars and 70 km/h for trucks. Two years later, limits were changed to 80 km/h and 60 km/h outside of settlements limits and 40 km/h in urban areas. In 1953, all restrictions were lifted to be re-introduced again in 1957.

Currently, the speed limit in the cities of Russia is set equal to 60 km/h. As the Administrative Code provides for a minimum fine of 500 rubles (€6.7) only for speeding by 20–40 km/h[1] the real speed of the traffic flow, including settlements, is 80 km/h.

The width of lanes in Russian cities is 3.5 m, while that for public transport is 4 m (not less than 3.75 m according to the standards), wide lanes are believed to encourage speeding (Fig. 5.1).

"What Russian does not love to drive fast?" This phrase from the first part of the "Dead Souls", a novel by Nikolai Gogol, very accurately describes the Russian soul. Even though it was written more than 170 years ago, now, in a time of high speeds, it is particularly relevant. Modern cars can go at tremendous speeds, and drivers often are tempted to go for a ride "quick as the wind".

Installation of speed limiting traffic signs is not enough. If the layout of the streets remains unchanged, car drivers will continue to drive at a high speed.

Therefore, to reduce traffic in the streets the most common and popular trick used in the modern global urban planning practice is the so-called "traffic calming"—the method combines technical and architectural/planning concepts. This type of measures was rarely reviewed in the Russian city planning literature and actually never used in our country. The reconstruction of Malaya Konyushennaya Street in St. Petersburg became the first project on traffic calming in the Russian city planning practice (Fig. 5.2).

Only some cities in Russia have introduced speed limits so far. For instance, restrictions are imposed on 70 streets setting the speed limit at 40 km/h in Kazan city. In St. Petersburg, there are plans to establish a single territory with the speed limit of 30 km/h.

---

[1]The fine for speeding by more than 10 km/h was cancelled in September 2013.

Fig. 5.1 Lenina Street (Ryazan, Moscow region) [*Source* Created by Maxim Shanin (2014)]

Fig. 5.2 Malaya Konyushennaya Street in Saint-Petersburg [*Source* Made by Vaganov (2014)]

Moscow got its first low speed zone in Pushechnaya Street along its entire length from the Lubyanka Square to Neglinnaya Street. The zone has only technical markings. All surfaces are made at the same level; the side-walk is not elevated above the road way. The speed limit in the low speed zone is 40 km/h.

In addition to that, another unofficial quiet movement zone—along the Bolshoy Tolmachevsky Lane, near the Tretyakovskaya metro station. Initially, a completely pedestrian street was established there, but eventually the residents asked to open it for traffic. As a result, it was decided not to rebuild it completely and simply to fence it with low poles instead.

Please note that in 2014 Moscow considered introducing speed limit of 40 km/h in the streets next to pedestrian zones. Voting was initiated at the «Active Citizen» website,[2] with more than 70 % of Muscovites voting against the measure.

Some traffic calming issues are further discussed in Chap. 8 within the analysis of the policy of road safety improvement.

[2]https://ag.mos.ru/.

5.2.2 *Pedestrian Areas*

The most radical means of reducing traffic density in urban centers are "car-free zones". They are characterized by a complete ban on traffic, with the exception of special vehicles (ambulance, police, fire and public utilities, service for shops). Such zones are set up in smaller areas.

No research on the urban pedestrian spaces planning in Russia was done over the period 1990–2005; no there was any strategy for their development.

Nevertheless, some first projects were implemented to restore pedestrian environment. Thus, In St. Petersburg, they restored a pedestrian area in the historic city center with intense functional saturation and historical flair. The area of 6–7th lines on the Vasilevsky Island is completely pedestrian. The number of pedestrian streets in St. Petersburg is higher compared to any other city in Russia—there are 10 of them there.

Many cities currently revive their historical areas introducing pedestrian zones combined with the citywide pedestrian network (for example, in Irkutsk, Podolsk, Ryazan, Samara, Kaliningrad, Vladimir). There were 16 pedestrian zones in the Moscow region in 2015, their number is supposed to be increased up to 20 by the end of 2016.

Old Arbat Street remains the benchmark of pedestrian streets. Pedestrian streets in many cities look like this street (Fig. 5.3).

Arbat (1200 m) became a pedestrian street back in the 1990s, following that pedestrian zones began to be actively established in Moscow since 2012. Kamergersky Lane, Stoleshnikov Lane, Kuznetsky Most Street and Rozhdestvenka

(a) (b)

Fig. 5.3 Pedestrian streets of **a** Moscow, Arbat Street [*Source* Made by Deror avi (2009)], **b** Yekaterinburg, Vainer Street (*Source* Author)

Street got their pedestrian zone status. The longest European pedestrian area was opened in Moscow two years ago. It stretches from the Gagarin Square to the Kievsky railway station Square, and its length is 6.5 km.

Competitions for pedestrian zone projects started to be organized in Russia. Thus, the best project for turning Pochtovaya into a pedestrian street in Ryazan city was selected through a competition. In Kaliningrad, they hosted an open international architectural and urban planning competition in search for the most promising concept of the historical center development (including the development of the pedestrian zone).

Thus, pedestrian streets have been the most common type of planning tools of personal car demand management in the last 20 years. Over the past 10 years, beautified pedestrian spaces appeared in many small and medium-sized towns, especially those with the historical centers in them. Commerce and public catering are the core functions of the pedestrian urban spaces, less frequent are office real estate. The demand for pedestrian spaces depends on the level of development of the services sector in the city.

5.2.2.1 Elements of Forced Speed Reduction

One of the few measures for traffic calming used everywhere in Russia is speed humps. The audit of such humps was held in Moscow in 2012. There were 2148 of them located in 980 places according to official data. Unregistered humps were installed at 707 locations. After checking, 49 humps at 109 addresses were deemed illegal and were dismantled. Speed humps at 658 locations were deemed necessary and were included into the city asset list. Sometimes, instead of real speed humps 3D-drawings on a flat surface are used to give a visual sense of the hump presence (Ilicheva 2012).

Photo-and-video cameras installed for recording violations of speed limits may also be considered as elements of forced speed reduction in Russia.

According to the data from the traffic police, there are currently 6.2 thousand of stationary and 4.1 thousand of mobile control zones in Russia. More than 50 million court rulings on the cases of administrative offenses (presenting 67 % of the total number of fines given to the vehicle owners and drivers) were based on the data collected from the devices in 2015. To compare, the figures stood at 40.9 million and 63 %, respectively in 2014, according to General Administration for Traffic Safety of Russia. The bulk of the total number of fines—44.9 million (89 % of the total)—are speeding ones.

So far no uniform national requirements applied to the automatic record detection devices have been developed in Russia. Besides, cameras sometimes are installed in places of mass violations of traffic rules, which do not eventually lead to accidents (i.e., where the maximum number of fines is collected) rather than where they are really needed to ensure the safety of traffic.

The "National safety improvement programme 2013–2020" increases the number of stationary photo-and-video recording systems. Local programs implying the installation of cameras on roads are also becoming increasingly popular. By and large, funding issues are the main barriers to these measures being implemented. Municipal budgets incur expenses for installation and maintenance of cameras, while the relevant revenues from fines associated are allocated to regional budgets.

Therefore, stationary traffic cameras are mainly installed in large cities and on federal highways. The leaders are Moscow, Kazan, and Chelyabinsk.

When the national photo-and-video recording cameras system was introduced a big problem the authorities encountered with was that ambulances, fire trucks and police vehicles began to receive fines. For that reason, an exemption list for all the special services vehicles with the usual license plates and the plates of public transport had to be introduced. But when it happened, the number of cars that freely used dedicated public transport lanes increased dramatically. Moreover, as noted by many drivers, these were often vehicles used by various officials.

Note that drivers use a variety of ways to "fight" cameras, ranging from covering their license plates with mud to sophisticated technical devices for their concealment (flipper flips, electromagnets behind the bumper to hold the metal plate on top of the license plate hiding numbers, etc.). Application and even installation of any device that complicates the identification of license plates entails extremely severe sanctions in case of detection. In particular, Art. 12.2 Part 2 of the Administrative Code provides for a fine of 5000 rubles (65.8 €) or driver's license revocation for 1–3 months. Driving a vehicle with license plates known to be forged shall be punished with driver's license revocation for 6–12 months (Art. 12.2, Part 4, of the Administrative Code).

Drivers legally are allowed to use radar detectors (anti-radars suppressing radars operation are prohibited in Russia), as well as navigators[3] and dashboard cameras with GPS-modules and relevant database systems installed. The range of such devices is very wide today, but outside Russia any of that may be found as banned.

5.2.3 *Street Patterns Forcing Drivers to Slow Down*

A number of new traffic reduction methods were applied over the recent years. The "My Street" project was launched in Moscow in July 2014. In line with the project, 47 streets were landscaped in 2015, another 59 streets are to be landscaped in 2016. The project also involves sidewalks repairs, the creation of bicycle lanes, construction of parking lots. Technical measures of traffic calming were applied, pedestrian crossings were raised flush with pedestrian sidewalks, the carriageway was significantly narrowed (Fig. 5.4).

[3]The "Yandex.Navigator" Service can now warn the drivers that they are speeding. The application knows about the speed limits on different roads and in various sections of the same road giving an audible warning signal when the speed limit is exceeded. In addition, the speed limit in the area is now displayed on the screen. The official website is at https://driver.yandex/news.

Fig. 5.4 Myasnitskaya Street (Moscow) after reconstruction (*Source* Author)

Sidewalks were widened by an average of 2.5–3.5 m. As a result, the number of pedestrians has increased between two and seven times on those streets that were landscaped, according to the Director of the Moscow center for traffic management. For instance, the flow has doubled from the initial 400 pedestrians per hour at the Myasnitskaya street.

One of the main program principles is taking the residents' opinions into account when working on city improvement. For that purpose, voting takes place at the «Active Citizen» website in order to approve street improvement projects that are included in the program.

In addition, they began to build safety spots in Moscow. Here, for example, is one of them on the Turgenev Square (Fig. 5.5).

They also started to build diagonal pedestrian crossings. Japan is particularly known for such crossings; the Japanese, much like Europeans and Americans, have achieved a lot in slowing traffic.

The cities began to replace crossroads with traffic lights with roundabouts (Fig. 5.6).

Fig. 5.5 Safety spot (Moscow) (*Source* Author)

Fig. 5.6 Intersection in the city of Perm [*Source* Yakimov (2014)]

Unfortunately, most of the traffic calming (and organization) measures in Russia can be applied only after making amendments to the relevant building codes and state standards.

5.3 Regulatory Demand Management Measures

In Russia, a wide range of regulatory demand management measures is used: improving transport services, improving public transport, developing pedestrian paths and bicycles, dedicated lanes for public transport, car sharing, improving taxi services.

Public transport is examined in separate sections, so let us consider other measures first.

5.3.1 Cycling

The assessment of the development of the cycling in the cities of Russia is an interesting endeavor. The features of the development of cycling in Russia are summarized in the Table 5.2.

In general, many factors influence the development of cycling infrastructure; for instance: the lack of regulatory framework for the design of cycling infrastructure facilities, lack of racks for bicycles, the level of development of cycling as a sport, mentality and temperament of the citizens, the minimum cost of a bicycle, etc.

Existing building codes do not always regulate the design decisions. Most of the streets were designed before the mandatory regulations requiring planners to include cycling infrastructure were introduced. Difficulties arise with the current profile of the road, historical development and the presence of the expressed terrain when it is necessary to integrate functional areas for cyclists into functional automobile, pedestrian, engineering and recreational areas. And it is hard to avoid a conflict of interests here because under the regulations the bicycle lane requires a minimum of 1.5 m of space for the single-lane traffic and 2.7 m for the two-lane traffic.

Often, designers add a bicycle lane when planning a reconstruction of a certain part of the road. Nobody seems to care that it starts from nowhere and goes nowhere. Due to such approach city maps get "dead" cycle paths with inefficient low traffic.

Public rent-a-bike systems are absent in most cities, there are no parking lots for bikes. Urban residents use bicycles spontaneously, more for sports and recreation rather than as an alternative means of transportation within the urban settlement. Therefore, the development of this type of transport has just began and does not yet have as a visible impact on passenger flows.

Table 5.2 Comparative analysis of domestic and foreign practice in cycling development

Feature	Other countries	Russia[a]
Status of bicycles (town planning and transport)	Three-stage: federal, regional, municipal (local). National projects of development of cycling have been developed. Bicycles are a full-fledged part of an integrated transport, recreation and tourism systems of countries and cities	Single-stage—regional Bicycles are not mentioned in the Transport strategy up to 2030. Bicycles are an element of beautification of local recreational areas
Definitions (international standard)	A cycle lane is any of the longitudinal lanes of the roadway designated by markings and having a width sufficient for the safe movement of bicycles in a row. A cycle lane is distinguished from the rest of the roadway by longitudinal road markings. In accordance with Article 26 bis of the Vienna Convention	Lane for cyclists—a lane of the roadway, designed for cyclists and mopeds separated from the rest of the roadway by horizontal markings and indicated by road sign 4.4.1 combined with plate 8.14 located above the lane
Traffic regulations	Adopted signs of the Vienna Convention indicating shared bike-pedestrian paths	Adopted
Approaches to designing cycling infrastructure	It begins at the level of the master plan of a settlement	Developed at the stage of beautification of some recreational areas
Methods of traffic management	Counter cycling on one-way streets is allowed. Left turn is organized. Trailers are allowed. Children 8 years and older can cycle on the roads	Drivers are required to give way to cyclists crossing the road when turning. Cyclists are to be over 14 years old to use roads
Sources and procedures of financing	Federal, regional, and local budgets	Regional budget
Admission to traffic	Children from 8 years of age are allowed to cycle on roads if accompanied by an adult. Adult passengers are allowed to be on the pedicabs	Persons of at least 14 years of age are allowed to ride bikes on the road. No passengers are allowed except 1 child up to 7 years of age
Procedures for the implementation of proposals (ideas, innovations) to develop cycling	Start by the initiative of NGOs on the ground to rise to the status of district, city, country, involving businesses, local authorities and MPs	Start following the decision of the governing bodies
Approaches to the division of responsibilities in case of accident (insurance, etc.)	Law, legal and financial responsibility is higher for the protected participants in a road accident	All participants in an accident are equal

(continued)

Table 5.2 (continued)

Feature	Other countries	Russia[a]
Securing property rights for infrastructure	Ownership rights, lease, road concession are stipulated by laws	The owner of the motor road is also the owner of the bicycle path located within the red line of the road. In other cases, the owner of the territory is considered to be the owner
Economic incentives for cycling	Tax breaks to businessmen whose employees use bicycles Subsidies to cyclists	Missing
Approaches to indicators of assessing the level of development of bicycle transport	BYPAD audit system Assessment by 9 indicators	Missing

[a]Included into traffic regulations pursuant to the RF Government Decree #221 of 03.22.2014

The guidelines for the planning and design of infrastructure for cyclists and pedestrians in the UK indicate that designers should always consider the possibility to reduce the traffic and set speed limits first. They can change the configuration of intersections and roadway only if such options are not available. They may consider building bicycle lanes and paths (Department for Transport 2004) only as a last resort measure.

In Russia, unfortunately they started from building lanes and paths first. Their length is taken as an efficiency indicator of the city transport policy. Here are the statistics on the length of improved cycle routes: Naberezhnye Chelny—130 km, Izhevsk—26 km, Ufa—25 km (69 km to be constructed in 2016), St. Petersburg—150 km, Moscow—250 km (Fig. 5.7).

Using "big data"[4] and various relevant research results the strategy for the bicycle network in Moscow was developed.

The cyclist routes map partly resemble that of the underground—it has circular and radial lines including an 80-km long "Green Ring" connecting parks and encircling the entire city, paths along the Garden and Boulevard Rings, radial highways and the chord lines (Departament transporta goroda Moskvy 2016).

The project will commence from landscaping of the Moscow River embankments, then barrier-free access for cyclists shall be provided to the neighboring areas; chord local routes shall be added at the final stage of the project implementation. According to the best-case scenario the project is expected to be completed within 6 years.

[4]Co-authors of the project—Federiko Paralotto together with a team of his Mobility in Chain bureau, with the active participation Michael Zeiler, an Austrian planner and founder of Rosinak and Partner.

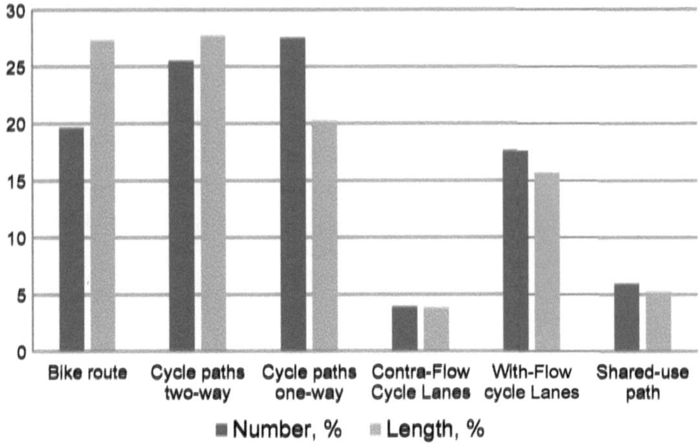

Fig. 5.7 Bicycle networks in Moscow [*Source* Open Data Moscow City Government (2016)]

The strategy involves conversion of about 700 km of roads into bicycle paths—it is approximately equal to the length of the M10 highway connecting Moscow and St. Petersburg.

On June 1, 2013 a bike sharing system was launched. It now consists of 300 docking stations and 2700 bicycles. As a result, the number of trips people cycle went 12.6 times up from 70 thousand in 2013 to 879.4 thousand in 2015.

5.3.2 *Dedicated Lanes*

In Russia bus lanes are part of the road reserved for certain vehicles (according to the local Traffic Rules and Regulations), and are called "lanes for route vehicles". The lane may be arranged to be placed anywhere in the road, except for at the curb on the side of the main direction of movement (on the right for the right-handed traffic).

In Kazan city, dedicated lanes were introduced at most of the streets having minimum three lanes in one direction in two stages: in 2008 and 2010. Motorists repeatedly tried to challenge their emergence in courts. Mainly bus routes use the lanes. The number of roads with dedicated lanes increased specially for the time when Universiade 2013 and the water sports championship in 2015 took place. The "sports" lanes were cancelled soon after the end of these events.

Since 2010 dedicated lanes have been considered as a method to deal with traffic congestion in Moscow. Their establishment was part of the "Livable City" plan developed by Moscow Mayor Sergey Sobyanin. 21 dedicated lanes are currently in use in the city (Fig. 5.8).

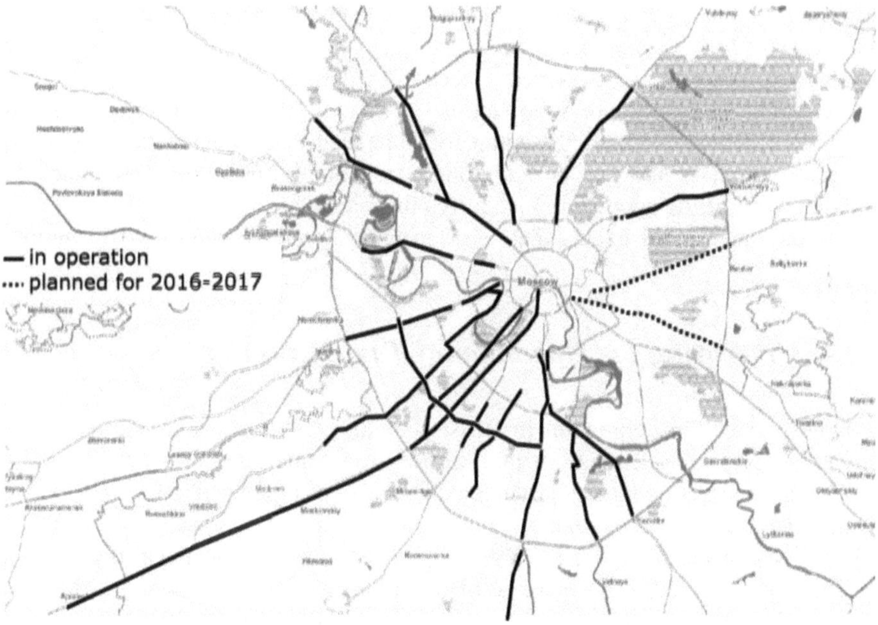

Fig. 5.8 Dedicated lanes for public transport in Moscow [*Source* Created by authors using the data from Mosgortrans (2016) and images by Stamen design (2016)]

Construction of dedicated lanes for public transport helped to partly solve traffic problems, particularly, the speed of buses, trams, and trolleybuses went up by 15–30 %. Passenger traffic on dedicated lanes is now more than 1.5 million people a day.

Main issues associated with organizing the dedicated lanes in Moscow are the following: dedicated lane ends once the road where it is arranged intersects with another road; early ending of a dedicated lane; the excessive width of lanes, that eventually encourages not only driving but even parking at them; the absence of a single concept of a road design, insufficient isolation of dedicated lanes from other lanes of a carriageway.

Another way to ensure public transport priority is the construction of isolated tram tracks that are common in our country.

5.3.3 Car-Sharing

In 2015, short-term car rental services became available in Moscow, such as Delimobil (with Italian company named General Invest as an investor), Anytime, YouDrive. Ca5 service became the last to be launched. Three out of four companies allow customers to use their cars only within the Moscow Ring Road (YouDrive + 1–5 km from Third transport ring). Payment is deducted from a customer's bank card linked to a mobile application. The rate is 6–8 rubles per minute. The fourth company requires its customers to sign a contract with the company in order to use the service (except YouDrive).

Delimobil company got involved in a scandal in 2016. After an accident involving the rented vehicle the company demanded that the users compensate for the cost of repairs and pay an additional fine (total of 600 thousand rubles). In addition, cars of the service are not fully covered by an insurance.

Moscow authorities expect that car-sharing will compete with taxis in future.

5.3.4 Improving the Operation of Taxi Networks

A large part of the population in Russia could not afford a taxi ride during the Soviet period so the demand for taxi services was quite low.

The taxi market undergone dramatic changes over the recent years with regards to its organization and regulation. Following passing of the relevant federal law taxis services were liberalized in 2011 in order to account for illegal taxis.

This approach led to an increase of the number of licenses issued in Moscow alone to 55,000 in 2015 (i.e. an 8 times increase compared to 2010). Thus, taxis have become very popular in the cities of Russia and affordable for a large segment of the population.

A number of newly registered taxi companies have grown in recent years, companies began to introduce and offer new services (mostly in large cities) including tow trucks, "sober driver". In some cities of Russia, as in many countries around the world, they offer car rental services as well as meeting passengers at the airport and picking up students from school (if their parents cannot do it themselves).

5.4 Price-Based Measures to Control the Demand for the Road Network

5.4.1 Parking Charge

The basic price-based demand management tools deployed are: an additional tax when buying a car, a motor vehicle tax; congestion pricing, area licensing, cordon pricing, vignettes schemes, electronic payment system (Electronic Road Pricing); parking charge; tax for automobile fuels. The objective of all transport taxes is to make car owners to pay the full cost associated with the movement of vehicles for the society encouraging drivers to reduce the use of private cars and encourage the use of sustainable transport such as public transport and cycling.

Russian and Western studies has shown the lower estimate of external effects is about 3 rubles per car-km, i.e. a 1 km trip by car in Moscow costs the community at least 3 rubles more than the motorist. To date, taking into account all existing transport taxes, less than a quarter of this value is internalized in Moscow (taxes cover only 0.62 rubles/car-km) (Hovavko 2012).

The pricing mechanisms are based on a fundamental economic principle stipulating that price signals may shift the mode choice from an a priori desired trip to alternative routes, modes of transportation, time or may even force an individual to abandon the trip eventually.

Among the first price-based travel demand management tools implemented in Russia was the parking charge.

Back in 1794 the order of the St. Petersburg Chief of Police established the first Russian rules for parking carriages: "There are large stones set near houses, you are not to park on these stones or near the walls where pedestrians walk not to be an obstacle for them".

During the Soviet times, there were no particular ethical barriers that would prevent enforcing a law so cruel, but there was no necessity to do so, either: the amount of chariots was minimal, and the drivers of military and other kinds of government-owned vehicle fleet would know perfectly well where they could or could not park their vehicles.

In the 1990–2000, in the cities of Russia (while the motorization level was growing rapidly) the regime of free parking dominated. Under these conditions, a daily car trip became the cheapest way to get from home to work: out-of-pocket cost of using a private car was limited to paying for gasoline needed for a 10–15 km trip and, as a rule, was no more expensive than a bus or subway fare. The price advantage and a relative comfort of a car trip accompanied by the considerations of prestige were extremely important factors for the first generation of the post-Soviet motorists who constituted 70–80 % of all the Russian car owners.

As soon as the motorization level in Russian cities grew from the Soviet level of 50–60 to the European figures of 300–400 cars per 1000 residents, the shortage of urban space became increasingly more acute.

As a result, the centers and business districts of Moscow, St. Petersburg, and all the other major cities of Russia from Sochi to Khabarovsk were filled with parked cars. Cars were parked everywhere: on sidewalks, lawns, public transport stops, in the courtyards and passages between the houses, and under all sorts of prohibitory signs. On the streets of more or less considerable width, cars would park in 2–3 rows: someone would be fortunate enough to occupy a free space, while someone would have to turn the emergency lights on while waiting for a free space.

The opinion polls showed that the mood of a typical city resident was dominated by unhappiness due to the hardships such as:

Getting away from the apartment building entrance: all the free space was occupied by parked cars;

Driving through the streets from home to work: nearly half of the roadway was occupied by the daily car parking;

Finding a place to park near the workplace: all the free space is occupied by parked cars.

The discussion of this problem in the media and blogosphere was reminiscent of the discussions that took place in the United States at the turn of the 1920–1930s:

"Aside from the weather, there is no question more discussed in our cities today than that of automobile parking". (Proceedings of the United States Conference of Cities 1928);
"The driver who engages in it shows little regard for the rights of others". "Parking reservations are as important as your hotel reservations". (Proceedings of the National Safety Council under the heading "Where Shall They Park?", Chicago. 1933).

The municipal and federal authorities were not inclined to interfere in the solution of this problem:

The growth of car sales was one of the major signs of the real improvement of living standards of a Russian citizen during the era of high oil prices;
The growth in the number and proportion of car trips seemed as a quite successful (and profitable for local budgets!) compensation for the public transport stagnation processes that were ongoing in the majority of the Russian cities.

Both Russian and invited foreign experts tried to attract the attention of the authorities to the obvious and extremely serious consequences of preserving the regime of "free parking" in the major cities of the country. It is appropriate to refer to the article by M. Blinkin, titled "The place for parking" (Blinkin 2007). A significant influence on the abolition of the "free parking" regime can be attributed to some of the well-known foreign monographs translated into Russian, namely "Transportation for Livable Cities"[5] by Vucan R. Vuchic and "The High Cost of Free Parking"[6] by Donald Shoup.

The severity of the problem was recognized at the political level in the early 2010s when the Moscow traffic jams (and traffic jams in many other cities of Russia) became so notorious that they eventually gained international renown. A good reference is the article "Stuck" published in a prestigious "The New Yorker" magazine (Gessen 2010).

The federal law contained the following provision: "the public areas (including squares, streets, driveways, embankments, squares, boulevards) are to be used free of charge by anyone". References to this provision allowed to declare that any local fees for parking on the road networks of cities would be considered as illegal.

The situation changed only in 2011, when two main federal laws "On the roads and road activities …"[7] and "On the road safety …" provided for the road owner's

[5]Russian issue 2011 edited by M. Blinkin.

[6]Russian issue 2014 edited by E. Muleev.

[7]Federal Law #257-FZ of 08.11.2007 "On automobile roads и road activities in the Russian Federation and on making amendments to some legislative instruments of the Russian Federation".

right to impose parking restrictions in the area of the road. In cases when road network was city-owned, the law gave a mayor an opportunity to introduce fees for parking along sidewalks.

The Government of Moscow took the first steps in this direction. Mayor S. Sobyanin launched a pilot project on November 1, 2012, which introduced a paid parking zone in the city center.

The subprogram "Establishment of an integrated parking system" of the city of Moscow state program "Development of transport system for 2012–2016" was established.

However, the paid parking system was finally established only by 2015; other cities in Russia started to adopt parking policies (Table 5.3).

Parking is free on Sundays and holidays. It is possible to pay for parking through an application on a mobile phone or via text message; credit card payments through street parking meters are possible; cash payments are not allowed.[8]

The analysis of consumer payment preferences in Russia indicates that they vary substantially across cities with text messages being the second most common payment method (Table 5.4).

Payment methods are being constantly improved, thus voice technologies (voice recognition) and voice enroll phone payment system are planned to be tested at the end of 2016.

The payment control is carried out solely by instrumental means (photo and video capture) handled by parking inspectors aided by "Parkon" cars. The penalty for unpaid parking ranges from €19 to €33 depending on a city.

The issue of paid parking systems was a subject of heated debates over the implementation period: social movements arose, rallies were held against paid parking, protest petitions signed.

According to surveys of the perception of paid parking in Russian cities, 76 % of social media users still do not accept the idea of being charged for the use of street and road network comparing this to a so-called "breathing tax".

The social media data[9] was collected between July 20 and July 30, 2015.[10] The idea was to measure the ratio of positive and negative attitudes of Novosibirsk residents to paid parking. In order to monitor, collect and analyze extremely large arrays of unstructured text information from the social media, Brand Analytics application was deployed.

The application detected 822 mentions from 703 individuals on social media during the period under study. Negative connotation was attributed to 38 % of

---

[8]Except Krasnodar. The change is returned to a parking card, user's personal account or in cash through the company's pay office. 661 people have used this opportunity since the start of the project collecting change of 420,000 rubles.

[9]The study was done for a thesis (Higher School of Economics) by L.P. Tufanova on "Factors of long-term efficiency of the transport levies system to fight traffic jams".

[10]On July 23, 2015 for the first time Mayor of Novosibirsk announced a pilot project of paid parking.

Table 5.3 Features of municipal paid parking in Russian cities

City	When introduced	Cost[a], € per hour	Number of existing parking spaces	Planned number of parking spaces	Administrator Organization
Nizhny Novgorod	April 14, 2016	€0.3[b]	301 (in summer 2016 65 more spaces)	3681 (400 by the end of 2016)	Municipal public institution
Kazan	February 9, 2015	€0.7	1314 Average occupancy —50 %, in some areas—95 %	2009 (by the end of 2016)	Municipal budgetary institution
Penza	January 18, 2016	Drama Theatre—0.4 €—from 8.00 to 22.00, €0.3—from 22.00 to 8.00, €2—for the entire period from 22.00 to 8.00 Multi-level parking lot at the intersection of Kuraeva and Volodarskaya—€0.1 first hour, €0.3 after that	-30 theater	–	Municipal Unitary Enterprise
Yekaterinburg	May 26, 2014[c]	€0.4	2,354	2387	Municipal budgetary institution
St. Petersburg	September 3, 2015	Category A and M—€0.4; Category B—€0.8; Category C—€1.6	2,907	46,000 by 2018	State public institution
Moscow	June 1, 2013	Differentiated tariff from €0.5 to €1.7	44,111	There is a selective extension of paid parking zone	State public institution
Krasnodar	2013	€0.4	5521 (17 % of the total number of parking spaces)	8000 by 2017	Limited Liability Company
Krasnoyarsk	April 2, 2015	€0.4	1422[d]	1500	Limited Liability Company

(continued)

Table 5.3 (continued)

City	When introduced	Cost[a], € per hour	Number of existing parking spaces	Planned number of parking spaces	Administrator Organization
Belgorod	November 24, 2014	€0.4[e]	2519		Municipal public institution
Tula	October 15, 2015	In the historical district €0.3, €0.1 the rest	5276	–	Municipal public institutions
Kaluga	September 15, 2014	€0.5	852	2000 till 2019	Municipal public institutions
Sochi	May 15, 2015[f]	€0.7 per hour or €6.6	1015	Up to 5000	Municipal public institutions
Rostov-on-Don	End of 2016[g]	€0.4[h]	1210	Until the end of 2016 it is planned to create about 6.5 thous. spaces	Limited Liability Company
Tver	July 1, 2015	€0.4	822	1500–2000	Municipal budgetary institution
Tyumen	May 12, 2016	€0.3	182	3500	Municipal public institutions
Vologda	February 1 2016[i]	€0.3	2323	–	Municipal budgetary institution
Perm	August 15, 2016[j]	€0.2	–	2500	Regional branch of a public joint stock company
Novosibirsk	Tender to be declared in 2016	€0.4	–	10,800 (1.500 in 2016)	Municipal budgetary institution

[a]Hereinafter financial parameters of the parking systems are calculated based on the rate of 76 rubles per €1
[b]In summer 2016 the second parking for 65 spaces will open at €0.7
[c]Fines for illegal parking are imposed only from 2016. Revenues from parking began to dropped when citizens realized they were not fined. If in 2014 80 % of drivers paid for parking, in December 2015 this share went down to 19 %
[d]In addition to paid parking organized by the project investor, 5 free of charge intercept parking lots for 1327 car spaces were built in the center of Krasnoyarsk
[e]First 30 min free
[f]In December 2013 a system of paid parking along the curb was launched in the test mode
[g]Currently the system is working in the test mode without payment collection
[h]So far the system is working in the test mode, no payment is collected
[i]In Vologda the paid parking project has been put on hold by the decision of the city mayor
[j]The commissioning was postponed 6 times in two years

Table 5.4 Parking payment methods by city, %

	Text messages	Parkomat	Special mobile applications[a]
Moscow	24.5	10	65
Ekateriburg	–	70	–
St. Petersburg	36.61	51.33	12.07
Nizhny Novgorod	29	58	13
Tyumen	–	90[b]	–
Rostov-on-Don[c]	40	40	20

[a]In all cities there are special web sites
[b]By a bank card
[c]When operating in the test mode

posts, while 81 % of the total number individuals analyzed voted against the new tariffs and only 19 % expressed positive attitudes towards the project. Generally, the arguments against the project were not constructed logically. Latent semantic analysis (LSA) performed unveiled the underlying reasons for negative attitudes expressed:

- 42 %—illicit enrichment of authorities;
- 28 %—vague theoretical concept;
- 17 %—illicit enrichment of Muscovites;
- 13 %—non-effective implementation of the measures.

Unlike the group where negative attitudes towards paid parking were prevalent, the supporters of the project built up logical arguments and reasoning pointing out as that if implemented the project eventually would ease congestion.

Thus, the main reason for the protests expressed is the absence of transparency in public spending. Therefore, now the city authorities take efforts to raise public support for parking projects.

For example, the Moscow's budget revenues from parking fees and tickets are allocated to projects aiming to improve city areas, in proportion to the amount collected in those areas.[11] It has been decided to provide project financing to the relevant bodies in 2016 prior to the revenues being collected.

At the same time, the district administration presents its own projects, based on the wishes of local residents, for approval by the municipal councilors. Here are some examples of such projects:

- Improvement and planting of greenery in courtyards;
- Improvement and planting of greenery around buildings;
- Improvement of streets and maintenance of street furniture;
- Improvement and maintenance of parks, squares and boulevards;
- Measures to create a barrier-free environment.

[11]Government of Moscow Resolution #849-PP of December 26, 2012.

Starting from 2014, up to 30 % of the city budget revenues from parking payments and fines for non-payment can also be invested in local initiatives to improve the transport infrastructure.[12]

There is yet another area of improvement closely associated with the introduction of paid parking regime. The fact is that wherever the paid parking was introduced, many motorists began to look for free parking in the courtyards of neighboring houses. As a result, the local residents started installing automatic barriers blocking the entry for unauthorized cars. The Moscow Government decided to co-finance the installation of barriers.[13]

At the same time Russian cities' authorities point out that revenues collection is not the main purpose of the projects to provide parking space.

The parking rates, as well as revenues associated vary greatly by city (Table 5.5).

A large share of revenues collected come from hourly parking payment. For instance, the share accounted for 88 % of the total revenue from paid parking in St. Petersburg over the first eight months after the paid parking was introduced, i.e. from September 3, 2015 to May 2, 2016. Budget also gets revenue from issuing residential, monthly and yearly parking permits.

They introduced parking lots intended for permits owners only in 22 streets in Moscow starting from March 15, 2016. Parking permits are issued to carsharing clients as well.

Subscription with monthly parking option for residents was introduced in Moscow on July 1, 2016. Seven parking lots of this kind will become available in the peripheral areas of the city.

In general, there exists a risk of losing existing facilities for long-term car storage, known in Russian cities as "garage co-operatives" without providing alternative parking locations. Also, garage parking lots were established for non-profit motorist organizations in the past (for example, there were 1483 parking lots of the Moscow City Union of Motorists). Their lease agreements are not to be extended. Following the decisions made by local courts, garages are now being often removed.

A survey showed that 70 % of car owners buy a garage for fear of theft and vandalism, 20 % use it to perform repairs, store winter tires, consumables and maintenance materials, and 10 % believe it is a method of investing money and making money on leases. For many, the garage is a place for pleasant pastime and socializing with friends. In addition, under certain conditions, it is possible to store preserved vegetables—the results of their dacha activities.

As an alternative to individual garages, Russian cities began to develop a variety of programs aimed at establishing multi-level parking.

---

[12]Moscow Government Decree N 88-PP dated March 3, 2014 "On Amendments to the Decree of Moscow Government dated December 26, 2012 N 849-PP".

[13]Moscow Government Decree dated July 2, 2013 N 428-PP.

Table 5.5 Key financial indicators of paid parking projects

City	Cost of the project, thousand €	Budget revenues from the project, thousand €	Occupancy of parking spaces, %
Nizhny Novgorod	140	During the first three weeks of the parking operation—0.2	–
Kazan	499.6	143.4 86.8—fines (2015)	50
Yekaterinburg	347.5	January–October 2015–154.3	–
St. Petersburg	3947.4	2015–626.3 (plan—1907.9) From January to May 2016—460.5	33.2 (2015)
Moscow	98684.2 (210526.3 taking into account the purchase of tow trucks)	2nd half of 2013—46052.6 2014—205263.2 2015–5.7	78.8 (2014) 75 (2015)
Krasnodar	157.9	During the first 9 months of 2015—64.3 (24.4 % of the plan) From May 2014 to January 2016—118.4	–
Krasnoyarsk	921	2 (per year)	–
Belgorod	513.2 (2015)	From January 1 to June 1, 2016 paid parking along the streets—56.4; covered plane parking—55.2	–
Tula	349.5 annually for maintenance	October–December 2015—104.2 January–March 2016—89.8	–
Kaluga	263.2, 171 for expansion	175 (2015)	–
Tver	131.6 the first parking lot (200 places)	6.6 (month) 65.8 (fines in 2015)	Uneven: at some streets occupancy is 85 % higher than at others
Tyumen	153.9	3 (For 1 month)	–
Perm	115.8 (2016) 344.5 (2017) and 393.4 over three years	–	–
Sochi	–	As of October 5, 2015—105.9	–
Rostov-on-Don	7013.2	–	–

For instance, the launch of the "People's Garage" program in Moscow in 2009 involved building a multi-story car park (usually from two to eight floors high). Parking lots sales started on February 15, 2010 with the average price of a parking space set at €4,600. By the end of 2015, it became obvious to the city authorities

Table 5.6 Number of parking spaces established near the residential dwellings in Moscow

	2011	2012	2013	2014	2015
Number of parking spaces, car-spaces	453,763	11,425	15,205	12,269	18,231

Source Data from Official portal of the Moscow Mayor and Moscow Government (2016)

that demand for was far below that had been previously expected because of the high cost and inconvenient location of parking. The authorities now wish to convert them into shopping malls and service centers.

As a result, they believe that planar-type parking lots may now resolve the issue of residential parking (Table 5.6).

The effectiveness of the paid parking is affected by a number of institutional factors.

Currently buying and selling of parking spaces are carried out in the informal sector and is not regulated by the state. In accordance with the current regulations, parking spaces located on the semi-basement floors of apartment buildings are an integral part of those buildings and regarded as shared property. However, in practice, developers actively sell parking spaces to all those who want to buy them. They cession of rights agreements are used here and parking spaces are classified as non-residential premises. The Ministry of Economic Development has proposed to amend the existing legislation to allow official registration of parking spaces located in the basements of apartment buildings as real estate.

It has to be noted that at the federal level there is no single legal act comprehensively regulating all issues related to the organization of parking spaces.

The establishment and development of parking are regulated by different types of the Russian legislation, including civil, administrative, land, tax, urban planning and budget laws.

With regards to the regulation of road activities, road safety and the establishment and operation of parking (parking lots) on roads, the separation of powers between the public authorities of the Russian Federation entities and municipalities and local governments is laid down in the relevant Federal statutory instruments. Thus, regional authorities exercise their powers in designated areas in relation to the public roads a the inter-municipal and regional level while local authorities do that in relation to local public roads.[14]

In addition, it should be noted that there is no single approach to developing regional standards of urban design by Russian entities. In some cases, these enactments are advisory in their nature and do not establish any mandatory requirements or values that must be taken into account when creating parking environment in cities.

[14]Federal Law #184-FZ of 06.10.1999 "On general organizational principles of legislative (representative) and executive bodies of state power of the Russian Federation entities"; Federal Law #131-FZ of 06.10.2003 "On general organizational principles of local self-governance in the Russian Federation".

Regional regulations provide that rates and parking charges should be set in line with conditions outlined in relevant service contracts. In legal practice, paid parking use agreements are also classified as the contracts for the provision pf services.[15]

Currently, the cities of the Russian Federation only begin to form the regulatory framework for various issues in the field of development and operation of parking lots. As a rule, the resolution of the municipal administration on the approval of the order of creation and use of car parks and parking spaces in the municipality outlines the rules for creation and use of parking (parking lots), including paid parking.

Municipal regulations also contain provisions setting specific benefits requirements applicable to certain categories of paid parking users including disabled people, WWII veterans, etc.

Russian cities use different conceptual framework when passing their relevant legal documents. In addition, sometimes the conceptual framework is not even outlined in their regulations.

State-financed institutions in most regions perform paid parking administrator (operator) functions: "AMPP" in Moscow, "City Parking Service"—in Yekaterinburg, "Zelenstroy"—in Tver, "Service of Uniform Order of Municipal Economy"—in Kaluga, etc.

In fact, commercial organizations don't have access to management of paid parking. An exception is given only to a few cities where the parking operator is selected on the basis of competitive bidding.

Krasnodar is the first city in Russia where parking space has been established without using public funds (Berdnikov 2016). All works and services performed are in accordance with the investment agreement signed with the company "City Parking" (the selection was done on the basis of competition).

Upon the expiry of the investment agreement the entire infrastructure (parking equipment, hardware and software systems, software) ownership will be transferred to the city of Krasnodar.

Ninety percent of revenues received from the use of parking spaces during the investment agreement period go to the investor as compensation of expenses he incurred by supplying equipment and its maintenance, and ten percent of the revenues are allocated to the city budget. Revenues raised from parking tickets go to the municipal budgets in full.

In order to attract private business to the construction of parking lots investors commonly receive tax privileges and preferences. Some cities set lower tax rates for organizations engaged in providing parking services and qualifying for paying a single tax on imputed earnings (for example, St. Petersburg, Kaluga, Chelyabinsk). In Kazan, there are income property tax incentives, as well as the option to use a Simplified Taxation System setting the differentiated tax rates.

[15]Among ways to regulate civil-law relations regarding the use of paid parking we can mention lease contract, paid services agreement or custodial agreement.

Parking tariffs together with the method for calculating parking tariffs are set only in a small number of cities.

At the same time, in Moscow and St. Petersburg the prices are set to achieve optimal occupancy rate of parking lots. In Kaluga, Ryazan and Chelyabinsk the base rate is set relative to the costs associated with the establishment and operation of a parking, i.e. it is focused primarily on the break-even operation which is a fundamentally different approach to the tariff policy. As a result, most cities use a single tariff is used, and only some—a differentiated scale.

Studies of international parking pricing practices indicate that optimal occupancy rate should stand around 60 during off-peak time and up to 85 % during peak hours (Shoup 2005). Staying within this range you get a reduction in the number in the so-called "wandering" traffic (drivers traveling on the road network looking for a place to park at a speed which is slower than that of the average flow) and, as a consequence, an average speed increase.

This effect does not occur if occupancy rates keep higher than those indicated above, the lower rates signify that parking space is poorly organized (supply of parking lots exceeds the demand for them), which leads to underutilization of expensive urban space. The actual occupancy rate of parking spaces within the Garden Ring in Moscow at the end of 2013 stood similar to that indicated above but, nevertheless, was deemed unsatisfactory (Figs. 5.9 and 5.10). As a result, it has become one of the arguments for implementation of a differentiation policy and price increase by the Department of Transport in Moscow.

Therefore, differentiated parking rates were implemented according to parking zones of the city:

- Within the Boulevard Ring, that is, in the Moscow Down Town, a parking lot charges 1.1 for the first hour of parking, and €1.7 for each additional hour;
- In the area between the Boulevard and the Garden Ring—€0.8 regardless of the duration of the parking session;
- Within the Third Transport Ring—€0.5.

In addition, the responsibility for parking violation differs across the entities of the Russian Federation. In general, regional laws establish liability for failure to pay

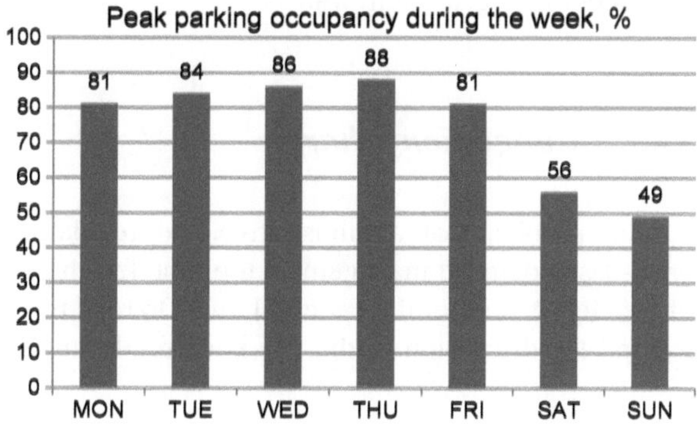

Fig. 5.9 Peak occupancy of parking in Moscow during the week at the end of 2013 [*Source* Reproduced by author from Moskovskoe parkovochnoe prostranstvo (2013)]

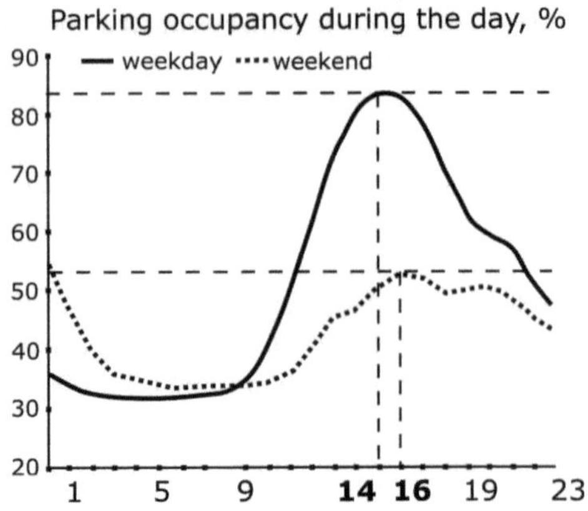

Fig. 5.10 Peak occupancy of parking during the day in Moscow at the end of 2013 [*Source* Reproduced by author from Moskovskoe parkovochnoe prostranstvo (2013)]

for fee-based parking (parking spaces), located on the public roads of local importance; the liability amount is established by municipal regulations. In some regions, liability is established for parking vehicles, including dismantled ones, at children's playgrounds and sports fields, lawns, green areas, outside of parking boxes and elsewhere outside the other places designated for parking within the territory between residential buildings.

A particularly urgent problem for many cities implementing parking projects and introducing paid parking is obtaining information about the owner or operator of the vehicle who has violated any provision of parking regulations as to impose relevant administrative sanctions against them.

For example, 56.1 thousand draft resolutions on an administrative offense were issued for parking vehicles without paying the parking fee in the first nine months of the operation of paid parking zones system in St. Petersburg. However, requests to provide personal data were turned down by the Main Directorate of the Interior Ministry in St. Petersburg and Leningrad region. In Yekaterinburg, the Russian Interior Ministry also refused to provide the data requested to regional authorities. In Kazan city traffic police was obliged by the court decision to share the database with the city administration.

Personal data of car owners may be shared between the Russian Interior Ministry and the authorities of the Russian Federation entity only if there is a relevant agreement between them in place.

5.4.2 Congestion Charge

Similar problems of administrative and regulatory nature apply to the second price-based demand management tool: the fee charged for entering a particular area of the city. A draft of the Federal Law "On road traffic management" introduced the fee and the discussion of the law is currently underway.

On one hand, fees for entering certain urban areas are a demand-management tool, on the other, they generate revenues that city receives.

A growing number of those large Russian cities (Moscow, Kazan) that develop and implement TDM strategies began to seriously consider adopting the foreign concept of "controlled access". The concept implies introduction of car-free zones in certain areas of the city (mainly, the historical part, areas of business activity and high-density residential developments). Technical, organizational and legal aspects of the congestion charge are under study now.

People react mostly negatively to the introduction of the charge. For example, Kazan residents' opinion survey[16] results collected through a social network revealed the following: despite the fact, that almost all respondents agree that Kazan face environmental pressures and traffic congestion, *80 % of respondents believe that charging for travelling within the city center zone is not the solution to traffic and environmental problems*.

The underlying reason for such responses is skepticism associated with the policy implementation rather than people having doubts about the effectiveness of the instrument itself.

Any additional charges are regarded as new taxes that do not necessarily lead to any significant changes. Moreover, Russian population believes that corruption remains prominent at all levels of the public sector, thus there are concerns that the revenues received from collecting the congestion charge will not solve any problems and will simply be stolen. Public perceptions of the congestion charge is that it is not a part of transport policy and just a new attempt to get additional fees from the population.

Let us consider the factors affecting the feasibility of congestion charge introduction, using the city of Kazan as an example.

The specific feature of the Kazan transport system is that the city road network is heterogeneous and consist of two tiers. The structure of the macro level is presented as the core network of radial and ring roads, while staggered layout is widespread on the micro level of administrative districts.

The total length of road network remained practically unchanged in the last five years (an increase of only 2.2 %). The area of the carriageways significantly changed only recently, when lanes were added to individual streets in preparation for the Universiade-2013.

The Kazanka river, Kaban lakes divide the Kazan city into separate parts with rather limited connection between them. The city transport system features are high and growing motorization rates and the high level of freedom of choice for transport users. Transport demand is almost unlimited; to a large extent it is met by individual passenger transport.

---

[16]This survey was conducted by the Institute of Transport Economy and Policy of Higher School of Economics within the framework of developing a congestion charge system in Kazan.

At the same time the following driving restriction are in place: cargo trucks are banned from entering and passing the city, one-way traffic, parking restrictions, restriction on performing certain driving maneuvers.

Compared to other Russian cities the traffic is better organized in Kazan, but it is still far from that in European cities.

In this context, performance improvements implemented in Kazan are behind those in European cities; while there is plenty of technical equipment for traffic control, management level and regulatory framework require improvement.

The total length of dedicated lanes for public transport is 125 km, which is 8.4 % of the total length of the street and road network.

A major issue of concern when introducing the congestion charge is the presence of the main road crossings of the River Kazanka in the central part of the city. The changes in the distribution of the traffic between the existing bridges in case any of them becomes overloaded are difficult to make as the river embankments shall be strengthened beforehand.

In addition, the number of traffic connections of the center with the periphery is insufficient to meet the existing demand. The issue is further complicated by the fact that there are only a few crossings over the main rail line going through the city.

On the other hand, as there is a limited number of arrival points into the city center, the administration of the system involving congestion charge implementation becomes easier.

In order to work out the amount of fee payable by users to enter the city center of Kazan, an estimation of the generalized travel cost was used.

Generalized travel cost includes the following elements:

- Direct cash costs for the user ("Out of Pocket Price"). If travelling by car, the cost is equal to that of fuel for a road trip;
- Value of Travel Time (VOTT) calculated as travel time multiplied by "the value of time".
- Fee charged for entering the city center (in this case, it is the congestion charge).

A key point here is to estimate the VOTT distribution among the total number of potential users. VOTT estimations for a particular user can either be based on the level of his monthly income, or the travel option that he may choose. Of course, both of the hypotheses ("the higher the wages—the higher value of time", "the higher the willingness to pay for the trip—the higher value of time") are not quite true. Foreign studies show that there is a large variation in VOTT among people and that the value of the estimate actually depends on purpose and circumstances of a particular trip. In addition, VOTT (as well as the value of spare time) stands above the average hourly earnings in high-income countries. VOTT is always significantly lower than hourly wage of a car owner in low-income countries (including Russia).

As for sociological research on VOTT estimates for an average Russian motorist (in terms of or value of free time), none of this kind has been done so far.

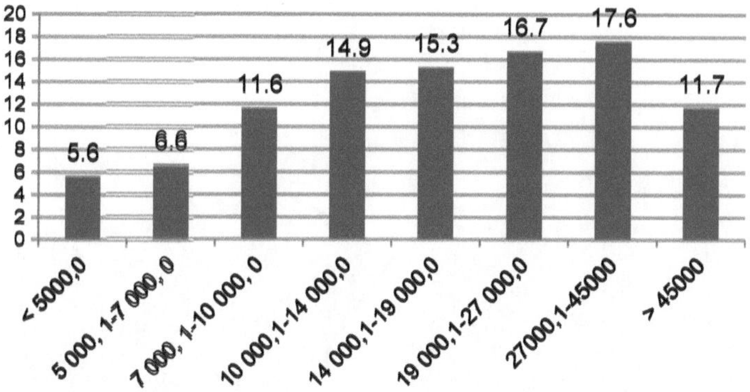

Fig. 5.11 Distribution of per capita income, Kazan, 2012 [*Source* Created by authors using the data from Russian Federation Federal State Statistics Service (2016)]

Fig. 5.12 Distribution of value of 1 h of Kazan car owners time (*Source* Created by author)

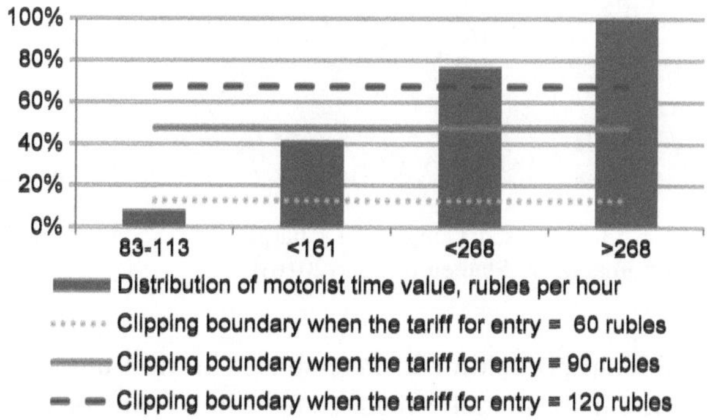

In this case, in order to calculate possible charge rates and traffic volumes for the projects implying a congestion charge introduction the only option left is to recourse to the available statistical data on the size of monthly wages.

Income distribution in Kazan city in 2012 served as the major input data for our analysis (Fig. 5.11).

Next, we make two conventional and plausible assumptions:

- Car owners are in the top 50 percent of the income distribution;
- VOTT is 75 percent of the hourly average income.

The approximate distribution of "time value" of Kazan car owners looks as presented below (Fig. 5.12).

If we further assume that the time required getting to the city center increases by 0.5 h provided restrictions imposed, then 12.8 % of drivers will abandon their idea of driving to the center of the city if the tariff is set at 60 rubles.

The assessment of changes in traffic levels on the street and road network if the congestion charge is introduced in Kazan was done based on the transport model developed by the Scientific Research and Design Institute of Regional Development

Fig. 5.13 Proposed Congestion Charge zone boundary in Kazan [*Source* Created by authors using images by Stamen design (2016)]

and Transportation (St. Petersburg) in the framework of the development of integrated map of traffic management in Kazan for 2014–2020 with a view to 2030. The Kazan Transport Committee has provided data on the current state of their SRN (in the .shp file format).

Using the ArcGIS system the data files were converted into the map of roads that was further transferred to the transport modeling suite "AimSun" (Spain).

An up-to-date origin-destination matrix (ODM) was estimated. According to study performed the following conclusions were drawn [The congestion charging zone is highlighted (Fig. 5.13)]:

- The traffic load within the congestion-charging zone reduces by 15–20 % on average.
- The traffic load within the neighboring to congestion charging zones increases by 12–18 %.

In addition, the international best practices for congestion charge suggest that a decrease (of about 15.8 % from the current level) in traffic load is expected to occur within the city center once a congestion charge is introduced. Thus, the most congested areas within the congestion charge zone shall be eliminated.

In this case, assuming the demand for travelling to city center contracts by 10 %, which is less than the number of those using the detour routes, then the additional traffic load on roads next to the city center, would rise by 2–8 %. The rise associated could actually have a negative impact on road traffic in the areas indicated.

In the end, we believe it is worth considering alternative ways to reduce passenger car trips to the center of Kazan city without resorting to measures aiming at reducing transit traffic flows.

Thus, we have come up with several alternatives, namely:

- the development and provision of parking facilities in the downtown area;
- the introduction of restricted environmental zones implying pollution restrictions being imposed on driving in city center; thus vehicles do not complying with certain requirements shall be rolled out.

It should be noted that the current Law № 257-FZ gives the right to the bodies of state power of the Russian Federation the authority to make decisions over the introduction (and termination) of a fee-based system on public roads of regional or inter-municipal importance.

However, such decisions may be implemented only in the case if there exists an alternative route that may be used by drivers charge-free (clause 37 of the Law) or such route should be provided prior to the commissioning of a new or renovated toll road or part of a road. At the same time, the Law provides that the total length of charge-free alternative route should not be more than three times higher than the length of toll road. Eventually the clause outlined above makes it impossible to impose a toll (de facto, but not de jure) on public roads.

Besides, along with roads of regional or inter-municipal importance within the proposed toll collection area public roads of local importance can also be located, where local authorities are to make decisions on tolls themselves.

In this regard, the introduction of administrative responsibility for toll violation is seen as a necessary step to prevent non-payments by drivers of vehicles.

Note that the administrative responsibility for failure to pay a toll on toll roads has not been established yet (even though current regulations provide for such).

Note that the federal regulations provide that parking tickets shall be issued for parking in places where parking is prohibited (it is regarded as a violation of traffic rules, the responsibility is envisaged in the Code of Administrative Offences, the minimum amount of a fine is set at 500 rubles.), as well as for parking within dedicated disabled parking spaces (amount of a fine varies from 3 to 5 thousand rubles); however, the fines associated with failure to pay for parking are introduced by the authorities of the federation entity.

However, before introducing a toll, it is necessary to establish a well-functioning pay-to-park system and make sure people get used to the fact that driving to certain areas might not be cheap at all. It took London, Singapore, Milan and others decades to do that. High fees charged for parking in centers of these cities paved the way for tolls and congestion charges being introduced afterwards.

The introduction of economic disincentives to the use of a car during peak hours shall take place only if an acceptable alternative such as high quality public transport is provided for. In addition, individual transport users' behavior has to be analyzed supplied with the forecast of their readiness to pay tolls and a breakdown of users by travel modes.

Thus, the organization of paid access to the city center is an effective traffic management tool but the right time for its implementation in the Russian cities has not come yet ...

Transport strategy of Russian cities still favors private cars compared to that existing in such cities as London, Stockholm, Milan, Singapore, Beijing, Bogota, where measures to reduce the number of private cars are much stricter.

The largest cities in Russia have only recently started implementing demand management tools for the use of personal vehicles for daily work trips from the periphery to the center.

The redistribution of road network in favor of public transport, the development of walking and cycling infrastructure and at the same time decreasing the capacity of streets within city centers; a reduction of the number of on-street parking spaces along sidewalks are currently among the most popular non-price-based methods. In this context, it is worth mentioning the number of activities undertaken (or planned to be implemented) over the recent years in Moscow and in some other major cities:

- The development of pedestrian-only streets, as well as street reconstruction works carried out to expand sidewalks;
- The provision of dedicated lanes for public transport (bus lanes);
- The development of cycling infrastructure and implementation of a bike-share system.

Charging certain polluting vehicles to discourage them from entering the centers of certain Russian cities (a delay of about 20–30 years compared to European practice) became one of the non-price demand management tools introduced.

Russian cities, being slightly behind European ones, have also began to introduce car sharing.

Parking fee is currently the main price-base demand management tool in the Russian cities.

References

Anderson ML (2014) Subways, strikes, and slowdowns: the impacts of public transit on traffic congestion. Am Econ Rev 104(9):2763–2796

APTA (2007) Public transportation: Benefits for the 21st century. American Public Transportation Association. http://www.apta.com/resources/reportsandpublications/Documents/twenty_first_century.pdf. Accessed 31 Aug 2016

Berdnikov A (2016) O sozdanii parkovochnogo prostranstva v Krasnodare (On creation of paid parking zone in Krasnodar). Urban Report 7. http://lanit.ru/press/smi/8909/. Accessed 31 Aug 2016

Blinkin M (2007) Mesto dlya parkovki (The place for parking) http://www.archnadzor.ru/2007/10/08/mesto-dlya-parkovki/. Accessed 31 Aug 2016
Cambridge Systematics (1999). Public transportation and the nation's economy: Economic development research group. http://www.apta.com/resources/reportsandpublications/Documents/vary.pdf. Accessed 31 Aug 2016
Cambridge Systematics. (2002) The benefits of public transportation: An overview. Reichman Frankle Inc. http://www.apta.com/resources/reportsandpublications/Documents/ben_overview.pdf. Accessed 31 Aug 2016
Davis T, Hale M (2007) Public transportation's contribution to U.S. greenhouse gas reduction: From science to solutions. http://www.apta.com/resources/reportsandpublications/Documents/climate_change.pdf. Accessed 31 Aug 2016
Departament transporta goroda Moskvy (2016) Bike lanes. http://transport.mos.ru/#!/page/bicycle/veloinf/index.php? Accessed 31 Aug 2016
Deror avi (2009) Arbat Street (Moscow). https://commons.wikimedia.org/w/index.php?curid=8444384 Accessed 31 Aug 2016
DfT (Department for Transport) (2004) LTN 1/04—Policy, planning and design for walking and cycling. http://www.ukroads.org/webfiles/LTN%201-04%20Policy,%20Planning%20and%20Design%20for%20Walking%20and%20Cycling.pdf. Accessed 31 Aug 2016
Duranton G, Turner MA (2011) The fundamental law of road congestion: evidence from US Cities. Am Econ Rev 101(6):2616–2652
Gärling T, Schuitema G (2007) Travel demand management targeting reduced private car use: effectiveness, public acceptability and political feasibility. J Soc Iss 63(1):139–153
Garrett-Peltier H (2011) Pedestrian and bicycle infrastructure: a national study of employment impacts. Political Economy Research Institute, University of Massachusetts, Amherst. http://www.peri.umass.edu/fileadmin/pdf/published_study/PERI_ABikes_October2011.pdf. Accessed 31 Aug 2016
Gessen K (2010) Stuck. The New Yorker. http://www.newyorker.com/magazine/2010/08/02/stuck-3. Accessed 31 Aug 2016
Hovavko IY (2012) Internalizatsiya vneshnikh ehffektov ot zagryazneniya okruzhayushhej sredy v RF: voprosy teorii i praktiki (Internalization of external effects of environmental pollution in the RF: issues of theory and practice). Teis, Moscow
Ilicheva I (2012) The first 3D speed braker to appear in Volgograd. https://ria.ru/society/20120529/660116583.html. Accessed 31 Aug 2016
Jaffe E (2013) Why people choose cars, even when mass transit would serve them better. http://www.theatlanticcities.com/commute/2013/02/whypeople-chose-cars-even-when-metro-would-be-faster/4566/. Accessed 31 Aug 2016
Litman T (2010) Quantifying the benefits of nonmotorized transportation for achieving mobility management objectives. Victoria Transport Policy Institute. http://mail.artshenkman.com/cs/groups/content/@webottawa/documents/pdf/mdaw/mdy3/~edisp/con056214.pdf. Accessed 31 Aug 2016
Litman T (2012) Evaluating public transportation health benefits. Victoria Transport Policy Institute. http://www.vtpi.org/tran_health.pdf. Accessed 31 Aug 2016
Litman T (2013) Evaluating public transit benefits and costs: best practices guidebook. Victoria Transport Policy Institute. http://www.vtpi.org/tranben.pdf. Accessed 31 Aug 2016
Litman T (2014) Evaluating active transport benefits and costs: Guide to valuing walking and cycling improvements and encouraging programs. Victoria Transport Policy Institute. http://www.vtpi.org/nmt-tdm.pdf. Accessed 31 Aug 2016
Mosgortrans (2016) Polosy dlya obshhestvennogo transporta (Dedicated lanes for public transport). http://www.mosgortrans.ru/?id=509. Accesses 31 Aug 2016
Moskovskoe parkovochnoe prostranstvo (2013) O tarifah za parkovku «s potolka» (On groundless fees for parking). http://parking-mos-ru.livejournal.com/1885.html. Accessed 31 Aug 2016
Nelson P et al (2007) Transit in Washington, DC: current benefits and optimal level of provision. J Urb Econ 62(2):231–251

Official portal of the Moscow Mayor and Moscow Government (2016) https://www.mos.ru. Accessed 31 Aug 2016
Online TDM Encyclopedia (2014) http://www.vtpi.org/tdm/. Accessed 31 Aug 2016
Open Data Moscow City Government (2016) http://data.mos.ru. Accessed 31 Aug 2016
Parry IWH, Small KA (2009) Should urban transit subsidies be reduced? Am Econ Rev 99(3):700–724
Russian Federation Federal State Statistics Service (2016) http://www.gks.ru/. Accessed 31 Aug 2016 (In Russian)
Shapiro RJ, Hassett KA, Arnold FS (2002) Conserving energy and preserving the environment: the role of public transit. American Public Transportation Association. http://www.gobrt.org/aptaenergyreport.pdf. Accessed 31 Aug 2016
Shanin M (2014) Lenina Street (Ryazan, Moscow region). https://commons.wikimedia.org/w/index.php?curid=32842846 Accessed 31 Aug 2016
Shoup D (2005) The high cost of free parking. Planners Press
Smart Growth America (2013) Building better budgets: a national examination of the fiscal benefits of smart growth development. Smart Growth America. http://www.smart-growthamerica.org/documents/building-better-budgets.pdf. Accessed 31 Aug 2016
Stamen Design (2016) http://maps.stamen.com/. Accessed 31 Aug 2016
Taylor MAP (2007) Voluntary travel behavior change programs in Australia: the carrot rather than the stick in travel demand management. Intl J Sustain Trans 1(3):173–192
Vaganov A (2014) Malaya Konyushennaya Street (Saint-Petersburg). https://commons.wikimedia.org/w/index.php?curid=35760287 Accessed 31 Aug 2016
Vuchic VR (1999) Transportation for livable cities. CUPR/Transaction, New Brunswick
Walker Parking Consultants (2011) Parking management plan Omaha Downtown Improvement District. Omaha, Nebraska. http://www.mapacog.org/images/stories/SSH_Publications/Final_Parking_Management_Plan_MAPA010712_Final.pdf. Accessed 31 Aug 2016
Weisbrod G, Reno A (2009) Economic impact of public transportation investment. American Public Transportation Association. http://www.apta.com/resources/reportsandpublications/Documents/economic_impact_of_public_transportation_investment.pdf. Accessed 31 Aug 2016
Winston C, Ashley L (2006) The effect of government highway spending on road users' congestion costs. J Urb Econ 60(3):463–483
Winters PL (2000) Transportation demand management. Transportation in the New Millennium. http://onlinepubs.trb.org/onlinepubs/millennium/00123.pdf. Accessed 31 Aug 2016
Yakimov M (2014) Gorod Perm, istoriya odnogo perekrestka (City of Perm, the story of one intersection). http://yakimovmihail.livejournal.com/139695.html Accessed 31 Aug 2016
Zali N, Abizadeh S, Bagherinia A (2013) New urbanism and urban design: Tools for changing behavioral patterns of the citizens. Int J Nat Eng Sci 7(1):31–36

Chapter 6
Formal and Informal Institutions for Urban Transport Management

Anton Vorobyev, Julia Shulika and Varvara Vasileva

Abstract After the total governmental control of the urban transport systems during the Soviet epoch came the 20 year long period of almost complete deregulation. Currently there is a trend towards the return to the practice of formal urban transport management, which represents a strange mixture of the remnants of the Soviet methods and the selective adoption of western urban planning practices. The chapter highlights the institutional aspects of transportation systems design and functioning. Using the neo-institutional approach, authors analyze urban transportation system management institutions as well as transportation policy of Russian authorities. The presented analysis consists of two levels: macro-level reveals trends at institution design, explains path dependency from the Soviet epoch. The micro-level put the light on the issue of decision design and the influence of certain actors. The clarification of the formal and informal urban transport management requires the overview of the following questions: the interests and the principles of authorities and private operators' interaction, practices of transport demand management implementation and public reaction, the evolution of public perception of private and public property. The chapter is organized as follows. The first, introductory part of the chapter is dedicated to explaining the approach and methodological framework used. The second part reveals peculiarities of Soviet transportation system heritage. The third part examines the challenges of 90s period—introduction of free market mechanisms and era of deregulation. Fourth part discusses the experience of the first decade of 21th century and relevant changes in transport system. In the final part authors analyze main institutions; both formal and informal which shape the modern transport system.

A. Vorobyev (✉) · J. Shulika
Institute for Transport Economics and Transport Policy Studies,
National Research University Higher School of Economics, Moscow, Russia
e-mail: avorobyev@hse.ru

V. Vasileva
School of Political Science, Faculty of social sciences,
National Research University Higher School of Economics, Moscow, Russia

M. Blinkin and E. Koncheva (eds.), *Transport Systems of Russian Cities*, Transportation Research, Economics and Policy,
DOI 10.1007/978-3-319-47800-5_6

6.1 Introduction

The search for an optimal configuration and management mechanisms of a transportation system is heavily dependent on the thorough analysis of the examined system and conditions under which it is functioning. Transportation system of modern Russia is a very complex one—based on Soviet fundament of centralized planning, after undergoing the "wild market" period in 1990s it is making steps towards era of sustainable mobility. Nevertheless, the system functioning and further development depends on many factors of a different kind—political, economic and social that define results and effectiveness of any reforms undertaken in the transportation sector.

Unfortunately, opposite to studies conducted abroad (Prybyla 1966; Grieco 2011; Kolik et al. 2015) the contemporary literature lacks studies of Russian transportation system in means of institutional approach. The approach allows to investigate key pillars that shape the system, reveal its possible points of development and problematic spots that needs attention. Big Russian cities like Moscow and Saint-Petersburg face transportation problems that some countries have already solved. But on the other hand, blind import of successful practices can be nonresultative due to specific context under which the systems exist. Still, the study of Russian experience can be useful for developing countries that are undergoing processes of reforming the existing system of transportation.

The presented chapter focuses on investigating formal and informal institutions that poses influence on Russian transportation system as a whole and local systems as well. In the first, introductory section we describe the theoretical approach used. The second section discusses the Soviet heritage that mainly shaped transportation system of modern Russia. In the third section we discuss issues of the transitional period—1990s, when market mechanisms started functioning under the new political and economic regime. The fourth section is dedicated to analysis of the modern period, after 2000, revealing recently formed trends that shape the state-of-art Russian transportation system. In the last, fifth section we present in-depth analysis of institutions and actors that pose influence on transportation system functioning, and give examples of their influence.

Theoretical framework Defining an institution is not a simple task due to a huge legacy of institutional studies. Despite the multidisciplinary nature of institutional studies, the most concentrated and thus fruitful research efforts (including adaptation) were made within political science. That is quite understandable as political institutions shape not only the political rules of the game, but indirectly economic (North 1990; Acemoglu et al. 2005; Hodgson et al. 2006) and social institutions.

The only reason to consider institutional impact on the public policy and its specific areas such as urban transport management is that institutions either shape 'the rules of the game' or are 'the rules of game' themselves. Having a thesis "institutions matter" as a starting point of this discussion, we would like to take an excursion into the institutional theory in order to answer the question "why do institutions matter?"

Why institutions matter? *First*, because institutions are established 'rules of the game' enforced and maintained by actors' behavior, identities, and repertoires. *Second*, because one incorrect institutional choice can put an end to the system's development or put clock back for years. *Third*, because institution determines the sequence of participation (e.g. voting) and procedure of decision-making process; and assigns power, resources and authority to certain positions and roles. Institutions indirectly determine an outcome of any given decision-making process by shaping incentives and opportunities for actor's behavior, and establishing and promoting evaluation criteria. *Fourth*, institution is not limited to formally structured rules of game, but also consists of ideas, norms, cultural conventions and cognitive frames that provide actors with meaningful information for interpreting the context. Even if the formal rules are changed, the ideas behind them might stay the same. Actors are trapped in the same logic of context interpretation, do not change their behavior and neglect the changes in formal rules. *Fifth*, institutions matter because they form, maintain and advance the relationship between actors, giving them advantageous positions for resource extraction, moderation of agenda setting and so on.

Theoretically described role of institutions in shaping actors' behavior and decision making process has a number of practical implications. The very first one is Hall's royal road to identification of institutions. Hall introduced a concept of Standard operation procedures (SOP) (Hall 1986) that are specific rules of behavior that are agreed upon and followed (in general) by agents, whether explicitly or tactically agreed (Rothstein 1996). Therefore, SOP unites notions of formal and informal institution, but distinguishes institutions from broader and more personally related categories like habits or customs.

Before describing how institutions work and how they manage to impose constraints on actors' behavior in details, we should briefly discuss the nature of these constraints. Generally institutions in public policy perform their action-changing role through rules, practices and narratives. Rules are the engine of formal institutions, they are officially constructed and sanctioned by coercive actions and formal rewards and punishments. Rules are 'stable means of making choices' (Peters et al. 2005) that mainly creates procedures of decision making. Rules implementation is the key issue determining the effectiveness of formal institution. If rules are routinely enforced and individuals generally comply with them, formal institutions are considered to be effective. Opposite case—where noncompliance appears frequently and is usually left unsanctioned—is a sign of weak institutions and a wakeup call of state having dysfunctional institutional system.

Practices constitute and maintain informal or unstructured institutions that lack means of coercive actions. They are 'ways things are done around here' that are not officially recorded of sanctioned. Practices are disseminated through demonstration of what is considered to be a proper conduct. Despite informal nature and lack of coercive means of implementation practices are powerful means of institutional pressure that can help maintaining formal institutions or contradict them (Helmke and Levitsky 2003).

Alongside with formal rules and informal practices, institutions constrain actors' behavior via narratives. Narrative is defined as a "sequence of events, experiences, or actions with a plot that ties together different parts into a meaningful whole" (Feldman et al. 2004). Narratives are about explanation, persuasion and legitimation (or delegitimization) of current rules and practices. They reinforce norms (both formal and informal) through cognitive framing that creates shared understandings of 'why the things are done this way'. Narratives play an important role in maintaining institutional design in the long term prospective.

How do institutions influence public policy? First of all, institutional design that shape incentives and opportunity and evaluation criteria for actor's performance also revises strategic goals set by the elite.

Second option for institutions to constrain public policy is by determining who and how can make decisions.

Thirdly, alongside with determining how decisions are taken, institutions shape public policy by establishing rules and practices of the 'influencing game'. Institutions provide incentives and opportunities for interest groups to influence the political decision making process. Institutions not only distribute public resources in a way to create and support certain interest groups, but also maintain political opportunity structure and veto points.

Summing up, institutions matter in the public policy formation and implementation, because they not only shape rules and procedures of any decision making process, but provide interest groups with resources and frame lobbying opportunities. In other words, lobbying success cannot be only attributed to the interest group's resources or solely to the current configuration of political institutions within political opportunity structure, open or closed to pressure efforts. Previously discussed 'veto points' determine how interest groups could influence policy outcomes. Shaped by the institutional design and current electoral results, veto points provide lobbying opportunities for interest groups (Immergut 1990).

How and when do institutions change? Since political institutions play an important role in shaping policy decisions in different areas, we should briefly discuss conditions of institutional changes. According to the rich research legacy of the new institutionalisms, institutions are prone to changes (1) under external shocks; (2) through the process of competitive selection; and (3) due to the institutional design (Goodin 1996).

First condition for institutional changes—external shocks—is typically described as 'critical junctures' interrupting long periods of path dependent institutional continuity in the standard model of punctuated equilibrium adopted by historical institutionalism. Less radical and more incremental approach to institutional changes suggests that not only external shocks can create disruption in institutional equilibrium, but also internal pressure. Internal pressure for change is caused by enduring gaps between institutional ideals and practices. An example of such internal pressure source is described by March and Olsen (1996): when people lose faith in institutional arrangements, they switch from rules to standard operation procedures and further more. This reallocation of resources reduces the ability of institutions to enforce the rules and increases the capacities of alternative structures, thus promoting institutional changes.

Competitive selection is a term coined to describe a competition of different institutional elements (rules, routines, etc.) for survival and reproduction. Competitive selection states that there are no 'immortal' institutions over time, instead an ever-changing mix of rules exists and structures actors behavior in any given moment.

The processes of change attributed to institutional design take place if actors specify design in order to achieve some objectives (for example, constitutional reform weakening the executive branch by redistributing powers to legislature). Another reason for this type of changes is conflict design, the one that reflects the result of a bargaining game between different actors pursuing conflicting goals; or learning, which perfectly fits the 'import reform' cases—when actors borrow from others or adapt design with the most appropriate feedback results (March and Olsen 2006).

6.2 Urban Transport in USSR in the 1980–1990s

The Soviet administrative system of the urban transport systems has passed through numerous and very contradictory transformations in the 1930–1980s, so by the time of the communistic system break-up it was organized in the following way.

The property complexes of all transport branches were in the state ownership, each transport entity was under direct administrative supervision of the relevant ministry or department. The entities were functioning on the basis of the state plan which was set in the natural indicators (the number of the transported passengers, the amount of the transported freights, the amount of transport work formulated in passenger- and tonne-kilometers). Such indicators like the income, the expenses, including the labor costs, the labor productivity and the output per unit of vehicles or per capacity unit (the load-carrying capacity) of public transport vehicles were also preplanned. At the same time the transportation tariffs were fixed in the centralized way, and the government carried out the deliveries of public transport vehicles according to the centralized formalized rules connected with the standard rates on the updating of the vehicle park and the accomplishment of the target figures.

The railroads were under authority of the Ministry of Railways of the USSR, in particular:

- the entities of the rail transport which carried out the transportation of life-support goods for the urban population and the operation of the industrial enterprises;
- the suburban railroads which served the suburbs of the majority of the large cities of the country.
- the subways.[1]

The decisions on the construction of the subways were prepared at the level of the party leaders of the corresponding republics or regions, as well as they were to be coordinated in the key allied departments (A State Planning Committee of the USSR and the State Committee for Construction of the USSR) and be finally approved at the level of the top party management.

The Soviet subway and first and foremost the Moscow subway were a classic for the Soviet practice example of contradiction between «in nominal» and «in real». «In nominal» the Moscow subway is an efficient transit system with the exceptionally high freight capacity and the speed of traffic. «In real» it was an exhibition of achievements of Soviet culture (the outdoor pavilions and underground subway lobbies were designed by the best architects and interiors were decorated by the first-rate artists) and, at the same time, the structures for mobilization purpose. The first circumstance led to the situation when the development of the subway network was always sacrificed to create the next installation of the monument of Soviet architecture. The second one led to the choice of decisions on design that are uncommon to the international practice of the subway construction. Moscow metro lines (that were constructed in 1930–1960s) are traced in deep tunnels. The distance between the stations is so great that the trips from the suburbs to the center, as a rule, are bimodal: the destination has to be reached by the land transport.

The tramway depots and the trolleybus parks were under authority of the republican ministries of housing and communal services (Minzhilkomkhoz), the largest of which was one in the Russian Soviet Federative Socialist Republic (RSFSR).

In the cities of Russia which are located on the big rivers, an essential role in the transportation of goods (first of all, the construction freights), and also in suburban passenger traffic was played by the river shipping companies which were under authority of the Ministry of the river fleet.

The public motor transport has been the key element for the most of the cities, and its enterprises were under authority of the profile republican ministries. In point of fact each ministry represented the large state corporation, which was aggregating the professional carriers in segments of the cargo and the passenger motor transport. The largest ministry in RSFSR was the Ministry of the motor transport (Minavtotrans) which was supervising the core management structures in all

[1]By the time of the collapse of the USSR the subways were functioning in the 6 cities (Moscow, St. Petersburg, Nizhny Novgorod, Novosibirsk, Samara, Yekaterinburg), and also in Kiev, Kharkov, Tbilisi, Baku, Tashkent, Yerevan, Minsk. The present-day names of the cities are used here and further.

regions of Russia, besides all the enterprises of the cargo and passenger motor transport, including the bus and taxi fleets were under authority of the Ministry. Concerning the urban areas (the cities) the key element of this system was presented by the bus fleets, the largest of which operated more then 800–1000 buses.

Because of the reason that the profile ministries of the Union were absent, the certain coordination powers on the level of the Union, first of all, regarding the technical and innovative policy, were assigned to the Ministry of the motor transport (Minavtotrans) of RSFSR and the Ministry housing and communal services (Minzhilkomkhoz) of RSFSR.

Some certain exceptions concerning the general management structure of the motor and urban electric transport have been made for Moscow and St. Petersburg. The bus and taxi fleets, the trolleybus parks, the tramway depots, which were united in a single structure—Mosgortrans, remained so far submitted to the city authorities of Moscow. The city authorities of St. Petersburg had the trolleybus parks and the tramway depots under control, while the local association of the passenger motor transport (Lenpassazhiravtotrans) remained in subordination of Minavtotrans of the RSFSR.

The system of the centralized directive planning of the gross measures promoted obviously a weak interest of the transport entities to satisfy the needs of the transport service consumers and an active interest in the «statistics fraud».

It is necessary to stress that transport planning (with the pure market grounds and the orientation on the demand) is arranged by the principle of *ex ante* while by its definition the centralized directive (in other words—socialist) planning is arranged according to the principle of *ex post*. The best of the passenger transport entities had quite good standards of the transport planning and were able to plan route schedules and to arrange the route network properly on the basis of the regular surveys of the transport demand. All of the passenger transport entities had the directive planned indicators on the amount of transport work expressed through passenger-kilometers: it is clear, that the directive indicators have to be simple and additive, in other words, suitable for the addition on the basis of the subordinate entities in total. It was difficult to combine the fine points of the transport planning with the directive indicators, and sometimes it was practically impossible. Meanwhile, the implementation of the directive planned targets was considered to be imperatively necessary: the resources, first of all, the motor fuel were allocated to the entities strictly «in line with the plan»; the deliveries of the new buses have been also connected with the accomplishment of the plan (the new vehicles instead of the disposed ones); the employee bonuses as well as the promotion track in the entities were also depended on the result of the plan's realization. It was only the number of the passenger claims to be dependent on the fulfilment of the transport planning parameters.

The tariff rates of the passenger traffic were strictly made-up and were maintained at the extremely low level due to the political populism. Thereafter, the financing system of the passenger transport entities was based on the centralized grants and/or the cross subsidizing (in other words, the income was redistributed from the profitable kinds of transport, like the cargo, the passenger long-distance and the charter transportation,—to unprofitable ones, like the urban and suburban transport).

The first option was typical for the bus fleets in the big cities where the main activity was presented by the transportations on regular city routes. The second option was used on a mass scale by the peripheral territorial associations including the carrier entities which were carrying out both cargo service and passenger traffic by all means of transport. Both the centralized grants, and the mutual settlements of accounts within the cross subsidizing have been subordinated to accomplish the directive plan.

The implementation of the directive plan for the bus transport entities was determined, firstly, by the unique objective parameter—the number of the transported passengers that was estimated by the number of the sold tickets, and secondly, by the two relative parameters—the number of the transported passengers groups entitled to the special benefits (free ride), as well as the average trip distance. In these conditions the entities had no serious incentive to respond to the demand, notably to carry out the route schedules properly, especially in hours of the passenger traffic recession. However there were extremely strong incentives to overstate the number of «preferential» passengers and the average range of a trip, this fact led to the regular falsification of the inspections results.

The impetuses to overstate the number of trips were also quiet strong on the lower level where the driver negotiated with the tariff controller for a reasonable fee. Both contracting parties were aware that the top-management won't demand the clear data (the implementation of the directive plan does not depend on the accomplishment/cancellation of the trips). Besides at the same time the statistics fraud of the number of the trip could provide the driver an extra award, as well as the remained fuel could be used to make a little money on the side. Usually the trips with a small passenger traffic during the interpeak time-periods were not performed in line with this algorithm (however they were included in the report). Moreover such trips were not able to be excluded from the transport planning document (the route schedule): the official of the town council while approving the schedules would always raise objections: «The communist party teaches us to increase the level of the transport service of the population!». The problems of the work disruption and the statistics fraud on the trips quantity became large-scale in the second half of the 80s, so that the Ministry of Transport of RSFSR has issued a special order demanding the 100 % fulfillment of trips from the subordinate carrier-entities (RSFSR Ministry of motor transport order 1987).

The statistics fraud concerning the cargo work was carried out even in a more simple way: for example, the transportation of the cargo which hasn't got any receiver in interest (garbage, snow) allowed «to improve» the figures in the reports at the end of the planned year or quarter.

These total falsified data was accepted for true on default and was used in order to form the wide and multilevel informal communication in line with the connection «an entity—the territorial association—the ministry—the budget departments of the RSFSR and the USSR». The matter of this communication was the exchange of the resources distributed in a centralized way for some career bonuses and some personal benefits of all the participants of this connection-chain: from the local «commanders of production», to the high-ranking officials distributors. However, the same or the similar «system of the hierarchical biddings» occurred in other industries of the economy (Gaidar 1990).

The most active actor of such communication was the establishment of the Communist Party of the Soviet Union (CPSU), especially on the upper levels of the bureaucratic ladder: the head of the regional transport management who was holding an influential post in the party hierarchy at the same time had good chances to receive the new buses than his colleague—the «ordinary» member of CPSU without any party position.

It is important to emphasize that apart from the public motor transport, there was a large park of cars and buses which was under authority of other institutions not presented by the transport ministries and departments. In the public documents the terms like «the department-level cars (buses)», «the cars (buses) of non-general use» were commonly used.

Till 1991 the number of park of the department-level cars and buses was considered as the information prohibited from publication in the open sources. In order to illustrate the state of things we will provide the table (Table 6.1) from the official statistical digest.

It is necessary to draw attention that the number of cars of the general use (trucks, pickups and passenger vans), and also the public buses appears in the official statistics accurate within the vehicle, while the number of «cars (buses) of non-general use» does not appear at all.

The experts made an indirect assessment of this data in terms of the number of cars (buses) which were produced in the Soviet factories. Also they took into account the dynamics of the allowances, this data was provided by the reports of the public enterprises. According to the Table 6.1, as of 1990 the number of public buses made in the USSR was about 310 thousand units. Judging by the indirect estimates, the number of so-called «departmental-level buses» was no less than 1 million units. The ratio between the nondefense trucks of the general and non-general use is approximately the same.[2]

Most of these «department-level» trucks and buses were possessed by the large industrial enterprises. The buses were used in order to serve the daily cooperative trips for the employees on a route «the residential community—the enterprise's entrance», as well as to provide for the numerous collective events (typical for the socio-political background of the Soviet period) like the collective farm works,[3] the

[2]The official data on quantity and structure of the car park in the USSR (as well as the Russian Federation) is still absent: the confidentiality was replaced with the long-term statistical confusion connected with disintegration of the country and mass privatization of industrial assets in the Russian Federation. The data on the number of buses and cars of non-general use are given on the basis of the subjectibe assessment of the science editor of this book who was one of the leading experts in the Ministry of motor transport of the RSFSR at the end of the 1980s.

[3]It was a typical practice in USSR to involve the city workers and employees in the harvesting and other events in agrarian sector. In the 1950–1960s this practice was compulsory; in the 1980s it

Table 6.1 Park of vehicles in the USSR in the 1980s

Transport vehicles	1980	1985	1990
The motor cars of general use, incl. the light vans and the estate cars	710,267	741,477	894,316
In the personal property	7	2	12379
The buses of general use	276,623	309,800	309,537
Motor taxi-cabs of general use	135,264	128,455	116,698
The motor service vehicles—totally	536,520	405,697	425,678
Of general use	17,526	17,160	18,316
Of non-general use	518,992	388,537	407,362
The motor cars in the personal property, thousands of units	8218	12,444	16,408
Trolleybuses	23,868	26,771	28,505
Subway cars	4567	5899	6758
Tramcars	20,559	21,647	22,101
Special vehicles (motor cars и autobuses)—total, thousands of units	1464	1800	1761
Of general use	25	32	20
Of non-general use	1439	1768	1741
The production of the motor vehicles, thousands of units	1327	1332	1259

Source created by author using data from Goskomstat (1991)

cultural, sports or political events. The department-level trucks have been included in the production process of the same industrial enterprises. These vehicles as well as the department-level buses had the two most important advantages over the cars and the public buses:

- They hadn't got to deal with the state plan on the tonne-kilometer or passenger-kilometer transport work, thereafter, they hadn't to make the extra transport work;
- they were always just in time at the disposal of the user in interest.

The administrative efforts to transfer the departmental park partly in the ownership of the system of the public transport and/or use it on the regular route transportations were made up to the end of the 1980s.[4]

These efforts were met with a latent, but tough resistance from the heads of large enterprises and the relevant industrial ministries. This confrontation has been connected with the fact that the leaders of the Soviet industry had a sufficient real notion of the quality of the Soviet public transport. The real state of the work was radically different from the ceremonial reports which were included in the official

(Footnote 3 continued)

became «semi-obligatory»: there was a principle "participation in exchange for bonuses", for example, additional days to annual vacation.

[4]The last attempt has been undertaken the day before «Gorbachev's reorganization».

reporting and, especially, in the sphere of public propaganda, as well as the materials which were regularly sent to UITP[5] by the Soviet institutions.

If one makes an objective comparison of the Soviet public transport with the best world analogs, the preference will be given to the latter. The data on the quality of public transport services on the Moscow routes was published soon after fall of «censorship curtain» at the end of the 1980s (Blinkin et al. 1988) (the data is provided in the Chap. 3). These data demonstrate a true ratio of the advantages of «socialist» and «bourgeois» public transport.

It is important to emphasize the following two circumstances. First, the order to carry out an independent and objective examination of the transportation conditions on the Moscow routes in 1986 proceeded from the main Russian reformer—Boris Yeltsin appointed at that time to a position of the party leader of Moscow. In other words, the declaring of the thesis «the emperor is naked» required permission of very high level.

So, the Soviet public transport was considered by John Kenneth Galbraith in his classical monograph (Galbraith 1973) to be one of the acceptable elements of the propagandized «new socialism». The famous scientist Vukan R. Vuchic, the author of the high-quality monograph «Transportation in Livable Cities», published in many countries, saw the Soviet public transport as a positive example of the solution of urban transport problems. Alas, the Soviet myths were much more durable, than the Soviet Union itself.

It is arguable that one of examples of the informal management practices in a transport complex were the running of the truck and bus fleets by the industrial enterprises, as well as the persistent unwillingness to devolve them to the general transport system. These informal practices were getting on with the Soviet centralized administrative system.

The segment of motor cars has been arranged in the most exotic way as compared with the modern reality. In this book we have already emphasized the extremely low level of car ownership in the Soviet cities. This was the result of the fact, that a «sector A» (the heavy and defense industry) was prior over a «sector B» (the consumer goods), and was supported in many respects due to the ideological reasons.

In the 1980s the level of car ownership has doubled due to the volumes of domestic production,[6] as well as due to the «grey» import market of the used cars which appeared after the adoption of «The law on cooperation». But, even taking into account the rapid automobilization growth, in 1990 there were 80 cars per 1000 Soviet residents that corresponds with the indicators of the Western European countries in the beginning of the 1950s, or the indicators of the USA of the end of the 1910s. It is clear, that in the mentioned conditions the segment of cars did not play a noticeable role in the structure of the urban mobility. Besides the activism of

---

[5]International Association of Public Transport.

[6]The AVTOVAZ entity was the main producer of cars with an output up to 1 million cars in a year (these were the remakes of the Italian FIAT car model—124).

the car drivers did not extend to the problems greater than the questions of the collective parkings.

The directive socialist planning extended even to a segment of the taxi transportations which had and still has strictly market ground. The Soviet taxi drivers were protected from the competition with the private carriers: in 1928–1988 in the USSR the self-employed taxi-driving was considered as an administrative violation. The behavior of the Soviet taxi driver was determined by the combination of directive planned targets (for example, coefficient of paid mileage) and the need to maximize the personal income in the conditions of a monopoly position in the market and the assured (non-market) provision with fuel.

It is to mention a segment of the «company-provided cars» as a very exotic practice. The nominal, or the reflected in official statistics number of the park of such cars made about 500 thousand units in the 1980s, that is 3–4 times more, than quantity of the taxi-cabs. Taking into account features of the statistical accounting of the department park of cars, there is a good reason to believe, that the number of such cars was not less than 1 million.[7] The term «official cars of uncommon use», represented the cars with a driver carrying the Soviet and party chiefs, and also heads of the large enterprises and the organizations.

The reason why this outmoded institute was saved in the country declaring the socialist ideals was clear not only to experts, but even to the top management. In 1988 the Council of ministers of the USSR issued the resolution «On regulating the use of official cars» (Postanovleniye Soveta Ministrov SSSR (USSR Council of Ministers order) 1988) which appealed to economy and replaced such kind of cars with the compensation payment. In the same 1988 the Ministry of Finance of the USSR issued the guidance letter with a certain amount of compensations (Ministry of Finance of the USSR 1988). However, these documents, as well as many other government documents in the 1980s had no significant effect on transport behavior of the Soviet elite. It must be noted, that nowadays after thirty years of «radical economic reforms» the transport behavior of the postsoviet elite remains just the same as at that time; the changes have concerned only brands and models of cars.

This complicated combination of the formal and the informal institutions, the nominal and the real goal-setting was keeping a certain stability due to the following circumstances.

In general the national economy had a sufficient resource to maintain the city transport systems in the condition acceptable for a very unspoilt Soviet citizen.

The resource allocation with all the «advantages» of the formal and informal practices described above, was adjusted and quite predictable; in any case, the chief

[7]The data on the number of official cars is given on the basis of the subjectibe assessment of the science editor of this book who was one of the leading experts in the Ministry of motor transport of the RSFSR at the end of the 1980s.

of any level was aware of what share he will receive. The skilled worker—the bus driver—also understood that in all cases he will get an acceptable salary.

During «Gorbachev reforms» there were attempts to improve this system in line with the ideas of the «market socialism». There are different points of view concerning whether it was possible to modernize the system of urban passenger transportation within the constant socio-political status quo. However, the happened events violated the first of the above-mentioned conditions: the country ran out of the resources, needed for maintenance of any life support systems. The situation was described by Gaidar (2010): «… the break-up of the socialist system was predetermined by basic characteristics of the Soviet economical political system: the institutes established at the end of 1920s—the beginning 1930s were too rigid and didn't allow the country to adapt to challenges of the world development at the end of the 20th century. The heritage of socialist industrialization, abnormal defense burden, a serious crisis of agricultural industry, a noncompetitiveness of the processing industries did the break-up of the regime inevitable. In 1970s—the beginning of the 1980s these problems could be regulated at the expense of the high oil prices. But it is unreliable ground to save the last empire».

6.3 The Urban Transport in 1990s

The fundamental changes which had happened in Russia at the beginning of the 1990s, concerned the urban transport systems in full.

The key aspects of these changes are reflected in the well-known presidential decree (Ukaz Prezidenta Rossijskoj Federacii (RF Presidential Decree) 1991) of B. Yeltsin which has been issued in December 1991, almost immediately after signing the Belavezha Accords which declared, that "the USSR, as a subject of international law and a geopolitical reality, is ceasing its existence".

The decree No. 341 has proclaimed, in particular, the obligatory privatization of the motor transportation entities in the second half-year 1992. In reality this regulation needed many years to be accomplished: even by the end of the 1990s about 30 % of the public road transport entities remained in the municipal property, or in the state-owned property of the federal subjects of the Russian Federation.

The privatization process of the taxi parks was carried out much faster; this process added up to the distribution of second-hand taxi-cars to the drivers as well as the land and the production rooms were restructured for the different business branches. Consequently the taxi transport as the centralized city service went practically out of existence as the subject of the urban reality. At the same time, this developed an almost non-regulated market of the private carting. Among its participants there were not only the former employees of the taxi parks, but numerous motorists of the different specializations (the private carting has replaced their former work, which was lost due to the closure or the restructuring of the Soviet entity).

The post-privatization destiny of the cargo entities depended on their location. The areas of the entities located in the nearest suburbs were much more valuable asset, than old cars and equipment for their servicing and maintenance. Therefore, such entities have been closed, and their land areas were used for construction. The new owners and managers of other cargo entities looked for suitable market niches; the success of these searches was various.

Later some Russian and foreign authors have criticized the process and the results of privatization in the 1990th: «the transformation of property didn't get the production moving, as the entities went for peanuts not for purpose of the production» (Luneev 2005), and «the main beneficiaries of privatization were presented by the rest of the old bureaucratic establishment and profiteers in order to make easy money» (Evans 2011).

Concerning the transport entities which were initially declared to be privatized in an obligatory way (the cargo carriers and the taxi parks), this criticism looks extremely naive. The planned land use led to the fact that the specified entities often occupied the vast pieces of extremely attractive urban or suburban lands. One was able to keep these areas for the same «*the production purposes*» like before the privatization only in the conditions, which allowed to maintain the Soviet model of the land use. Moreover, «*the production purposes*» of the Soviet entities of the cargo transport and the taxi parks could make any sense only in the conditions of the permanent made-up deficit of the transport services offer and the statutory bar, which canceled the private carting and the ownership of the cargo vehicles for private individuals. Thereafter, it is difficult to consider those new owners as «*profiteers*», who have decided to close such entities and to sell the parcels of land to representatives of violently growing developer business.

The large taxi parks were brought back to life only in the 2010s in a new quality: without any extra land, without their own service base, without any constant staff, besides, concerning the technological formats they were compatible in framework of the «Gett» and «UBER» century.

After the privatization the city entities of the cargo motor transport continued to function successfully, but only when the new owners understood that the scenario «to boost the efficiency of the entity» used to have the NPV (net present value) better compared with the scenario «the liquidation and selling of the land». The first scenario was achieved by the «trial-and-error» method while trying to comprehend the microeconomics and marketing fundamentals: the elasticity of demand to a tariff and the price policy optimization in the competitive conditions; the implementation of the «returns to scale» within the competition with the small transportation business; the expansion of the client base and the adaptation of the park structure to the special features of demand; the deliverance from the extra assets and the constant staff, the outsourcing of the technical maintenance, etc. In a word, the efficiency of the entity was increased at the expense of the quite rational business.

It is important to pay attention to one of the above mentioned circumstances: the competition from the small transportation business. It is easy to understand why this circumstance was regarded as sharp in terms of a simple comparison: in 1990 about 12,4 thousand cargo vehicles were in the private property, after 25 years of the exponential growth this indicator reached 3775 million units in 2014 (data from Main Directorate for Road Traffic Safety).

In these conditions, even if we take into account the indisputable desire of the new transport enterprises owners «to make fast money», as well as the quite reasonable suspicion about its «bureaucratic» origin, it is possible to call them «profiteers» only in connection with the Criminal code of RSFSR.[8]

Consequently, in the above described segments the officials pursued the policy of the rapid and almost total privatization: besides the privatization of the «old» entities of the Soviet type, both in the segment of the cargo motor transportation and in the taxi services an extremely huge group of the private carriers has appeared. Thereafter, the major shifts occurred: the offer deficit of the transport services has been completely liquidated; the system of the tariffs was made to be based on the balance of the demand and offer.

Certainly, this process was not conflict-free. Since 1990 in political and the expert circles there has been a broad opinion on «plundering of the public property», the selling of the «high-capacity technological base» for next to nothing, the disorganization of «the unitary motor transportation industry», the loss of the «social safeguards of employees», the replacement of the «skilled personnel» by profiteers and «effective managers», etc. In due time professor E. Yasin used to respond sternly: «Nothing was stolen from you, because you had nothing!».

Concerning the informal mechanisms, in the segment of the cargo motor transportations there was not only a normal competitive business, but also the normal labor union structures, which had nothing in common with the Soviet heritage.

The decree No. 341 has ranked the rail transport entities among the objects, which could be privatized only by the resolution of the government of the Russian Federation. The reform of the railroads has been dragged on for many years and is not finished so far. The Ministry of Railways of the Russian Federation (the former Ministry of Railways of the USSR) was reorganized only in 2004, though the privatization wasn't the key point of the reform, but the functions partition within the public administration (all of them were under the jurisdiction the Ministry of Transport of the Russian Federation) and the business activity which has become a prerogative of the state-owned JSC Russian Railways. The reform of the railway industry was added up to the immediate (since December, 1991) transfer of the underground railway (as well as all other assets of urban passenger transportation) into the ownership of the municipal government.

---

[8]Article 154 of the Criminal Code of RSFSR gave the following interpretation of a term «speculation»: «buying up and resale of goods and other objects for the purpose of a profit» also prescribed punishment for this «crime» up to seven years of imprisonment. It is interesting, that this definition repeats almost literally the definition of the term «Trade» from the Webster dictionary: «activity or process of buying, selling or exchanging goods or services».

Subsequently, by the end of the 1990s there was an arrangement of the specialized suburban joint-stock passenger companies with the shareholding of JSC Russian Railways, the regional administrations and (only in the few cases) the private investors.

The nontrivial scenario was carried out in respect to the entities of the land urban passenger transport. These entities were in a state of shock not due to the prospect of privatization,[9] but because of the absence of the supporting point: the management of any bus (trolleybus) fleet or the tram depot was aware that despite the importance to reach the mutual understanding with the local government, the top-management was found in the branch ministry in Moscow. This patronage actor arranged the regular work checks-ups, raised the level of the executive personnel's skill managing personnel and gave the directions on tariffs, the transportation rules and the instructions on the technical maintenance of the vehicles. But, the most important is that this patron provided them (on the basis of the mentioned above administrative haggling) with resources, first of all, the vehicles, and also with a regular operational grant.

The Resolution No. 3020-1 adopted by the Supreme Council of the Russian Federation in two days prior to the historical decree on privatization «...On the differentiation of the state-owned property» announced: the traditional branch ministries «as the subjects of economic reality» do not exist anymore. The decree No. 341 has fully confirmed this state of affairs: The ministries which used to be the key control centers of all the assets of national economy are mentioned very briefly and in a slighting way: «... In order to consider the features of the privatized entities the ministries and the departments of the Russian Federation take part in the development of the standard conditions of the conclusion of the privatization bargains approved by the State committee of the Russian Federation on the control of state-owned property. ... The ministries and the departments of the Russian Federation can send their representatives the committees on privatization of entities within their jurisdiction».

It is necessary to stress, that the first Government of newly established Russia (after signings of the Belavezha Accords in December, 1991) involved the Ministry of transport of the Russian Federation along with the aforementioned Ministry of Railways. It has been established on basis of the Ministry of the Road transport of RSFSR including the departments of the Ministry of the River fleet of RSFSR and the Ministries of the Highways of RSFSR. In December, 1991 the two former Union ministries of the civil aviation and the marine joined the Russian Ministry of Transport, thus losing their influence to a certain extent.

Nominally the new ministry has received extremely wide powers. In reality the situation was absolutely different. The case is not that the airlines and the large

---

[9]According to the decree No. 341, this was under jurisdiction of the public authorities of the territorial subjects of the Russian Federation or the municipal authorities, who owned these entities since December 1991.

steamship companies have been solving their issues with the government bypassing the new ministry.

On the assumption of the above, the most important thing was that the role of the ministry has been cut to the bone also in the branch of the road and city electric transport. The fundamental innovation was presented in a total absence of a subject of the administrative haggling: in the 1990s the ministry had no resources to support the urban passenger entities.

The relationship between the passenger entities and the new owners (the local administrations) emerged on the following grounds:

(a) The privatization is not allowed: the quarter of the century after the adoption of the decree No. 341, the decisions on privatization have been made in relation to a few number of bus fleets only, while the entities of the city electric transport remained state-owned;
(b) The are the made-up unconditional low tariffs and also the free trips for all the groups entitled to such benefits. As it was already mentioned in Chap. 3, the right of the preferential (free) journey was enjoyed by 27.3 % of all population of the country, in some cities this figure reached 35 % (Rodionov 2005);
(c) The grants on a covering of losses (at least of the preferential transportations) are given «within the limits of the possible»;
(d) In city budgets the investment on the re-equipment of the vehicles are provided rather as an exception, than a standing rule.

The last two items reflect the budget realities of the vast majority of the Russian cities: in the conditions of the permanent budget deficit the support of the municipal carriers wasn't among the priorities of the city authorities. Moscow and some other largest cities, including St. Petersburg and Kazan, were an exception to the rule.

The processes which were resulted from such strategy are described in detail in Chap. 3:

- The all-round stagnation of the municipal transport entities;
- The expansion of the paratransit services provided by the private carriers in the formats like the African jitneys;
- The radical shift from public transport to private cars.

The main competitive advantages of the Russian paratransit depended on the factors which had nothing to do with the market fair play. The private carriers had no liability to transport the groups of passengers entitled to benefits, but there were good opportunities to avoid taxes at the expense of the unrecorded revenue as well as the savings due to the disregard of the elementary regulations on the technical maintenance and the work conditions of the drivers.

All these questionable advantages were not a secret for the city administrations, however there was a forcible argument in favor of the private carriers: the jitneys (the minibuses) were carrying out the passenger transportations properly and without any municipal grants. The unrecorded revenue also prepared the grounds for the «corruption element».

The educated city officials were aware of the fact that the increasing commuting by the private cars fits badly into the planning parameters of the Russian cities,[10] besides it is fraught with the permanent jams. However, this was also argued well: the car owners got to work independently and at the same time did not require any budget subsidies. The standard counter-evidence used in the discussions abroad lies in the fact that for the municipal budget it is more profitable to support public transportations, than to respond to the increasing number of the private cars due to the high-cost of road construction. Unfortunately, this counter-evidence was absolutely non-valid for the Russian provincial realities: the road construction was suffering from the deficit of fund just like the support of the municipal carriers.

As a matter of fact the authorities took a laissez-faire attitude towards the situation with the urban mobility: the municipal administrations had no resources to pursue a distinct transport policy, the central power was adhering to the position of the «positive non-interference». This state of affairs can be described as «fragile state», the term used in the political science.[11]

The issue concerning the division of power and responsibility between the central authorities and the local governments deserves more thorough review.

The Constitution of the Russian Federation adopted in 1993 has declared independence of the local government from the state authorities. Their areas of jurisdiction were determined by the Federal Law «On Local Government» (28.08.1995, No. 154-FZ). In relation to the transport policy the law supposed numerous so-called "unfunded mandates" or the expenditure commitments of the local government determined by the federal or regional legislation, but without any sources of their covering.

The area of responsibility of the local government included «the organization of transport service for the population», and also «the municipal road construction and the maintenance of the local roads». The issue of the funding sources, required for implementation of these mandates, has been formulated in a florid way:

«The authorities of the local government provide satisfaction of the basic needs of the population in the responsibility areas referred to the municipal divisions at the level above the minimum state social standards. It is guaranteed by the state, federal authorities and authorities of subordinate entities of the Federation by consolidating allocations from federal taxes and dues collected in subordinate entities of the Federation into revenues of local budgets.

[10]This theme is described in Chap. 2.

[11]Since 2005 the organization Fund for Peace measures an index of the state fragility. The moment of disintegration of the Soviet Union could be characterized as «failed state» (the failed state not capable to pursue more complete policy and management), but it is possible to regard the Russian Federation as new formation in the time perion of 1991–1993 and as the state with high degree of «fragility» (a fragile state).

The question «What do the minimum state social standards mean?» in terms of the street network and the transport service of the urban population wasn't clarified during the whole validity period of the abovementioned law. Over the years there was no practice to contribute the federal and regional taxes to the local budgets in connection to the commitments of the local governments to satisfy transport needs of the population.

In spite of the fact, that the ambiguities as well as the evasive language of law were the features of this law, it reflected the political guideline on decentralization in the 1990s and particularly on the relative autonomy of the local government from the federal center.

Concerning the transport issues the law No. 154 provided quite proper ideas:

- The local government authorities are responsible for a street and road network of the cities and for the provision of the transport service (as well as for some other questions of life support);
- The federal government authorities set certain minimal social standards of life support;
- if the city has not enough funds to accomplish these minimum social standards, the authorities of the federal center and the federal subjects of the Russian Federation transfer a certain part of revenue from the federal and regional taxes to the city.

Unfortunately, in reality this system wasn't working properly, but at least it was considered as quite convincing at the conceptual level.

6.4 The Experience of the First Decade of the 21st Century

By the middle of the Noughties, during the second term of the President Vladimir Putin, the political regime came back to the quite centralized and authoritarian forms of the governance. It was clear that the transition from the Soviet totalitarianism to democracy isn't an irreversible process (Ottaway 2003); in this regard Russia has shown an example of an «unstable» transformation. The authors which studied this political phenomenon paid attention to such factors like absence of the steady formal institutes (as well as civil informal one) in the transition phase, a sustainable economic development (Evans 2011), as well as a political willpower (Carothers 2002).

The political scientist Th. Carothers describes both the structural and the procedural (relating to the actors' activity) factors which determine the success and the irreversibility of the transition from autocracy to democracy. The structural factors include: the level of the economic development, the level of the ethnic heterogeneity, the sociocultural factors and the path dependence phenomenon, which is extremely important for Russia concerning the formation of the state and civil society institutes. As regards the procedural approach, the main idea is that, despite the importance of the above-mentioned factors, there is a number of cases when the countries followed the democracy road while lacking for the due structural preconditions. The reason is the political willpower of the elite to move towards the democratic development. When it comes to democracy, "anyone can do it" (Carothers 2002). Unfortunately, the political establishment in Russia hasn't shown such political will.

Another important issue we need to focus on is the influence of the democratic institutions or their absence (or centralization/decentralization) on the economic development and, particularly, on the development of the urban transport systems.

The Russian experience in the Noughties shows that this influence is very ambiguous. The matter is that in most of the Russian cities on any elections and in any public events the «controlling interest» belongs to the government-financed stratum of the population—such as pensioners, employees of the budget sphere, and employees of the municipal entities. Their political views are determined by the paternalistic values of the Soviet type mixed up with the ideas of the fair division of the income from the oil and gas export.

In this value system there is no place for an idea that the government is no more than a hired manager whose existence depends on the funds of the taxpayers. In addition, there is no principle of the private responsibility of the citizen as a voter and a taxpayer, as well as the understanding on the division of power, and also of limits of the authority on the federal, regional and local levels of the government. Moreover, the representatives of the political establishment are lacking understanding of how the whole system functions.

Experience of the first decade has provided several convincing cases on the subjects designated above. The first one is connected with an attempt of the government to follow the practice of the direct payments, particularly, the monetization of the free trips.

The Federal law No. 122-FZ (August 22, 2004) provided quite reasonable measures to replace the right for the free travel with the direct monetary adjustment. The law was supported not only by the arguments from the liberal economy articles, but also by the common sense:

- in the beginning of the 2000s the municipal transport entities were functioning with 30–40 % Farebox Recovery Ratio, in addition to that the subsidies from the city budget weren't able to cover even their straight losses. Respectively, each trip performed on a city route wasn't able to bring in anything to the carrier, but a dead loss; besides there weren't any economic incentives to set the buses (or trolleybuses and trams) on routes;
- Despite the bad financial performance of the municipal transport enterprises, mass corruption was one of the main features of the branch. This was the result of the lack of the credible information on the quantity of passengers with free-trip benefits, and, thereby, there was a complete uncertainty about the pecuniary obligations of the city as a transport service customer;

– On the merits the urban transport benefits represented a discrimination in relation to the residents of the other settlements which didn't have any public transport routes.

The law No. 122-FZ made a division of the financial responsibility for the compensations payment to passenger group entitled to the benefits. The persons who received the benefits under the federal legislation shall receive the compensation payments at the expense of the federal budget. Although this regulation seemed quite logical, there was a traditional practice to delegate these commitments from the federal level to the regional and the local ones, which were out of the budget sources.

According to the federal legislation, the passenger group entitled to special benefits included the participants and disabled veterans of World War II, the armed forces personnel, the persons with disabilities and other small population groups. Hereby, the monetization of these benefits didn't cause some kind of problems for the federal budget. At the same time the regional benefits applied to the veterans of labour and the pensioners that was up to 30 million people. The most of the federal subjects' budgets had obviously no opportunity to compensate these benefits within reasonable limits.

The public response to the law on monetization of social benefits was unprecedentedly violent. The level of participation in the protest actions was comparable with the revolutionary events in 1991. The protests movement concerned wide sections of the population including pensioners, students and officers. The protesters have blocked the railroads and the highways; they demanded the resignation of the mayors, the governors and the ministers of the economic development. The monetization of social benefits was openly opposed by the heads of regions, the Mayor of Moscow Yu. Luzhkov wasn't an exception.

Consequently a quite reasonable reform turned into increased budget expenses, a downward rating (however, for short time period) of the federal authority and giving up some key points of this reform. For example, in Moscow there are still free trips with so-called «social cards» for veterans of work, pensioners and some other population groups.

The second case was connected with the introduction in a number of the cities of the country, first of all in Moscow, of the institute of paid parking. This case relating in 2013–2014 is in detail sorted in Chap. 5.

It is important to emphasize, that this quite reasonable conventional measure caused the strong protest moods from the capital car owners, who were presented by the people younger and financially more secure, than the Russian electorate at the average. The public activity of the civil society was reduced by a number of laws and measures of the law-enforcement practice, which were adopted after the rallies against the falsification of the election results. So, the protest against the paid parkings was expressed generally in the social networks and in the form of mass use of various technical devices allowing to avoid the parking charge.

The Moscow administration with its sufficient political stability haven't made any concessions: the practice of paid parking has been extended to the most of the city areas.

Two more cases which are also described in Chap. 7 are important to be mentioned:

- the introduction of a toll on federal highways regarding the trucks with an axial load more than 12 tons. The amount of the toll was estimated by the so-called system «Platon», which was an analog of the German system «LKW Maut»;
- the introduction of the toll road M11 «Moscow-St. Petersburg».

In the first case the protests of drivers of the heavy trucks were so mass and violent that the government was forced to reach a compromise, which reduced the budget effect of the new payment system to zero.

In the second case the protest activity was expressed by the residents of Zelenograd[12] and the cities of Moscow Area, who had plans to use the section of the new road for their daily labor trips. This activity was latent and didn't seem as a threat for the federal and Moscow authorities. Nevertheless, the response of the empowered politicians was immediate and especially populist: the government adopted a resolution, which determined the limit size of tariff rates on the toll roads.

All these cases, despite all the economic distinctions and the response of the authorities, reflect the public moods dominating in Russian society. The electorate majority is quite ready to submit any restriction of rights and civil liberties, but will persistently stand up for the free riding right.

Concerning the perspective of the local government, the abovementioned law No. 154-FZ was replaced with the more completed legal act «On the General Principles of the Organization of Local Government in the Russian Federation» (October 06, 2003 No. 131-FZ with further amendments up to 2016).

The main difference between the two versions of the law lies rather in the sphere of the legal language and casuistry. The law No. 131-FZ referred the following items to the «questions of local importance of the urban district»:

«… the establishment of the conditions for the transport services provision to the population within the city district», and also
«… the road activities in relation to the roads of local importance within the city district …».

It has to be clarified that the term «road activities in relation to the roads of local importance within the city district» makes just little sense, as the thesis on «construction, repair works and maintenance of the urban street and road network». How it is possible «to organize transport servicing of the population» without any preliminary conditions—this is the question for the Russian lawmakers.

[12]The city is located outside the Moscow Ring Road and enjoys the status of the district of the city of Moscow.

Certainly, the problem is not only in formulations. The main thing is that a question of the funding sources, needed for implementation of these mandates, is mentioned in new law even more indistinct, than in the previous version:

- The local budgets get subsidies in compliance with the Budget Code of the Russian Federation and the relevant legal acts of the federal subjects of the Russian Federation
- The local budgets can obtain other interbudget transfers from the budget of the federal subject as provided by the budget code of the Russian Federation.

As may be seen there are no references of «the minimum state social standards» or promises to give the local government any «assignments from the federal taxes and taxes of the federal subjects of the Russian Federation». But it is quite clear, that the city can receive money from the federal or the regional budget only on the basis of co-financing.

In major cities all these legal regulations were reduced to a simple principle: if the city's mayor accurately submits to the governor, or doesn't have any conflict with the regional power, there will be some funds for the support of the Urban transport system in the regional budget, and vice versa.

One more key point in the relationship of the federal public authorities and the local government is connected with the adoption of the amendments to the Urban planning code[13] of the Russian Federation in 2014.

According to these amendments, in the field of urban planning activities the federal authorities were obliged to «determine the requirements to the complex development programs of the transport infrastructure of settlements, city districts». In compliance with the adopted requirements (the resolution of the Government No. 1440, December 25, 2015), the local government were ordered, that the «Programs of complex development …» have to provide «the development of the transport infrastructure in relation to the needs of the population in transportation, as well as the entities provide the transportation of passengers and freights within the settlements and city districts». It is difficult to challenge such common truths. It is necessary to stress, that, first, just the same theses were in the Soviet documents in 1930–1940s, secondly, such requirements from the federal center mean the assignment of a new set of the non-financed mandates on the local government.

In general there are all reasons to consider that the efficiency of the urban transport policy corresponds to the composite governance indicators, estimated within the Global Competitiveness Index of the World Economic Forum (Table 6.2).

As we see, the indicator of the political institutes efficiency in Russia reaches below average value within last 8 years: the range is between 2.91 and 3.16. The efficiency of the budget spending also remains in the same values, but unfortunately, this indicator wasn't measured in the first decade of the Noughties therefore it isn't possible to compare the periods of centralization and decentralization of the

[13]Amendments to Town-planning Code of Russia were made by Federal law No. 80-FZ (20th April, 2014).

Table 6.2 Global competitiveness index by World Economic Forum on Russian Federation

	2007–2008	2008–2009	2009–2010	2010–2011	2011–2012	2012–2013	2013–2014	2014–2015
Institutions 1–7 (best)	2.92	3.13	3.11	3.11	2.91	3.05	3.16	3.16
Wastefulness of government spending (1 = extremely inefficient; 7 = extremely efficient in providing goods and services)	2.83	3.24	3.13	3.06	2.78	2.81	2.83	2.83
Transparency of government policymaking 1–7 (best)	3.05	3.17	3.5	3.84	3.7	3.63	3.77	3.97

Source Data from World Economic Forum (2016)

public administration. In spite of the fact that the indicator of transparency of government policymaking improves every year (3.05 in 2007 and 3.97 in 2014), it reflects the development in the area of e-government and the free-access to information, but isn't able fix the issues of the informal institutes activity.

6.5 Main Actors of Urban Transport Policy and Their Interests

In this part of the chapter, we will concentrate on the main (minus the city authorities) actors influencing or potentially capable to influence the development and the functioning of the city transport systems, and also to reveal the consequences of their activity.

The big business The main actors in this segment are the building companies and the producers of the vehicles. They have a considerable potential to influence not only on transport policy, but also on town-planning one.

The *building companies* are interested in those regions and the cities where (on a fixed or temporary basis) there are sufficient financial opportunities to implement great projects on transport infrastructure. During the whole decade of the Noughties such potential was kept by Moscow where the large-scale transport construction was carried out generally at the expense of the city, which also enjoys the status of the federal subject. St. Petersburg was also attractive for the transport construction: along with the city budget, the financing was provided by the federal funds and funds of private investors within PPP. The list of the cities with the active transport infrastructure development includes also Sochi (due to hosting of the winter

Olympic Games-2014), Kazan (in connection with the celebration of the 1000 anniversary of the city and hosting of the world Universiade-2015), Vladivostok (because of the hosting of the APEC summit), some other the cities.

The most interesting case for the analysis is the experience of Moscow and the Moscow Area, because in some years 60–70 % of investments in the transport construction objects were concentrated there.

In 2012–2015 in line with the «Development of Transport System of Moscow»[14] program the investments into the road and subway construction made about €2 billion annually for each direction, whereas the budget costs on the land public transport made about €200 million a year. At the same time in structure of population mobility of Moscow the public transport makes 80 %, whereas the private cars make only 20 %.

The comparison of these indicators shows: first, there is an inconsistency between the structure of the budget investments and the mobility structure, second, there is an obvious predominance of programs on the transport construction over other forms of the transport system support.

On this basis it is possible to claim that the companies (while being the leaders of the infrastructure construction segment) defend their interests in Moscow and dictate the priorities of the budget expenses.

Despite these circumstances, it should be noted that the S. Sobyanin's administration managed to weaken the influence of a construction lobby. Although the structural issues on the distribution of budgetary funds still remain, it is possible to claim that the practice of the ineffective projects (initiated by companies close to the state authorities) in the city programs has become a thing of the past.

An illustration in point is the construction of a so-called Alabyano-Baltic tunnel. This deep and complex tunnel was built across Leningradsky Avenue below the level of the Zamoskvoretskaya subway line. It was about 10 years under construction and the project costs made about €1.3 billion.[15] The expedience of this construction was questionable in terms of traffic. As a matter of fact, there was a selection of a construction object which had to serve the interests of the certain contractor (the scientific-production association «KOSMOS») and was close to administration of then Moscow Mayor Yu. Luzhkov. This example is typical for the crony capitalism. The president of the contractor-company which failed to complete this tunnel was included in the Forbes list of billionaires and is wanted by Interpol. The construction was finished by the new contractor, which had sought for the new project decisions and additional budget costs in order to make the project more reasonable.

It is important to mention the companies producing the vehicles among the influential actors of the urban transport policy. Their interests are also concentrated generally in Moscow, St. Petersburg and other big cities where there are sufficient

[14]Financial indicators are calculated at the exchange rate 76 rubles for €1.

[15]The costs of the project are recalculated taking into account exchange rate fluctuations of 2006–2015.

financial opportunities for the large-scale wholesale purchases of the passenger vehicles.

The standard example is presented by the tenders in 2014–2015 on the purchase of 1600 cars for the Moscow subway. Not only the Russian machine-building concerns (Uralvagonzavod (UVZ), Sinara and CAF Group) got interested in these tenders, but also such leading global manufacturers like Bombardier.

The winner of both tenders was "Transmashholding" concern. Among the shareholders are JSC "Russian railways" (25 %), "Alstom" (25 %) and business-structures controlled by I. Makhmudov and A. Bokarev. Tenders resulted in two contracts—supply of 832 rail cars in amount of 1.88 billion euros and 768 rail cars amounting to 1.72 billion euros.

The case itself clearly represents relations between Russian government and large business.

Conclusion of these deals raised a wave of public criticism—people thought that M. Liksutov, Moscow Vice-mayor responsible for transportation policy, had connections with the supplier.

On the other hand, these deals ensured the conclusion of the contract with 30-year life cycle, allowing the long-term stable supply of high-tech rolling stock. Conditions given allowed the ordering party to receive guarantee for maintenance of the rolling stock in stable technical condition by the manufacturer for the whole period of their operation.

At the same time, competing companies filed a complaint to the Russian antimonopoly agency, paying attention to the fact that the initial tender price was too low for them, and made participation unreasonable in economic sense. The nominal buyer—Moscow Metro and the real payer—the Moscow budget were at an advantage over the conditions of these contracts.

We can find some more specific examples of the rolling stock manufacturers participation in urban transport policy.

In the 1990s main active transport market players in Russian cities (except, of course, Moscow and St. Petersburg) were suppliers of used buses, trolleybuses and trams, not even the producers of new equipment. Old rolling stock decommissioned from the EU countries due to the exhaustion of the warranty mileage, or non-compliance to existing environmental standards has fulfilled the supply side. Local officials and transportation managers actively participated in conclusion of the supply contracts, which had "gray" character and often involved corrupt schemes. Brokers bought old vehicles from the cities where latter were decommissioned at give-away prices similar to the prices for scrap metal. The margin of the transaction appeared to be more than comfortable, even with the low prices at which Russian cities were willing to buy old rolling stock.

By any criteria—political, moral, environmental—these deals looked quite doubtful. However, the purchase of second hand equipment has allowed the peripheral cities of Russia to update the fleet of vehicles and, therefore, to maintain municipal carriers "afloat".

Another case of 2014–2016 period is associated with the production of NGV buses. That is equal to for approximately 10 % of buses manufactured in Russia. The demand for this technique was not very active: peripheral Russian cities have very scarce resources for the renewal of municipal carriers' fleet and are extremely reluctant to buy vehicles requiring additional infrastructure for maintenance and fueling.

NGV buses manufacturers get the lobbying support from Gasprom gas-engine fuel[16] and have succeeded in adopting the decision of the Russian Government (RF Government Decree (Postanovlenie Pravitel'stva RF) 2014), allowing "the provision of subsidies for bus purchase ... working on gas motor fuel, under the subprogram "Automobile industry"" of Russian state program "Development of Industry and increase of its competitiveness". After the lobbying demand from cities and regions to participate in the program significantly exceeded planned budgetary provision. It turned out that conditions for NGV buses purchase under federal program were more profitable for cities than buying buses built on conventional diesel technology.

Another illustrative case refers to trolleybus manufacturers, including leading Russian producer—JSC "TrolZa".

Here it is necessary to make an excursus. In our opinion, the trolleybus—land transport that runs directly on the city streets and gets energy from the overhead wiring, has no future in sustainable urban passenger transport systems. The "sustainable mobility" ideology had provoked strong demand for high-quality rolling stock and, accordingly, an active supply of new technology by the world's leading market players in the transport engineering field.

The main trends in this process are quite clearly defined nowadays. The first of them—rail cars, HRT and LRT. Second—all state-of-the-art buses: diesel buses of Euro-5 and Euro-6 environmental classes; vehicles with an engine running CNG (compressed natural gas) or LNG (liquefied natural gas), various type "hybrids" (gas-diesel, diesel-electric, gas-electric). Finally, the electric buses with longer capacity of autonomous run. In all segments fierce market competition involves many of the leading market players—manufacturers of cars and rolling stock. This leads to radical technological superiority of the vechicles produced over old exemplars in terms of energy efficiency, comfort, speed, ease of entry/exit, operational reliability, design advantages, and, of course, adaptability to the needs of people with restricted mobility.

Unfortunately, this process affected the trolleybus sector in a very small extent: today about 70 % of the trolley buses world fleet is still concentrated in the ex-USSR cities. As far as we can judge from the outside, urban managers and transport planners in developed countries, do not consider this transport mode to be competitive and promising.

For example, in Belgium, homeland of Alain Flausch, the current UITP (International Association of PublicTransport) Secretary General, authorities have

[16]Associated company of the Public Joint Stock Company Gazprom.

turn down trolleybus systems. The same trend one can observe in France, Pierre Laconte motherland, previous longstanding UITP Secretary General. Trolleybuses have disappeared or almost disappeared in all other developed countries, which are major "fashionable" trend-setters in public transport over the last fifty years: Austria, Germany, Netherlands and Japan.

A number of reputable experts, one way or another involved in the UITP Trolleybus Committee activities, retain the belief in a short renaissance of this urban transport type. They appeal to the highest environmental, technical and economic advantages of the trolleys, as well as to new opportunities associated with an increase of autonomous run distance. With all due respect to colleagues, we cannot share their optimism: progress in the trolleybus segment, alas, never keep up with innovations in car and bus segments, backed by massive investment and engineering potential of leading market players. For example, mass production of electric buses that are able to compete with conventional diesel buses at the life cycle price and the ease of use—is a question of next few years. Obviously, the appearance of such electric buses will have finally close the chapter on prospects of the traditional trolleybus tied to the overhead contact wires.

Let us now return to domestic realities. Trolley crisis began in Russia in the 1990s and since then has not ended. The flagship of the Soviet trolleybus industry —ZiU (Uritsky factory, now JSC "TrolZa") was configured to production up to 4000 trolleybuses per year and deliver them to the cities of the USSR and "fraternal countries".

Motives for trolleybuses mass use were far from the present-day realities. The reasons were laying on the surface—artificially low electricity prices for public utilities, key peculiarity of the planned economy; a chronic deficit of large capacity buses; employment of female labor, etc.

JSC "TrolZa" initially bore the burdensome weight of excess space, obsolete equipment, archaic lineup and production assets calculated according to planned production indicators, which seems unreal under the present conditions. Despite all the joint efforts of owners and management "TrolZa" was trapped in a vicious cycle —"the stagnation of demand—lack of investment to upgrade production capacity and model range—deeper stagnation of demand." That resulted in significant drop in production—in recent years it did not exceed 100–120 units per year.

All these products, starting with the modification of obsolete models ZiU-9 (project of 1966, last trolleybus rolled off the production line in 2015, in total more than 42,000 trolleys were produced) and ZiU-10 (1978) and ending with lowered floor "Optima" (1999) and low floor "Megapolis" (2006) can barely be considered to be the result of hi-tech transport engineering.

At the end of 2015, SUE "Mosgortrans" (main Moscow land transport operator) conducted the analysis of the relative complexity of the operation and maintenance of buses and trolleybuses. The enterprise has the largest fleet of buses (about 6500 units) and trolleybuses (about 1500 units) in the world. The results of big panel data analysis clearly shows that contrary to the book learning trolley buses drastically lose on the complexity of operation and maintenance.

The actual data refute the «book» ideas about low levels of noise made by trolleys—ZiU-9 are equipped with the old type of compressor units, a very sonorous ones.

The same situation concerns the environmental hazard. It is clear that estimation of emissions should be made not on «passport» values only, but include negative externalities public transport creates. According to technical documentation trolley has zero emission and the bus of «Euro 5» standard has little more than zero. However, the total value of the exhaust generated by the total flow of vehicles, in which there is, respectively, a trolley or bus is obviously higher. Research on the trolley as a point of turbulence (and thus a source of externalities) in the transport stream were very actively carried out in 1970–80s, right at a time when Western European cities started to liquidate this type of transport. Later it turned out that trolley, due to the acceleration dynamics and lock on contact wiring, is, alas, a very powerful generator of negative externalities.

Nevertheless, the main argument trolley advocates put forward is the environmental damage associated with the replacement of trolleybuses by buses. Here we must note that under Moscow's conditions (as well as any Russian big cities) such arguments have clearly demagogic and populistic character.

The problem is in different sector. Up to 4 million passenger cars and about 400 thousand vehicles are used in Moscow. Approximately 20 % of them are cars of Euro-2 and lower ecological class. However, it is impossible to remove these environmentally unfriendly vehicles from the city: their owners are usually unwealthy retirees, and poor representatives of the self-employment segment. Any city government, seeking voters support, would never start this confrontation.

The total annual mileage of Moscow car park of is about 50 billion kilometers. The same 20 % are made by Euro-2 lower class vehicles. The total annual mileage of "Mosgortrans" buses is about 0.4 billion kilometers, but 80 % of them are of Euro-5 class. It is easy to calculate that the contribution of buses in the air pollution combined with carbon monoxide, nitrogen oxides and other normalized hazard is statistically insignificant.

So, in the strategic perspective the future of Moscow trolleybus system is, in fact, predetermined. Trolleybus transport will go down in history in the next 10–15 years. As mentioned above, under the current conditions Moscow trolley fleet will decrease according to the dynamics of write-offs, adjusted for modest numbers of new supplies. Accordingly, we expect gradual closure of certain routes, and the remains of an aging rolling stock—to be redistributed between the remnants of the wiring network. Trolleybus will first leave the city center, and then the whole city.

It goes without saying that the vehicles of new generation, will gradually replace the trolleybuses, and certainly not worsen the ecological situation in the city and the conditions of passenger transportation.

Carrier companies In the Russian practice carrier companies are divided into two highly distinct segments. Among the first one are state-owned (or municipal) unitary enterprises, which continue to be the main operators in the country's major cities. Second group is represented by private carriers.

The state (or municipal) passenger unitary enterprises are fully dependent on the city authorities, both financially and administratively. As a rule, they cannot be considered to be independent actors of urban transport policy.

A certain exception in this regard up to the last few years were the largest Moscow transport enterprises—SUE «Moscow Metro» and SUE «Mosgortrans». Until Sergei Sobyanin took his office, directors of these enterprises were independent decision-makers, while maintaining almost complete independence from the city transport authority in policy formulation and assuming themselves accountable to personally the mayor of Moscow Yuri Luzhkov. The result of the independence is the fact that two major transport operators had different and autonomous from each other fare collection systems up to 2013.

Today, this exceptional status can be considered lost: a real actor of the Moscow transport policy is the Department of Transport of the Moscow Government, not subordinate transport enterprises.

Private carriers Enterprises operating in the paratransit format (jitneys or minibuses) played a very important, sometimes a key role not only in the local, but in the federal transportation policy. Transport, technological and economic aspects of this phenomenon are described in Chap. 3.

Private carriers appeared in Russia in the early 1990s and formed a vast paratransit transportation segment. It has been and remains extremely popular, not only in provincial towns, where the stagnation of the public transport gave it opportunity to flourish, but even in Moscow and St. Petersburg. A clear symbol of this segment became RAF minibuses from Riga car factory, remaining in operation since Soviet times, and then, after 1994—GAZ-3221 «Gazelle». The latter became the personification of the term «marshrutka», the alias of private operated minibus. Reasonable price, the possibility of DIY-repair, cheap and available parts made the car (and its subsequent modifications) the backbone of rising private companies' fleet.

The above-mentioned competitive advantages of this transportation mode (legal and barely legal) allowed private transportation business community to gain the political influence. Moscow Transport Union became the main lobbying center expressing interests of the community, the main lobbyist, respectively—Yury Sveshnikov, the Union executive director. Private transportation lobbyist gained strong support in the Duma (in all convocations), as well as in the legislative assemblies of Russian cities, including Moscow City Duma. They succeeded in achieving the adoption of a number of legal provisions in the spirit of public transport deregulation promoted in the UK when Margaret Thatcher held her office (Pucher and Lefevre 2014).

The adopted Federal Law 259-FZ «Statute of road transport and urban land-electric vehicles» (8th November, 2007) had the following key points:

- Absence of mentioning the «city route network» concept and, moreover, any requirements on the network design and its approval procedure by municipal authorities;
- The law provided a procedure of a declarative character for opening the route by the initiative of any carrier: private companies or even private persons;

- The law allowed for any form of cash collection, even «money in the driver's pocket»;
- Legitimization of the «competition for passengers on the route», polar opposite to the recommended by UITP «competition for the route in the tender».

This state of affairs persisted for many years, until the adoption of the Federal Law 220-FZ «On the organization of regular transport of passengers and luggage by road transport and urban land-electric vehicles ...» (13th July, 2015). For the first time it established rules and norms of the organization of passenger carrying on urban public transport routes. The norms complied with the UITP recommendations and modern European urban practices: unified route network and a common tariff and ticketing systems; city as the contractor; gross fixed-route transportation contracts; tender competition for city routes.

The passage of the law lasted for several years due to long conflict between Moscow and other major Russian cities authorities and lobbyists from Moscow Transport Union. Authorities insisted on immediate adoption of the bill and MTU stood for preservation of the former legislation №259-FZ using their connections in State Duma.

Moscow administration had to use the administrative resource to prevail in the confrontation. One can treat this episode as a defeat of the liberal concept of the passenger transportation organization. However, it is unlikely that the African type of paratransit, which was supported by Moscow Transport Union, corresponds to the canons of liberal economy.

Let us note that the largest and most successful private companies put before the fact of the new "rules of the game" introduction have quickly shifted their attitudes towards cooperation with a new organization—Moscow agglomeration transport associations (TAMA). Tama is a typical GONGO,[17] emerged with the active participation of the Moscow Government Department of Transport.

Civil society The impact of civil society on the transport system is situational and quite specific. This is partly due to the underdevelopment of civil society, as well as the peculiarities of the relationship between society and government in Russia.

Among the social movements and organizations that have existed in recent years, we can outline few types:

(a) *The protest movement of local residents*

The best known local residents protest is the movement «In Defense of Khimki Forest», led by Eugene Chirikova. The emergence of this movement and its long-term activity (2007–2011) was caused by the siting plans of M11 high-speed highway «Moscow—St. Petersburg». The highway advance section should pass through forest land plot in the vicinity of the town of Khimki (Moscow region), called by locals Khimki forest.

[17]GONGO—a government-organized non-governmental organization.

By the time of the construction works beginning the ecological value of this area has been jeopardized in many ways: in 2000s authorities traced regional road through this site, active construction of cottages around the perimeter started. In addition, since 1990s, regular forest protection and reforestation measures have not been taken. Regional Forest Service has experienced a chronic shortage of budget financing and no voluntary forest care movements existed in Khimki.

According to the project, cleared strip, required for the construction of M11 advance section had length of 3 km and a width of 80–100 m. There is no doubt that the project would cause significant damage to a forest area. However, taking into account very poor condition, of Khimki forest as well as significant amount of environmental compensation measures envisaged by the project, the problem clearly was not one of the most acute and intractable.

The movement «In Defense of Khimki Forest» began as a typical NIMBY.[18] The founders were residents of houses and cottages located in the immediate vicinity of the future M11 route. The rest of Khimki population had positive opinion about the construction of a new road. The old federal route M10 «Moscow—St. Petersburg» with its daily traffic of up to 100 thousand cars is a source of Khimki ecological trouble. M10 is traced directly through the densely built-up areas of the city, literally under the windows of the high-rise development area erected in 1980s.

The conflict arose and went in the acute phase, primarily due to the inability and unwillingness of Moscow region and Khimki authorities to play the role of moderator in the dispute between local population, the grantor and the concessionaire of the M11 project.

M11 project has been implemented in the form of public-private partnership, and planned to become a toll road. The grantor until 2010 was FSI (federal state institutions) «Roads of Russia», then state-owned company AVTODOR; the concessionaire—«North-West Concession Company», whose major shareholders were structures controlled by A. Rotenberg and Vinci group (Concession & Construction Group, France).

Khimki authorities[19] preferred to communicate with members of «In Defense of Khimki Forest» indirectly, only through private security companies with doubtful and criminal reputation. It resulted in immediate politicization of the «Movement …» and its expansion. Moscow environmental activists, marginal left-wing and anarchist groups have joined the movement; E. Chirikova became public figure, well-known in Russia and abroad. The conflict has led to a number of extremist manifestations (including arson of the city administration building by left activists), as well as criminal incidents (number of members of the «Movement …» including city editor of the independent newspaper, were severely beaten and maimed). The

[18]NIMBY (Not in the My Backyard) term consolidate the ad hoc protest social movements, which are associated with dissatisfaction of local residents by any transport projects planned for realization near their places of habitation.

[19]Two Khimki administration heads, holding their office during M11 construction, were consistently prosecuted for corruption activities under the Criminal Code.

road through the Khimki forest became well known to PACE, the World Wildlife Fund, and even the Pope's office long before the construction started.

The only real result of the «Movement …» activities was a three-year delay in putting the M11 advance section in the operation, as well as a high public awareness of its chairman—Evgeniya Chirikova.

It is interesting to note that almost simultaneously with M11 project, under the same conditions authorities started another project—Odintsovo bypass.[20] A considerable part of the road was traced through forest areas of Odintsovo district with a significantly higher bonitet and ecological value than the Khimki forest Social status of the residents of Odintsovo district is significantly higher than of the other residents living in Moscow region and, in particular, Khimki district. However, the concessionaire company (JSC «Main Road») and the district administration were successful in organizing a constructive dialogue with the locals. After a series of adjustments to tracing plan and adequate compensation package of measures, acceptable and compromise solutions were elaborated for both sides. No scandals and, even more, no incidents of criminal character did not happen during its implementation.

In 2003, Moscow authorities have made attempt to close (with further demolition) the Apakov tram depot. This depot serves southern routes of Moscow and is one of the oldest in the city. It occupies a vast land, near the outer boundaries of the Garden Ring, where the land value is extremely high. This led to an idea of previous Moscow Mayor Y.M. Luzhkov to sell the land for private development without regard to possible transportation problems. Arisen active opposition to this plan had purely informal character and consisted of local residents and supporters of the tram transport preservation. In 2011, against the background of years of civil protests, The Urban Planning and Land Commission,[21] headed by the new mayor S. S. Sobyanin put an end to the redevelopment project, saving the depot and, therefore, tram routes in the southern part of Moscow.

Another example of the purely informal activity of local population is connected with the reconstruction of Leninskiy Prospect. Citizens stood against authorities initiative to transform the avenue into a kind of urban freeway.[22] Similar projects of city arterial streets reconstruction (avenues or freeways in the Moscow's terminology) have repeatedly been implemented in Moscow in the last 20 years. Generally, these expensive projects led to reduction of property prices and the quality of life in the surrounding buildings, without creating a significant positive effect on street's traffic capacity. When city government sounded the initiative to

[20]New exit from federal road M1 "Belarussia" (Moscow-Minsk).

[21]The Urban Planning and Land Commission is an advisory body to the mayor of the city, which addresses key issues of planning, construction and land-use of citywide importance.

[22]The previous chapters have repeatedly drawn attention to the fact that the Russian planning standards do not provide for the separation of the elements of the road network into the streets (streets, avenues, arterials) and roads (freeways, expressways).

reconstruct Leninskiy Proskpect, the traditional place of residence of the Moscow academic and cultural elite, latter stood against it.

The process started with public hearings on reconstruction project, conducted with accordance to current legislation. But the voices of opponents (among highly educated locals) were easily blocked by employees of municipal enterprises (usually migrant workers who do not speak Russian and are totally dependent on their employers). That represents a typical practice of controlled public hearings, where the concerned party tries to gain support and fully control the hearings results. The situation changed after putting the problem in public agenda and address of reputable group of architects and urban planners to mayor Sobyanin.

Treatment began with a statement of principle: «our estimate of the reconstruction project (redevelopment) Leninsky Prospekt is made solely on the professional field; We deliberately leave out of the equation any interests of business kind and political speculations». This appeal was the reason for the re-examination of the project in Moscow government and its further abolition.

(b) *Automobile Societies*

In recent years in Russia there were both the establishment and the dissolution of many organizations declaring themselves the representatives of the Russian motorists.

The Russian Automobile Society (RAS, the chairman—Valery Soldunov) is the oldest one. Formally speaking, this society refers to the category of GONGO. It was founded in 1973 due to the initiative of the Ministry of motor transport of RSFSR and the State automobile inspection of the Ministry of Internal Affairs of the USSR. The directorate is formed by a number of the high-ranking officials, including V. Kiryanov, the deputy minister of internal affairs of the Russian Federation, and V. Fedorov, who is the member of the Federation Council of the Russian Federation and the former chief of Main Directorate for Road Traffic Safety of the Ministry of Internal Affairs of the Russian Federation.

RAS wasn't involved in any socially significant actions and events during last years. Nevertheless, this society considerably surpasses other organizations of motorists in quantity and the real influence in the cities and regions of Russia. The number of RAS members makes more than three million motorists, while its local representative offices are open in 800 municipalities of Russia.

Its popularity and influence power is based on the material and technical resources inherited from Soviet period: first of all, the land areas for parkings and other objects of the automobile infrastructure which are in property or rented by the society.

At the beginning of the 2000s the deputy of the State Duma of the Russian Federation Victor Pokhmelkin made an attempt to establish an independent and liberal All-Russian social movement «Movement of Motorists of Russia». During the State Duma elections in 2003 Pokhmelkin headed the electoral bloc «A new course—the automobile Russia».

However, there was no success as no progress was reached: it became clear that the car owners, who belonged mostly to the Russian middle class, are attracted not by the liberal values, but by the left-populist slogans on the price reduction concerning the ownership of the car.

Although V. Pokhmelkin was formally regarded as a politician of the liberal wing, he tried to support these slogans, but, probably, this niche has been already occupied. The core associations of this organization are V. Pokhmelkin himself and his deputy and the notorious lawyer L. Olshansky. The activity of the movement is very low, except for couple of interviews to the second-grade media.

The interregional public organization of motorists «Freedom of choice» was created in 2005 by Vyacheslovy Lysakov, who was a public activist, a defender of the owners of right-hand cars at that time.

The «Freedom of choice» was mainly a protest movement during its first years. According to the official site it arranged «10 resonant All-Russian actions» in defense of the owners of the Japanese right-hand drive cars; in defense of the owners of the American cars with so-called «red turn signals»; in support of Oleg Shcherbinsky under the slogan «Down with the blinkers»; against increasing prices on fuel.

Despite the public fame of the organization, its influence power and the opportunities remained very poor. In 2011 the situation changed when V. Lysakov became the member of the Coordinating Council of the All-Russia People's Front (ONF), and then the deputy of the 6th State Duma of the Russian Federation due to the ONF quota in the party lists of the United Russia. As a delegate of the State Duma V. Lysakov took up a post of the first deputy chairman of the Committee for Constitutional Legislation and State Building.

From this point on the organization «Freedom of choice» finally transforms from the protest movement into the GONGO.

V. Lysakov was the author of a number of important legislative initiatives, for example, one of them was to insert the concept of the «possible total error of measurements» when determining the state of alcoholic intoxication of the driver in the legal ground. This initiative has allowed to cancel the adopted standards about zero alcohol per mille in the driver's blood.

At the same time, he initiated a number of populist initiatives with the aim to discredit and block the measures held by the city administration of Moscow concerning the parking regulation in the city. Another populist idea referred to the sphere of traffic safety, it was an initiative to fix 20 km/h minimum limit of overspeed.

The further prospects of the «Freedom of choice» depend mainly on Lysakov's success to pass to the 7th State Duma of the Russian Federation being a candidate of a single member electoral district of Moscow.

The interregional public organization «Federation of Russian Car Owners» (FRCO) was found in 2006. The protest actions of motorists on the same above-mentioned occasions (in defense of Oleg Shcherbinsky's and the owners of the right wheel cars) became an impetus to its establishment.

In contrast to the «Freedom of choice», the Federation of the Russian Car Owners is quite successful interregional movement which units dozens of independent and active regional organizations.

The Federation positions itself as «community of the free people», who pursue the principles of «legality, justice and equality» and admit that «they and the society in general are responsible for freedom». Its declarations contain both liberal and left-populist ideas: «... the Federation considers safety, equality and mutual respect on the roads to be one of the main priorities of activity on an equal basis with the problem of the car owning cost».

The Federation provides the car owners with the juridical education, the legal aid and implements the anti-corruption projects.

Sergey Kanayev, the head of the Federation, positions himself on equal terms with other leaders of the organization. He is a founder and the organizer of the Russia's first National award in the sphere of the organization and traffic safety «Safety is everybody's business». The award has a set of the nominations and attracts a significant number of participants from many regions of Russia. To all appearances, the management of the Federation interacts effectively with business society and is capable to raise considerable sponsors funds.

- «Society of Blue buckets» (the coordinator is Pyotr Shkumatov), the informal Moscow community created in 2010. Its initial aim is to fight against the archaic practice of privileges in traffic due to the official status. The symbol of these privileges are the blue emergency vehicle lightings (flashing beacons), in common parlance—the flashers. Such blue color signals are installed in Russia not only on emergency cars, but also on numerous cars of law enforcement agencies, special services and the high-ranking officials. The number of the Russian officials having traffic privileges is unprecedented in comparison with world figures.

The humorous publication on the official website of the «Blue buckets» explains an origin of the name and the purpose of this community:

«... Once upon a time a driver asked his five-year-old son for a blue sand-box bucket. Its size was just same as a standard flasher on a roof of a regular official car. He has attached this bucket on a roof of the car. Consequently, everything resulted in a mass action of civil unrest, so the state authorities were in a state of shock. And so it turned into a grass-roots movement against the state rudeness on the roads of our country».

The main target of the «Blue buckets'» activity is the monitoring of the law offenses by officials—the owners of the traffic privileges. The recorded violations are made public, first of all, in the social networks; besides the organization lodges complaints in the relevant departments.

The «Blue buckets», as well as many other public organizations of motorists wasn't standing only for the equality on roads, but also it has joined the populist actions against the paid parkings and other measures of the Moscow government.

- The public organization «Probok.net» positions itself as «a crowdsourcing project based on the useful information on the transport situation received from the inhabitants. The best ideas on the solution of the transport issues are directed in the governments of Moscow and other regions. Initially the project was engaged only in the motor transport. Nowadays the organization has an aim to make a convenient environment for life and traffic of all groups of citizens regardless of the transport mode». This organization was founded in Moscow and headed by Alexander Shumsky in Moscow in 2010. It enjoys the financial support from the city administration being in status of GONGO.

It combines the functions of a public organization and engineering center for young employees. By receiving regular information from a variety of interested drivers and other citizens, as well as carrying out their own form of transport surveys, specialists make a suggestion for local activities aimed at improving traffic management. They managed to find a good business niche: their projects are not big enough for the professional companies, but certainly not available on the engineering complexity to «activists from the street». The mechanism of promotion projects is as follows: some initiatives are put forward, the site users are voting for them, and then they are sent to the Government of Moscow. Many of these initiatives have been successfully implemented in practice.

However, the status of GONGO charges the organization with the task: concerning the collision of opinions on the reconstruction of the Leninsky Avenue the organization Probok.net followed strictly «the overall strategy», challenging (sometimes in a political incorrect way) the position of the academic critics of this project.

Alexander Shumsky has become a significant public personality in recent years. In 2015 he ran for the post of deputy of the Moscow Duma under very ambiguous slogan: «We took back Crimea, so we will take back free parking too!». The mention of the Crimea is clearly not part of the Moscow agenda, the promise of free parking was clearly inappropriate from the public man who is close to the Moscow authorities and is a head of qualified engineering center. They launched an ambitious and clearly not cheap election campaign; but then he withdrew as a candidate.

In 2016 Alexander Shumsky became the member of the Moscow Civic Chamber.

(c) *Passenger and Pedestrian Societies*

In this segment there is a number of the organizations which participate actively in urban transport policy:

- The interregional public movement for the rights of pedestrians «The union of pedestrians» headed by Vladimir Sokolov;
- «The union of passengers» was established in 2012 as an independent public organization, which aim was to protect the rights of transport services consumers. The organization is headed by the retired government official Kirill Yankov, the member of the Public Council of the Russian Ministry of Transport

- the public movement «Muscovites for the Tram», headed by Alexander Morozov, etc.

In comparison with the above-mentioned public organizations of motorists, these associations and the organizations similar to them represent rather prototypes of future real structures and are recognizable in the public space generally due to personalities of the leaders.

6.6 Conclusion

In conclusion we state that formal and informal institutions of Russian cities transport systems management represent a very picturesque mixture of heterogeneous elements succeeded from the Soviet legacy and newly formed elements as to overcome this legacy including fragmentary assimilation of diverse international experience. Formal institutions since 1991 went through the real «roller coaster».

Transport management institutions have made the long way from total centralized control of the city transport systems, peculiar for the Soviet period, and almost complete «laissez-faire», observed in the 1990s. In 2000s we observe the comeback of the formal government regulation by a number of centralized legislative provisions. But still, in institutional terms, these regulations are quite fragile. They are not sufficiently maintained by complex ideology and budget.

Informal institutions and stable practices still hold a strong position in transportation management process due to low check and control capacities of underdeveloped civil society structures are still in the embryonic stage of development. Existing institutionalized structures of civil society mostly are trivial GONGOs or act as a tool for the promotion of their leaders. Latter tend to accumulate political gain using populist rhetoric, eagerly supported by the population.

The mechanisms of «direct democracy» such as public hearings, or online surveys are largely fake and used as formal legitimation procedures. Their results are easily rigged and, in all cases, has minimal influence on the adoption of large-scale solutions in the field of transport and urban policy. Results are often used to point on quazi inadequacy of the public demand in conflict situations.

Local communities demonstrate «signs of life» mainly in the NIMBY format. But still, people do not estimate positive or negative external effects of the whole project. The solution for it requires more intensive dialogue initiated by local authorities. Existing gap between authorities and people results in deep distrust by latter in former. It means that people will have negative attitude towards any government initiative in the sphere of transport—reorganization of parkings, replacement of trolleybuses, raising transport taxes, etc.

Citizens' collective unconscious is based on a free-rider pattern. In particular, people hope that the city authorities may and must solve all transport problems within resources they have, providing car owners with roads and free parking lots, and other residents with a high-quality public transport. At the same time standard and internationally approved road pricing measures are perceived as directed against the people. As a result, the willingness to pay for parking, toll roads (and, especially, for the paid entry) is still low even among owners of cars with a cost of €50–100 thousand and more.

Due to these problems, politicians of federal, regional and municipal levels, usually do not touch the transportation issues while proposing programs to their voters. But if they do, the promises take form of populistic ones, pleasant to the ear of potential voter.

With all the problems mentioned above, it should be emphasized that the reasons for optimism remain. In recent years, in the federal legislation, as well as in municipal legal acts have been introduced a number of quite reasonable innovations in the field of transport policy. Civil society of the mid 2010s period (with all its shortcomings) is much more mature than their predecessors 10–15 years ago. So, the authors retain the hope that the second post-Soviet generation will live in Livable Cities of Sustainable Mobility era.

References

Acemoglu D, Johnson S, Robinson J (2005) Institutions as a fundamental cause of long-run growth. In: Aghion P, Durlauf S (eds) Handbook of economic growth, vol 1, Part A, 1st edn. North Holland, Amsterdam, pp 385–472

Blinkin MY, Gurevich GA, Sarychev AV (1988) Avtomatizirovannye sistemy transportnogo planirovaniya (Automated transport planning systems)

Carothers T (2002) The end of the transition paradigm. J Democracy 13:5–21

Evans AB (2011) The failure of democratization in Russia: a comparative perspective. J Eurasian Stud 2:40–51

Feldman MS, Sköldberg K, Brown RN (2004) Making sense of stories: a rhetorical approach to narrative analysis. J Public 14:147–170

Gaidar YT (2010) Collapse of an empire: lessons for modern Russia. Brookings Institution Press, Washington

Gaidar YT (1990) E'konomicheskie reformy i ierarxicheskie struktury (Economic reforms and hierarchical structures). Nauka, Moscow

Galbraith JK (1973) Economics and the public purpose. Houghton Mifflin, Boston

Goodin RE (ed) (1996) The theory of institutional design. Cambridge University Press, Cambridge

Goskomstat (1991) Narodnoe hozjajstvo v SSSR v 1990 g (The national economy of the USSR in 1990). Finansy i statistika, Moscow

Grieco M (2011) Transportation in African cities: an Institutional Perspective. In: Brooks N, Donaghy K, Knaap G-J (eds) The Oxford handbook of urban economics and planning. Oxford University Press, Oxford, pp 834–844

Hall P (1986) Governing the economy: the politics of state intervention in Britain and France. Oxford University Press, Oxford

Helmke G, Levitsky S (2003) Informal institutions and comparative politics: a research agenda. Notre Dame

Hodgson GM, Archer M, Castro AC et al (2006) What are institutions? J Econ Issues XL 1–25

Immergut EM (1990) Institutions, veto points, and policy results: a comparative analysis of health care. J Public Policy 10:391–416

Instruktivnoe pis'mo Ministerstvo finansov SSSR (Ministry of Finance of the USSR instructive letter) (1988) O kompensacii za ispol'zovanie lichnyx avtomobilej dlya sluzhebnyx poezdok (About the compensation for the use of private cars for business trips). USSR

Kolik A, Radziwill A, Turdyeva N (2015) Improving transport infrastructure in Russia. Paris

Luneev VV (2005) Prestupnost' XX veka : mirovye, regional'nye i rossijskie tendencii (Crime of the 20th century: global, regional and Russian tendencies). Wolters Kluwer Russia, Moscow

March JG, Olsen JP (1996) Institutional perspectives on political institutions. Governance 9:247–264

March JG, Olsen JP (2006) Elaborating the "new institutionalism". In: Rhodes RAW, Binder SA, Rockman BA (eds) The Oxford handbook of political. Oxford University Press, Oxford, pp 3–20

North D (1990) Institutions, institutional change and economic performance. Cambridge University Press, Cambridge

Ottaway M (2003) Democracy challenged: the rise of semi-authoritarianism. Carnegie Endowment for International Peace, Washington

Peters B, Pierre J, King D (2005) The politics of path dependency: political conflict in historical institutionalism. J Polit 67:1275–1300

Postanovleniye Soveta Ministrov SSSR (USSR Council of Ministers order) (1988) Ob uporyadochenii ispolzovaniya sluzhebnyx legkovyx avtomobilej (On regulating the use of official cars). USSR

Prybyla JS (1966) Transportation in Communist China. Land Econ 42:268–281. doi:10.2307/3145344

Pucher J, Lefevre C (2014) Urban transport crisis in Europe and north America. Palgrave Macmillan, London

RF Government Decree (Postanovlenie Pravitel'stva RF) (2014) On approval of rules for granting subsidies for the purchase of buses and equipment for housing and communal services, running on gas fuel, in the framework of sub-program "Automobile industry" of the state program of the Russian Federation. The development Russian Federation

Rodionov AY (2005) Methodical recommendations on the organization of the transport service of the population of municipalities. Fund "Regional studies", Moscow

Rothstein B (1996) The social democratic state: Swedish model and the bureaucratic problem. Pittsburgh University Press, Pittsburgh

RSFSR Ministry of motor transport order (1987) O vvedenii vremennoj instrukcii i rekomendacij po obespecheniyu vypolneniya rejsov na avtobusnyx marshrutax (On the introduction of the temporary instructions and recommendations for the implementation of service on bus routes)

World Economic Forum (2016) The global competitiveness report. http://reports.weforum.org/global-competitiveness-report-2015-2016/. Accessed 31 Aug 2016

Ukaz Prezidenta Rossijskoj Federacii (RF Presidential Decree) (1991) Ob uskorenii privatizacii gosudarstvennyx i municipal'nyx predpriyatij (On the acceleration of the privatization of state and municipal enterprises). Russian Federation

Chapter 7
Road Pricing as a Modern Mechanism for Road Sector Financing

Ekaterina Reshetova and Nikita Krupenskiy

Abstract Total government spending on road construction and maintenance in Russia are about 2.5 times lower than the funds required by the technical norms. This happens due to the lack of the interconnection between road funding and taxes and charges levied on road users. The mechanism of road funds has been introduced in 2011, but the problem still exists. Taxes, which are concerned with the car ownership and road usage (transport tax, fuel tax, utilization fees, custom dues) cover about 50 % of the total governmental spending on road infrastructure (and about 25 % in Moscow), the rest is covered by the general taxes. The road pricing is seen as a good alternative way of road funding. The involvement of the private investor into the construction of the toll roads was first implemented in the end of 1990s, when several pilot projects were realized. The large-scale implementation of the PPP principles is connected with the creation of The State Company Russian Highways (Avtodor). The mission of the company is to form and develop Russia's national toll highway network using the private investments. Toll roads are also constructed or discussed in the big cities, such as Moscow, St. Petersburg, Ryazan, etc. In cities where the majority of offices, plants, and shopping malls are concentrated the road pricing is not only the best source of funding, but also the transport demand management instrument. Some other sources of road funding are emerging in the largest cities (Moscow, St. Petersburg, Kazan): parking fees, cordon tolls, pollution charges. Moreover the heavy vehicles (more than 12 ton) have been obliged to pay for the distance travelled on federal road network since November 2015. This is expected to provide additional income of about 50 billion rubles per year for the federal road fund. This chapter provides a detailed overview and appraisal of the road pricing implementation in the Russian Federation.

E. Reshetova · N. Krupenskiy (✉)
Institute for Transport Economics and Transport Policy Studies,
National Research University Higher School of Economics, Moscow, Russia
e-mail: nkrupensky@hse.ru

M. Blinkin and E. Koncheva (eds.), *Transport Systems of Russian Cities*, Transportation Research, Economics and Policy,
DOI 10.1007/978-3-319-47800-5_7

7.1 Roads of Russia: History

In 1828, the great Russian poet Alexander Pushkin made the following forecast for the national road network development:

"When progress and amelioration
have pushed their frontiers further out,
in time (to quote the calculation
of philosophic brains, about
five hundred years) for sure our byways
will blossom into splendid highways:
paved roads will traverse Russia's length
bringing her unity and strength;
and iron bridges will go arching
over the waters in a sweep;
mountains will part; below the deep,
audacious tunnels will be marching:
Godfearing folk will institute
an inn at each stage of the route".[1]

188 years have passed since these lines were written. Pushkin's forecast has survived the era of horse-drawn transport and the age of mass motorization and remains relevant in the new era of "digital age transportation". There are 312 years to go until the deadline of this forecast. The dream of "paved roads" which "will traverse Russia's length bringing her unity and strength" remains elusive: the Engel[2] coefficient for the national road network in Russia (2.8) is one of the lowest in the world: of the lowest value among countries of BRICS and Eastern Europe (Fig. 7.1, Table 7.1).

The road development programs implemented in the territory of the Russian Federation in 20th century did not belong to the category of national priorities and have been neither large-scale nor systematic. These programs were based on the implicit hypothesis that roads were of minor importance compared to railways and waterways, playing the role of the reserve and/or auxiliary routes of travel. Road freight transport has mainly played a supportive role in the transportation system of the country. The role of private cars remained minimal until the end of the Soviet regime.

The main paved road network and urban streets had only been laid out by the end of 1980s, lagging many years behind the developed countries.

Until the recent times, private cars using these roads were deemed to belong to a kind of "second-rate" category: occasional traffic generated by recreational journeys on weekends and public holidays. Private cabbing, i.e. the implementation of

[1]Eugene Onegin. Chapter 7. XXXIII (translated by Ch. Johnston).

[2]Engel coefficient provides the estimate of the density of road network, adjusted to population density, thereby eliminating the presence of unpopulated areas.

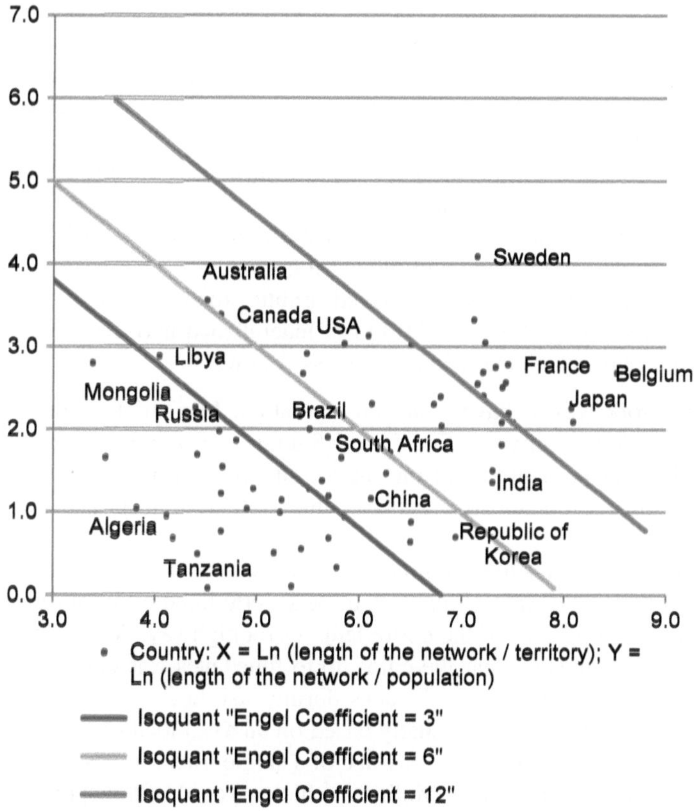

Fig. 7.1 The density of road network in various countries

Table 7.1 Country comparison based on Engel coefficient

Countries	Engel coefficient
BRICS	
India	7.32
SAR	4.46
Brazil	4.42
China	3.8
Russia	2.67
Eastern Europe	
Czech Republic	14.03
Poland	12.22
Belorussia	6.76
Romania	3.76
Ukraine	3.33

commercial transportation using private cars was considered an administrative offense till 1989. Even in the present-day Russian speech, some expressions referring to these second-class road users have been preserved, quite peculiar and untranslatable into other languages, such as a "motorist"/"motor-car enthusiast" or a "private car owner".

The urban road network in Russian cities has been designed and built in a similar fashion and guided by similar reasoning: the network density and its capacity were expected to accommodate the public transport, transportation of industrial and commercial goods, and special and official transport.

By 1991, the total length of paved public roads reached approximately 500,000 km, including about 45,000 km of federal road network.

The major network characteristics were as follows:

- Road network was to have a star configuration. In a strict sense, it could not even be called a network. Such a configuration was natural considering the political attitudes of the time: the main issue was not the inter-territorial connections but the connections between the strategic objects and the capital of country.
- The road network was to have no more than two lanes for most of its length. The length of the multilane sections was to be several hundred kilometers, that is, a few tenths of a percent of the entire length of a highway. At the same time, the widening of highways near the cities up to four lanes was carried out solely to provide capacity for buses and trucks during rush hours.
- The pavement strength was usually based on an axial load of 6 metric tons, that is, it was designed to allow for passage of a single truck. It was assumed that the majority of freights and heavy military vehicles would be transported by rail and inland waterways.

The shortage of paved roads that, for example, deprives some people of year-round access to railway stations or regional centers remains a problem even today. The decrease of this value from the appalling 10 % of the population living without access to roads to the current 1.8 % has to do with the depopulation of the peripheral territories rather than with the successful development of the road network.

The era of mass motorization began in Russia in the 1980–1990s, that is, 60–80 years later than to the developed countries. Thus, since the early 1990s, the heavy traffic generated by the rapidly growing vehicle fleet has been in discord with the poorly configured road network with insufficient capacity and permissible axial loads.

The vehicle fleet in Russia has grown by a factor of 4.5, that is, from 12 to 54 million cars over the past quarter of a century (1990–2015). The road network has also improved, but the imbalance between the scope of the vehicle fleet and the road network growth is evident. It should be noted that the issue of the nationwide road network development and enhancement of its performance, including its

Table 7.2 The transport system of Russian Federation in international rankings

Competitiveness indicators	Rank in Global Competitiveness Index (GCI)			
	2011/2012	2012/2013	2013/2014	2014/2015
Quality of overall infrastructure	100	101	93	74
Quality of roads	130	136	136	124

Source Data used from World Economic Forum (2012, 2013, 2014, 2015)

Table 7.3 Length of paved roads in the Russian Federation (2014)

Roads by ownership	Length of paved roads		
	Paved roads, in all	Freeways and expressways (%)	Other paved roads (%)
Federal roads, thousand km/%	50.66	5.29/10.4	45.37/89.6
Regional and intercity roads, thousand km/%	460.69	1.83/0.4	458.86/99.6
In total, thousand km/%	511.35	7.12/1.4	504.23/98.6

upgrade to meet the current requirements, remains highly relevant and essentially unsolved.

This conclusion is supported by the rankings of "The Global Competitiveness Report" annually published by the World Economic Forum (Table 7.2).

The quantitative aspects of the problem are presented in Table 7.3.

7.2 The History of Road Funds in Russia

The past 100 years in Russia have witnessed three attempts to arrange the financing of the road sector on the basis of target road funds. The first attempt was made in 1913 but failed because of the World War I and the subsequent collapse of the Russian Empire.

During the entire Soviet period, the road construction, renovation, and maintenance were conducted using three main sources, the ratio between them changing over time: the use of forced labor (GULAG) and/or road duties ("corvée") and budgetary financing.

(1) Forced labor. From 1936 to 1953, the basic road network was administered by the Main Highway Administration of NKVD of USSR. In addition to GULAG prisoners who were the main labor force used for construction of roads, the Main Highway Administration was allowed to enlist for this purpose the "rural population working on the basis of unpaid personal labor for six days a year during two periods—spring and autumn (when free from agricultural work)".

(2) Road duties ("corvée"). The most recent execution of the "corvée" regulations, needed for the maintenance of existing roads and construction of new ones, was established by the Decree of the Presidium of the Supreme Soviet of the Soviet Union in 1958, titled "On the participation of kolkhozes, sovkhozes, industrial, transportation, building and other enterprises and economic organizations in construction and maintenance of roads". This regulation existed with some amendments and additions until the early 1990s.
(3) Direct government funding. It was directed to "the roads of national importance", for the Moscow Genplan implementation, and a number of other projects of high political importance. The process of budgetary road financing became regular in the 1970–1980s; however, the respective items in USSR and RSFSR budgets were not given high priority in any period of the Soviet history.

The second attempt to create road funds was made in 1992 during Russia's economic transition period. The system established at that time included the Federal Road Fund and regional road funds. It was a kind of a mixture of old Soviet and new market institutions.

The predetermined sources of road funds included, in particular, fuels and lubricants tax (the analog of taxes included in the fuel price and customary around the world) and also vehicle ownership tax (the analog of the transport tax, also customary in various countries around the world).

The obvious insufficiency of these two sources (the level of the country's motorization did not exceed 80 vehicles per 1000 inhabitants!) led to the introduction of "road users tax". This tax was charged as 2.5 % of the revenue of all businesses, regardless of the size of vehicle fleet and actual road usage. In fact, it was a monetized road obligation, similar to the grants allocated to the road sector, as prescribed by the abovementioned Decree of the Supreme Soviet of the Soviet Union of 1958.

The Federal Road Fund was eliminated in line with the total consolidation of all budget revenues undertaken by the financial authorities after the default of 1998.

Currently, Russia has a system of road funds, restored in 2010–2011. In accordance with the legislation, the total tax burden on a vehicle owner consists of federal, regional and local taxes and also mandatory payments. The main fees are as follows: transport tax, fuel excise taxes, customs duties, and recycling.[3]

---

[3]The transport tax is regional, enforced by the regional laws of the Russian Federation, and required to be paid on the territory of a respective subject of the Russian Federation. All income flows into the Russian Federation's budget for regions. The transport tax payers are the persons who, in accordance with the Russian law, have a registered vehicle (Art. 357 of Russian Federation Tax Code of Law). The tax rate valid on the territory of each subject of the Russian Federation is established by each individual subject of federation.

Fuel excise duties are federal indirect taxes included in the price of fuels for motor vehicles (diesel, gasoline, motor oil) and are paid by the buyers. They are regulated by the Tax Code of Law (Chapter 22 "Excises"). Excise amount is set by the federal authorities. Recycling takes place on the territory of Russia since September 1, 2012, in accordance with the Federal Law of Russian Federation N 128 "On Amendments to the Federal Law N 89 "On waste from production and

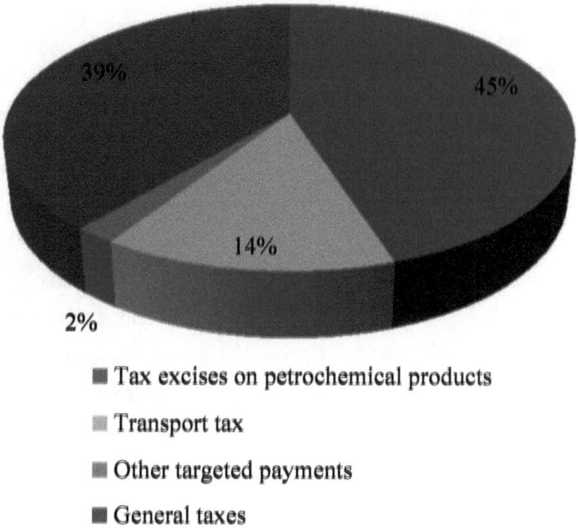

Fig. 7.2 Coverage of expenses for the road sector in the Russian Federation in 2014. *Source* Estimations made by authors

Formally, the current procedure basically complies with the experience of the USA ("The Highway Trust Fund") and other developed countries.

The issue, however, is that the total allocations for the road sector in the budgets of all levels are about 2–2.5 times below the minimum threshold cost required to prevent the road network degradation and ensure, at the very least, its moderate growth. However, even these obviously insufficient allocations are covered by the car owners' payments at about 60 % in the whole of Russia and at no more that 25 % in Moscow.[4] The rest is provided by general taxation (Figs. 7.2 and 7.3).

Nevertheless, the "collective unconscious"[5] of Russian car owners stays confident that they have paid the full price for using roads by buying "expensive" fuel and paying "breathtaking" transport tax. Accordingly, the lack of roads, low quality of roads, and systematic congestion in urban areas are viewed as a direct responsibility of the ruling elite.

(Footnote 3 continued)

consumption" dating from June 24, 1998, and Article 51 of the Budget Code of law of Russian Federation" dating from July 28, 2012. Recycling is a non-tax revenue forming part of the federal budget (P.1, Art. 51 of the Budget Code of law of Russian Federation) and, in fact, is a material warranty that a vehicle at the end of its life cycle will be disposed of rather than left to decay outdoors. Custom duties are established by the Customs Union Agreement of June 18, 2010 and shall be paid by the owners of foreign cars, the income flowing into the federal budget. The customs duty rate depends on multiple parameters that relate to the car as well as to the person importing it.

[4]The authors' evaluations are based on official tax budget statistics.

[5]The term of Jungian psychology.

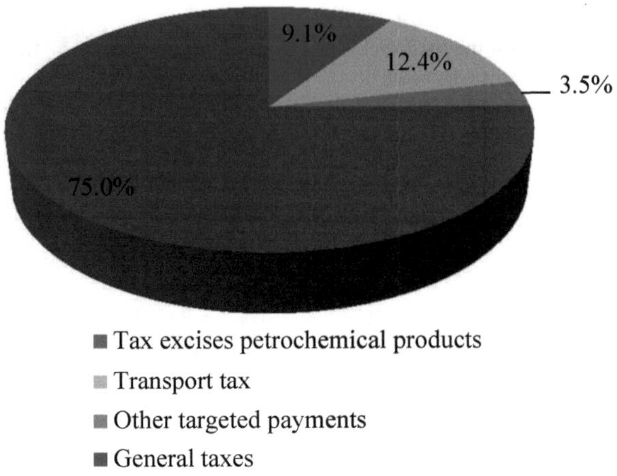

Fig. 7.3 Coverage of expenses for the road sector in Moscow in 2015 and planning period 2016 and 2017. *Source* Data from The City budget (2014)

As a result, the interconnection between the road financing system and taxes raised from road users is not clear enough either fiscally or in the socio-psychological sense.

This interconnection became even more ephemeral after the significant reduction of fuel excise taxes that was part of the so-called tax maneuver,[6] effective January 1, 2015, and even more so after the simultaneously made decision to transfer all revenues from fuel excise taxes to the regional road funds and fill the federal road fund with the revenues from general taxes.

At the time this book was being written, there was a regulation, according to which the federal road fund annually received a certain legally written amount of money from the general taxes, indexed according to inflation plus a certain share of fuel excise taxes. The potential of the federal road fund in GDP percentage is presented in the Table 7.4 showing the factual data for 2015 and governmental estimations for 2016–2018.

The federal road fund, by legislation, is used for renovation and maintenance of federal roads, subsidies for regional road funds, and, as a residual, on priority construction projects. The priority construction project for the next three years will be the bridge across the Kerch Strait. This politically motivated project will require about 8–10 % of the total federal spending on road sector.

[6]The tax maneuver includes the reduction of export duties on oil with a parallel increase of tax on resources extraction rates. The aim of the maneuver was to reduce the dependence of the budget of the Russian Federation from export duties which tend to dwindle when the oil price goes down. The increase of tax on resources extraction rates resulted in the increase of motor fuel price on the domestic market. To mitigate this effect, the maneuver envisaged the reduction of the excise tax on gasoline and diesel fuel.

Table 7.4 The reserves of federal road fund for 2015–2018

	2015	2016	2017	2018
Information resource	Factual data	Budget law	Ministry of finance estimates	
Federal road fund as percentage of GDP	0.83 %	0.76 %	0.72 %	0.71 %

The regional road funds will also receive the income from per kilometer payments collected since 2015 from the owners of heavy freight vehicles (12 metric tons or more).

The factual data on the recent years as well as the official estimates for the next few years give no reasons to expect any considerable recovery in the field of road construction in Russia.

Given the low, by the world standards, total allocations to the national road network, the structure of their distribution is most important. In this regard, attention should be paid to the extremely high inconsistence of the allocations in the regions. Thus, the allocations made for the road sector in Moscow and St. Petersburg are about 430 million rubles annually (per 10,000 cars), the average for Russia (without Moscow and St. Petersburg) is 240 million rubles, and in some subsidized regions of the country, less than 50 million rubles.

Calculations show that the projects constructed (or reconstructed) within the 50 km zone from the Moscow Kremlin at the expense of the budget of Moscow and Moscow oblast, as well as at the expense of the federal budget and other sources, account for more than half of the total allocations for all road construction projects in Russia.[7]

It is unlikely that such centralization of investments in road construction has been ever seen in the history of transport infrastructure development around the world.

7.3 Alternative Sources of Road Sector Financing: Public-Private Partnership (PPP) Projects

The idea of attracting private investments to the road sector as a supplement (or even replacement) of budget allocations has been in circulation in Russia since the beginning of the era of economic reforms. Boris Yeltsin's Presidential Decree "On

[7]Calculations were based on the analysis of the major program documents of the Russian Federation: Moscow State Program "Development of Transport System in 2012–2016", State Program of Russian Federation "Development of Transport System" and the Transport Strategy of the Russian Federation for the period up to 2030.

toll roads" appeared in 1992 and, to some extent, legalized PPP in the Russian road construction field.

The first PPP were implemented in the late 1990s: a viaduct in Saratov (1998), a bridge over the river Don in Voronezh oblast (1998), and a section of the federal road "Don" bypassing the town of Hlevnoe in the Lipetsk oblast (1999). Their introduction was marked by serious conflicts that had to do with flaws in the legislative framework as well as with the public rejection of the idea of toll roads as such. In particular, the collection of tolls on the bypass of the town Hlevnoe triggered regular protests by the regional Prosecutor's Office.

The participation of private investors in such projects is complicated by many factors, among which the following should be highlighted:

- Institutional factors: return on investment, even in the case of a most successful road project, is 10–15 years or more; consequently, the investor must have confidence in the stability of the "rules of the game" for the duration of extremely long, by Russian standards, time intervals;
- The cost of borrowed capital: at the current situation at financial markets, the cost for a private partner is 2–3 times higher than that for the government departments;
- Inflation and exchange rate risks: besides the obvious macroeconomic reasons behind the plan, unrelated to the specifics of the sector, there is also the factor of costs in euros and dollars. These costs include payments for engineering services and supply of high-tech materials, lease payments for road construction equipment etc.;
- High risk of not achieving the projected revenue, which is determined by working in the conditions of regulated tariffs, as well as high flexibility of demand (traffic) for the tariff and, consequently, higher traffic risks.

In these circumstances, Russian investors are ready to participate in PPP projects only under the conditions of significant equity contribution of budget investments, as well as the maximal possible transfer of these risks to the grantor. The most common practice pertaining to concession agreements in Russia is the provision of grantor payment for the completion of construction, as well as for the services provided during the period of service as set out in the agreement while fees are charged from users in favor of the grantor. This option actually exempts the concessionaire from any traffic risks.

Promotion of PPP in the road sector is associated with the adoption of the federal law "On Concession Agreements" in 2006 and the establishment in 2009 of the State Company "Avtodor" which became the analogue of PPP agencies, well-known in the world practice.

Through the efforts of this company, "most favorable conditions" have been formed that ensured the attraction of private funds to road construction and allowed for accumulation of certain experience in implementing large-scale projects from tendering to launching into regular operation on a toll basis. Also thanks to "Avtodor's" efforts, a number of large companies and banks were involved in PPP projects, allowing these companies and banks to master this format and be sure of its perspectives. A number of competing financial and building consortiums have been formed with the participation of leading financial institutions (Sberbank, VTB, Gazprombank, Russian Direct Investment Fund and etc.), forming a kind of a "top league" of road construction.

Currently, the State Company "Avtodor" has 2860 km of federal roads under asset management; by the end of 2030 it is planned to increase the length of the road network up to 12,000 km.[8] The total length of toll sections is 520 km as of January 1, 2016.

Table 7.5 presents the portfolio of major projects managed by "Avtodor", currently at the various stages of implementation.

The figures presented in the Table 4.5 reflect the situation as of the end of 2014. Taking into account the macroeconomic and budget realities of 2015–2018, a review of the timing of implementation and financial parameters of the highlighted projects should be expected. However, in all cases, the portfolio of the State Company "Avtodor" appears very impressive even by international standards as regards PPP-projects.

In 2014, the revenues from the toll roads controlled by State Company "Avtodor", their overall length at 405 km, was €31.4 million,[9] that is, about €79,000 per kilometer.

It should be noted that the average cost of construction of 1 km of a lane based on the 2014 prices is about €877,700 and €640,800[10] for a road of I and II technical categories respectively.

A toll road is built according to the standards of the abovementioned categories and has at least 4 lanes; thus, its construction cost is within the range of €2.6–3.0 million. In these conditions, the revenues from tolls are hardly comparable with the cost of construction and maintenance of the road.

The problem with the revenues rests on a set of adequate tariffs, which is always limited by the rigid factors of a socio-political plan. The point is that the majority of main roads in Russia serve not only the transit but also the local traffic. The long distance transit is characterized by the quite high "willingness to pay", and there is a possibility to set the market price. It is clear, however, that no system of payment administration on toll roads can be built based on the principle of "tariff discrimination": a tariff cannot vary depending on duration and aim of transportation. Accordingly, a tariff should be set with an eye to the "weak link", which is local traffic. Formally, the requirements for the availability of a free alternative route (Art. 37 of the Federal law N 257) are always met; however, in fact, a toll road in many cases is the only viable route for numerous local trips and commercial transportations. Under these conditions, each decision on tariff is associated with a distinct political reason, which necessitates looking for a compromise between the

[8]In accordance with the project of the Strategy development of State Company "Avtodor" to 2030.
[9]Exchange rate €1 = 76.4 rubles.
[10]The author's calculation is based on the statistical data by the State Company "Avtodor".

Table 7.5 Projects by the state company "Avtodor"

Section of the road and its length	PPP format	Period of construction/period of agreement	Project cost, € millions	Share of private investors (%)
M-1 «Belarus» (33–132 km)	The concession agreement with the pay of grantor	2015–2018/35 years	1225.0	22
M-3 «Ukraine» (124–194 km)	Long-term investment agreement	2014–2017/23 years	244.7	10
M-4 «Don» (1091–1319 km)	Long-term investment agreement	2015–2016/20–30 years	151.3	12
Moscow-Saint-Petersburg (58–149 km)	The concession agreement with the direct collection of fees	2015–2018/30 years	1,057.9	41
Moscow-Saint- Petersburg (208–258 km)	Long-term investment agreement	2015–2018/20–30 years	425.0	29
Central Ring Road (3rd launch complex)	The concession agreement	2016–2018	1206.6	41
Central Ring Road (4th launch complex)	The concession agreement	2016–2018	1185.5	36
Central Ring Road (5th launch complex)	Long-term investment agreement	2014–2018	639.5	6
Bypass of Balashikha and Noginsk	The concession agreement	2017–2020/30 years	798.7	48
Summary				
Total length of new sections—875 km	Project costs—€6934.2 million	Share of private investments—30.8 %		

Here and hereafter the financial parameters of the projects are calculated based on the rate €1 = 76 rubles
Source Report data by the State Company "Avtodor"

social acceptability of tariffs and the level of profitability of toll roads. Unfortunately, such compromises are not compatible with rapid transition to adequate returns on investments in toll roads. So, the principle "the user pays" included in the "Transport Strategy of Russian Federation until 2030" remains reserved for the future.

Let us take a look at urban segments of toll roads built on the basis of PPP mechanisms. These projects have not become widespread in modern-day Russia due to the following constraints:

- Deficit of regional and municipal road funds for governmental provision of co-financing;
- High flexibility of traffic on toll roads vs the tariff amount: in cities, there are more opportunities to choose alternative routes, therefore even a slight increase of tariff reduces the toll road usage;
- Difficulties associated with the land acquisition in the conditions of a dense built area.

Here are a few examples of the successful implementation of such projects:

- The Western High Speed Diameter in St. Petersburg (the project was supported by the state immediately after the adoption of the Federal Law "On Concession Agreements" in 2006);
- New entrance to the Moscow Ring Road from the federal road M-1 "Belarus" Moscow-Minsk (Odintsovo bypass);
- The main section (15–58 km) of the high speed toll road M-11 "Moscow-St. Petersburg";
- A toll overpass in Ryazan.

7.3.1 The Western High Speed Diameter in St. Petersburg

The Western High Speed Diameter, the first toll motorway in Russia, was designed to connect the northern and southern areas of St. Petersburg by a shortest route and reduce the transit load on urban street network. Its route corresponds to the planning decisions provided in the "General Plan for Leningrad development in 1956–1965".

The technical side of the project is characterized by the parameters of its three main sections (Table 7.6 and Fig. 7.4).

To date, the construction of the southern and northern sections has been completed; they are currently operating on a toll basis. Opening of the central section is expected by the end of 2016. The road will be operated under the current PPP contract until 2042. The total funding amounted to €2799 million, with the share of public and private investments at 50.7–49.3 %. Table 7.7 and Fig. 7.5 represent the current tariffs and the estimated traffic.

It is possible to pay with cash, a bank card, a contactless card, or using a transponder. A transponder allows passing the checkpoint on a separate lane and provides 20 % of the sales. A personalized contactless card provides additional 10 % of the sales. Thus, the technology of toll collection meets the standards of the contemporary foreign counterparts.

The combination of high enough (and reasonably differentiated) tariffs with high daily traffic creates favorable (and probably unique for Russia) conditions for investment return on the project as a whole.

Attention should also be paid to one important detail of the planning. The routing and project parameters of the Western High Speed Diameter meet the

Table 7.6 The parameters of three main sections of the western high speed diameter

The sections of the Western High Speed Diameter	The southern section (from the junction with the ring road in the area of the Predportovaya station, including the exit to Avtomobilnaya street, to the interchange near the Ekateringofka river embankment)	The central section (from the Ekateringofka river embankment to Bolshaya Nevka river embankment)	The northern section (From the Bolshaya Nevka river embankment to E-18 highway "Scandinavia" in the area of Beloostrov village)
Road category	I A		
Speed limit, km/h	110		
Traffic intensity, cars per 24 h	More than 100,000		
Length, km including:	11.5	8.95	26.15
Overpass sections, km	7.5	8.95	5.35
Ground roads, km	1.5	–	20.8
Number of lanes	8	8	4–8
Road interchanges	7	3	6
Purpose	Provision of freight flow from the Bolshoy sea port of St. Petersburg to the federal roads network, including M-10 "Russia" and toll road M-11 "Moscow-St. Petersburg" that is under construction	Removal of the transit load from the street network in the historical center of St. Petersburg	Provision of the shortest exit from the ring road to the federal road "Scandinavia" and also to the resort area of the city. Removal of the excessive load from Vyborg and Primorsk highways

highest international standards of urban freeways: the highway is completely separated from the residential building areas and does not combine the functions of a road and an urban street on any of its sections. Such positive examples are quite rare in Russia where the combined function of a road and an urban street became an unfortunately common phenomenon embodied by the regulatory formulation "high street of continuous traffic flow", quite exotic in the global context.

Fig. 7.4 The Western High Speed Diameter routing and implementation phases. *Source* Western High Speed Diameter (2016)

Table 7.7 Average basic tariffs on Western High Speed Diameter, € per 1 km

Vehicle category	Southern section	Northern section
Class 1 (cars)	0.08	0.058
Class 2 (heavy SUVs)	0.206	0.093
Class 3 (trucks)	0.309	0.128
Class 4 (heavy trucks)	0.186	0.186

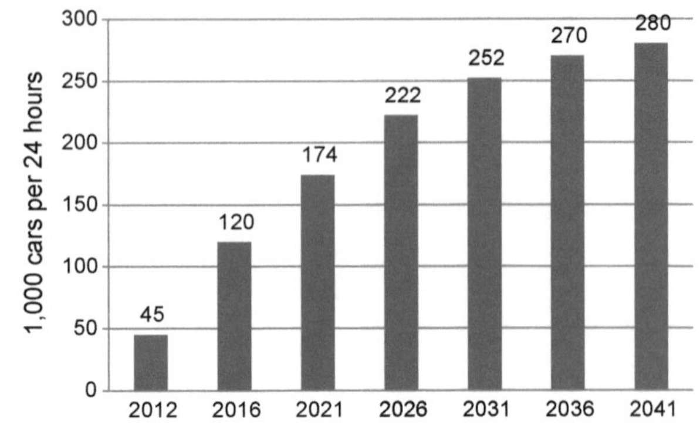

Fig. 7.5 The western high speed diameter traffic, 1000 cars per 24 h. 2012—factual, 2016–2041—forecast

7.3.2 New Entrance to the Moscow Ring Road from the Federal Road M-1 "Belarus" Moscow-Minsk (Odintsovo Bypass)

The construction project of the new entrance to the Moscow Ring Road from the federal road M-1 "Belarus" Moscow-Minsk (Odintsovo Bypass) is one of the first infrastructure projects of federal importance which was executed basing on a concession agreement. In addition, the routing of this section within the boundaries of the Moscow agglomeration provides a high-speed exit from the capital, which makes it virtually an urban project.

The concession agreement for construction and further operation of the 18.5 km-long toll road connecting the Moscow Ring Road in the area of the Molodogvardeyskaya junction to the 33rd km of the federal road M-1 was signed in 2009. The grantor functions are performed by the State Company "Avtodor", the concessionaire is JSC "Glavnaya doroga".

The total funding amounted to €426.3 million, and the ratio of private and public investments is 66:34. Since January 1, 2014 the road operates on a toll basis. Due to a significant number of potential entrance points, an open toll system is used. The toll booths are located so that the users pay only for the section of the road that was actually used (Fig. 7.6).

Payment can be made by cash, credit cards, smart cards, and a transponder mounted on the windshield of a vehicle.

The rates for cars not using transponders are within the range of €0.15–€0.23 per 1 km at daytime and within the range of €0.13–€0.22 per 1 km at night. A transponder provides a discount between 5 and 37 %.

Thus, the company encourages drivers to use automated devices for payment by making it possible to pass through checkpoints without stopping, that is, much faster (the checkpoint passage time is reduced by a factor of 4 on average as opposed to cash payment option and by a factor of 2 as opposed to credit card payment option).

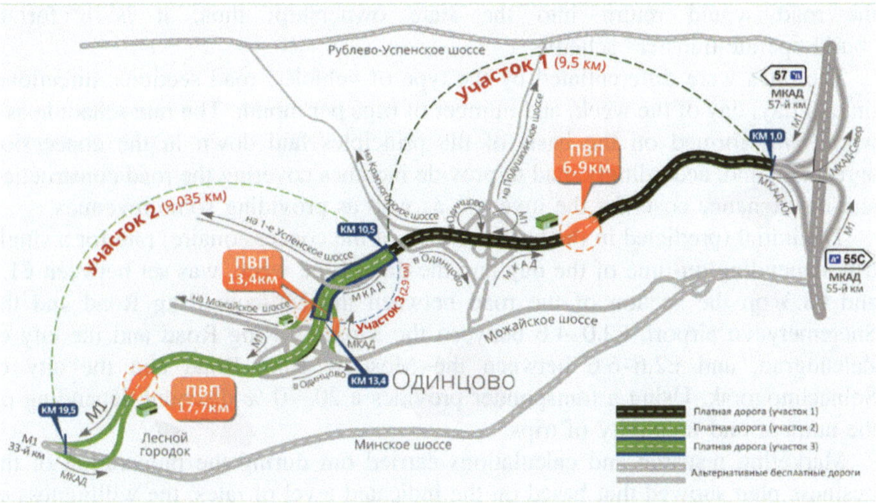

Fig. 7.6 The scheme of road by sections. *Source* Glavnaya Doroga (2016)

The project can be considered quite successful for the following reasons:

- The new section of the road is the first new entrance to the Moscow Ring Road from within the Moscow oblast built in the last 30 years.
- This is the only PPP project in road construction in Russia where the share of a private partner exceeded 2/3 of the total allocation.

7.3.3 The Main Section (15–58 km) of the High-Speed Toll Road M-11 "Moscow-St. Petersburg"

The main portion of the highway M-11 Moscow-St. Petersburg (15–58 km passing through the districts of Khimki and Solnechnogorsk of the Moscow oblast and bypassing the city of Khimki) started to operate in a toll mode in 2015 (Fig. 7.7). The grantor of the project is the State Company "Avtodor", and the concessionaire is the North-West Concession Company (NWCC) LLC.

The shareholders of the NWCC LLC are VINCI Concessions, a French operator company specializing in the field of road infrastructure, and MOSTOTREST, one of the largest Russian road construction companies. The total funding amounted to €947.4 million, while the share of private investments amounted to 54.1 %; a major portion of €513.2 million of private investments was provided by attracting €381.6 million in loans from Russian banks.

The concession agreement was signed for 30 years. It was assumed that during this time, the concessionaires would receive their return on investments, after which the road would return into the state ownership; thus, it is a formal "build-operate-transfer" scheme.

The rates were differentiated by the type of vehicles, road sections, directions, time of day, day of the week, and number of trips per month. The rate schedule as a whole was formed on the basis of the principles laid down in the concession agreement and, accordingly, had to provide incomes covering the road construction and maintenance costs for the investor, as well as providing some revenues.

The initial (predicted in the business plan of the concessionaire) rate for a single trip, depending on time of the day and the day of the week, was set between €1.3 and €3.3 on the section of the road between the Moscow Ring Road and the Sheremetyevo airport, €2.0–4.6 between the Moscow Ring Road and the city of Zelenograd, and €2.6–6.6 between the Moscow Ring Road and the city of Solnechnogorsk. Using a transponder provides a 20–70 % discount depending on the number and regularity of trips.

Marketing research and calculations carried out during the preparation of the business plan showed that based on the indicated level of rates, the willingness of drivers to pay and, accordingly, the daily traffic would be sufficient to return the private investments.

In November 2015, the main section of M-11 became a toll road. It became clear very soon that the optimistic calculations of the concessionaire proved to be unrealistic. This was largely explained by a radical change in the macroeconomic situation in Russia in 2012–2016 and, accordingly, a significant decrease of willingness to pay tolls even among the wealthiest residents of Moscow and the Moscow region.

Whatever was the reason, the rates announced by the concessionaire resulted in user dissatisfaction and fairly negative public reaction. The new toll road, despite the systematic congestion on the alternative, free route (the Leningrad highway) turned out to be redundant. The "inflated" rates were harshly criticized by representatives of the federal authority, the Moscow oblast administration, and even the General Prosecution of the Russian Federation. The concessionaire was obliged to reduce the basic rates up to 29–33 % depending on time of the day. It is expected that if the shortage of the expected traffic volumes persists, the rates will be lowered again.

Based on the conventional logic of a market economy, a rate adjustment should be expected to lead to a revision of the original concession agreement as well as the restructuring of the concessionaire's debts to the creditor banks. At the time this book was being written, the authors had no information on this matter.

In addition, the rate adjustments led to a curious paradox in the field of demand management: for two (and especially, for three) individual commuters from Zelenograd,[11] a carpool to the Belorusskiy railway station became cheaper than two

[11]The city of Zelenograd in the Moscow oblast is one of the administrative districts of Moscow; the distance from Zelenograd to Tverskaya Zastava where Belorusskiy railway station is located is about 50 km.

(or three, respectively) single trips by a suburban train. It is unlikely that this price effect will please the transport administration of Moscow, which in recent years has been looking for an opportunity to increase the demand for public transport.

However, the consequences of this situation around the main section of the M-11 highway turned out to be even more fundamental. In response to the request by the General Prosecution, the Government of the Russian Federation adopted a resolution that specifies the maximum rates for toll roads.[12] The biggest novelty of this decision was the establishment of maximum rates depending on the PPP format:

- If a toll road was constructed on the basis of a concession, then the rate (depending on the category of a vehicle) cannot exceed €0.158–0.507 per 1 km;
- In case of a long-term investment agreement, the same limits are set in the amount of €0.039–0.125 per 1 km.

Fortunately, this government decree is not retroactive, so the rates charged on the existing toll roads are not supposed to be reviewed. Also, it is important to take into consideration that the corrected rates on highway M-11 fall within these limits: the average price of travel on this road is €0.132 per 1 km.

The problem is that the very fact of the government setting limits to maximum rates is a very negative signal to potential investors. Essentially, an investor is challenged to replace the analysis of the willingness to pay, price flexibility on demand, opportunity cost of travel, amount of the borrowed funds etc. with the constant figures preset by the government. All of this, of course, does nothing to improve the investment climate in the country.

7.3.4 The Toll Overpass in Ryazan

Railroad crossings are ever the sources of chronic road congestion and places of high traffic accident rate. This problem is particularly severe near large cities where railway lines cross local roads (or even streets) with high-intensity traffic.

Construction of overpasses at such points is an acute, long-standing problem: investments in such facilities are non-earmarked for JSC "RZD"[13] and are often unaffordable for local budgets. The way out of this situation is constructing toll overpasses using private investments.

The toll overpass near the railway crossing at the 208 km of the Ryazan branch of the Moscow Railway (Fig. 7.8) was opened for traffic in December 2012.

Before its opening, the road between the districts of Sokolovka and Dashkovo-Pesochnaya (in fact, a city street) crossed the electrified railway tracks

[12]The Russian Federation Government Resolution N 47, dated 01.30.2016, "On congestion charge for vehicles on toll roads of federal importance, toll section of such roads (including cases when the toll section of road is a separate artificial road construction)".

[13]JSC "RZD" is a Russian state-owned company, the owner of public infrastructure and a significant part of the rolling-stock, and a major operator of the Russian railway network.

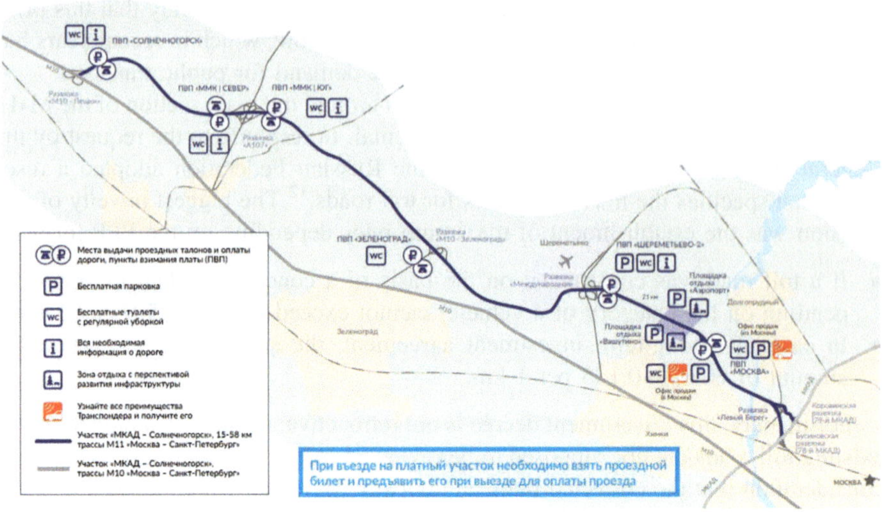

Fig. 7.7 The scheme of the toll section (15–58 km) of the M-11 "Moscow-St. Petersburg" highway. *Source* Severo-zapadnaya koncessionnaya kompaniya (2016)

Fig. 7.8 The scheme of the Ryazan overpass location. *Source* Authors

with high traffic frequency (up to 120 pairs of trains per day). The average time spent waiting at the crossing was 20 min, so a toll-based overpass became an alternative to losing time while waiting for the barrier to lift.

The cost of the overpass construction was €6.91 million. The sole investor was a non-government pension fund "Norilsk Nickel". The government of the Ryazan region and the administration of Ryazan provided only the legal and technical support.

The rate level is based on the cost of gasoline consumed during the 20 min operation of an engine in the idle mode while waiting in front of a barrier, or a 5 km detour through an alternative route. Currently, the rate charged for using the overpass, if paid with cash, is €0.26 for vehicles with height of up to 2.1 m and €0.53 for vehicles taller than 2.1 m. In the case of smart card payment, the rate is reduced to €0.21 and €0.42 respectively.

For public transport, emergency vehicles, and municipal services, the passage is free. Free passage is also allowed for vehicles serving budgetary institutions and two-wheelers.

It is noteworthy that the Ryazan overpass offers a discount based on "excess capacity", which is the first precedent in Russia. This idea was first suggested by Alan Walters (Walters 1968) many years ago and presumes a free passage for cars from 11 p.m. to 6 a.m.

On the overpass, there are 3 toll stations in each direction. Also, there is a rule prescribing to open the barrier and let the cars pass free of charge if the line-up at the terminals is longer than 70 m.

The rate of €0.53 along with the intensity of traffic up to 6000 cars per 24 h on the road of the 3rd technical category provides an annual income of about €1.12 million. Based on these initial conditions, the project has all the chances to return the investments within reasonable time.

According to the concession agreement, the overpass will become a municipal property after 20 years of operation. Since the standard life cycle of the span is 56 years, the local administration will be able to operate the overpass in a free mode for the subsequent 36 years.

The idea of toll overpasses over railways has become quite popular in many regions of Russia. In particular, only in the Moscow oblast the construction of 50 toll overpasses is planned at the points where regional and municipal roads cross the Moscow railway tracks. The first toll overpasses are also planned to be built near the railway stations located within the new borders of Moscow.[14]

Summarizing the experience of projects implemented in the metropolitan and urban areas, it should be noted that repeating the success of the Western Speed

[14]The expansion of the municipal area of Moscow in 2011 (the so-called "New Moscow" or "Big Moscow") was the biggest in the history of the administrative-territorial division as part of the city expansion project. Moscow became approximately 2.4 times bigger at the expense of the Moscow region territory. The main objectives of the project were to dismantle the traditional monocentric structure of the Moscow agglomeration and also institute a town-planning zoning, assigning clearly formulated administrative and government profiles to the newly attached territories.

Diameter project in St. Petersburg is possible only in a few, very exceptional cases. In particular, the possibility of constructing a northern alternate route of the Kutuzov Avenue in Moscow is already being discussed. With a certain level of probability, toll bridges across rivers may appear in some cities, for example, the fourth bridge across the river Ob in Novosibirsk.

However, as a rule, Russian cities lack both the sufficient budget resources for participating in PPP projects and the sufficient user willingness to pay for implementation of such large-scale projects.

The situation became even more complicated with the adoption of the above-mentioned decree on the maximum level of rates charged on toll roads: attracting potential investors became even harder than before.

The projects such as Ryazan toll overpass have every chance of widespread replication. Required investments in such projects (from €6.6 to €19.8 million) are quite feasible for local investors and the rates that can provide return on investment in such projects are not critical for local population.

Heavy trucks fare (axle load of more than 12 metric tons) In Russia, there are about 1.5 million vehicles weighing more than 12 metric tons. Another 400,000 vehicles of this kind, registered in other countries, operate on the roads of Russia, executing export, import, and transit transportation.

As part of the standard estimates provided by AASHTO, a famous American methodological document (AASHTO 1993), damages to roads[15] afflicted by vehicles can be calculated as a quadrupled value of axle loads. Accordingly, the damage caused by a truck with an axial load of 10 metric tons is equivalent to the damage caused by 10,000–25,000 private cars. Meanwhile, the specific fuel consumption and, accordingly, the tax included in its price is 4–5 times more for a truck with an axial load of 10 metric tons than for a private car, at the very most. In these circumstances, the tax collection from 12-ton trucks proves to be a more equitable way of road fund replenishment in comparison to the vehicle tax calculated based on the engine power: it is bigger for some private cars than for trucks.

The idea of taxing heavy trucks was borrowed from international experience, primarily from the German experience with "LKW Maut" system introduction. The respective project was developed at the initiative of the Ministry of Transport of the Russian Federation and the Federal Road Agency (ROSAVTODOR); the project included both legal and technological components.

For the purpose of providing proper regulations related to the introduction of a new duty, additions and revisions to the Budget Code of Russian Federation, the Code of Administrative Offences, the Federal Law "On the roads and road activities in Russian Federation", as well as to a number of other federal laws have been done.

The Government has also approved the Decree N 504, dated June 14, 2013, "On charging of fees in respect of compensation for damages caused to the federal roads by the vehicles having a maximum permissible weight of over 12 metric tons".

[15]A road is a complex assembly which includes pavement and subgrade with drainage, retaining and reinforcing structural elements.

Technological solutions used in the implemented system that was named "Platon[16]" are in line with the best global practices, which include:

- Payment for the actual travel distance;
- Use of global satellite navigation systems GLONASS/GPS;
- Regular monitoring of the traffic and identification of vehicles that fall in the category of payers;
- Collection of payment in an automatic mode without stopping the vehicle;
- Automated registration of payment rules violations.

The concession agreement between the Federal Road Agency (ROSAVTODOR) and "RT-Invest Transport Systems" LLC was signed for 13 years. During this period, the concessionaire will execute the construction entirely at its own expense and operate the charging system, including automated collection, processing, storage, and transfer of traffic data on trucks weighing over 12 metric tons in a real time mode. The system itself is in the ownership of the Russian Federation. The "Platon" is designed to operate on the total length of 50,774 km of federal highways.

The operation of the "Platon" system created an important precedent of the administration of road charges (which is, in fact, a federal tax) delegated to a private company. In this regard, the Russian federal agency focused on the German experience: the system "Toll Collect" performs the same functions on German roads and is owned by a consortium of private partners, including "Daimler-Chrysler", "Deutsche Telecom", and "Cofiroute".

The investments in the installation and launching of the "Platon" system amounted to 29 billion rubles, of which 27 billion were paid by Gazprombank and 2 billion were provided from the funds of OOO "RT-Invest Transport Systems".

The funds raised by the "Platon" system are directed to the federal budget (to federal road fund accounts). The operator services are also paid from the federal budget. An annual award to the operator is 10.6 billion rubles; 46 % of this amount is subject to annual indexation to the inflation level. The reward has a fixed structure: 59 % is used to cover the current expenses, 35 % is spent on the return of loan, 8 % for equipment replacement (once in 12 years), and 6 % is the operator profit.

The law taxing heavy vehicles travelling on the federal roads of the country came into force on November 15, 2015. The scheme of payment presumes either payment in advance or upon trip. Payment can be made in various ways such as: a bank transfer to the operator account, payment by credit card through terminals, QIWI,[17] in person in the "Platon" offices, through a user account on the service website, by E100 fuel cards, and by a DKV-card.[18]

---

[16]The name of the system is an abbreviation of Russian words "payment for tons" and is unrelated to the name of the Greek philosopher.

[17]Electronic payment system in Russia.

[18]E100 is the international fuel card with a microchip which allows to perform non-cash transactions for a full range of services related to the international and local cargo transportation. DKV Card is the main European non-cash fuel card.

The price for the travelled mileage is calculated automatically by an onboard device or by a route card on a personal account through the mobile app or in the "Platon" offices. The route map can be obtained for a single trip or for a month.

According to the governmental decision, the funds raised by the "Platon" system and directed to the federal road fund are used for subsidizing the regional road funds and are intended for road construction, renovation, and maintenance.

The initial rate was supposed to be set at 3.7 rubles per kilometer. The introduction of charges for 12-ton vehicles, as well as the value of rates caused vehement protests on behalf of the companies operating on the freight market, especially small entrepreneurs, that is, single-truck owners. The indignation of these companies was exacerbated by the technological and organizational shortcomings of the new system.

Protests took place in many parts of the country and forced the government to make compromises: the rate was reduced to 1.53 rubles per kilometer. The decision on the expiration date of the preferential rate will be made on the basis of the system's operation during the first 9 months, i.e. in October 2016. In addition, the tax laws have been revised, according to which the owners of 12-ton vehicles are granted a deduction from the sum of the transport tax, equal to the amount payed during the tax period through the "Platon" system.

The technological and organizational shortcomings of the new system were quickly resolved. These measures allowed for allaying the conflict with the carriers and introducing a new payment collection system as part of the routine operation.

It should be noted that the originally introduced rate (and certainly, the reduced one) is one of the lowest in Europe (Fig. 7.9).

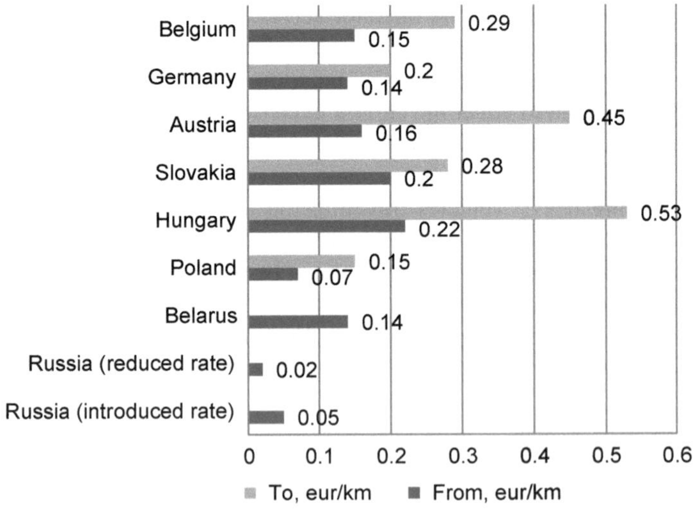

Fig. 7.9 The rates in European countries, €/km

The establishment of the fare for 12-ton trucks was a very informative case with regard to the implementation of the "user pays" principle in Russia and the prospects of transition to the road taxation in the "Pay-As-You-Go" tax format.

(1) In Russia, there are no technological obstacles to the implementation of the "pay-as-you-go" format, not only for 12-ton trucks but also for the entire vehicle fleet.
(2) The Russian legislation and practice in this sphere successfully adopts the respective achievements of the developed world.
(3) As in the case of toll roads, a user's willingness to pay is very low. Moreover, any innovations in the field of road pricing associated with the increase of fiscal burden are fraught with active protests on behalf of potential contributors.
(4) In the case of introducing the fees for 12-ton cargo vehicles, as well as in the case of toll roads, the Russian authorities are ready to suffer economic losses (i.e., shortfalls in payments) rather than pay the price of social discontent (i.e., protests from the disgruntled payers).

7.4 Perspectives of the Transition to the "Electronic Road Pricing" on the Basis of the "Pay-as-You-Go" Tax Formula

The decrease of income from fuel taxes that took place in the last few years in most developed countries has triggered a growing interest in the transition to taxation on the basis of the actual vehicle mileage—"Pay-as-You-Go" tax. The technical term proposed for this method is VKT-payments or VTM-payments,[19] with rates differentiated by vehicle category, roads category and also by the place and time of a trip.

The interest in the transition to VKT-Payments was amplified by the fact that the implementation of the "Road Pricing" policy has been limited to the introduction of multiple diverse payments[20] which, in spite of their seemingly obvious usefulness and validity, do not provide a fundamental solution to the problem of financing the national road network, and at the same time, often become a cause of public indignation.

In the Russian academic discourse, the "Road Pricing" related ideas initially appeared as purely theoretical constructions. Of course, the Russian economic theorists were familiar with the Pigovian Tax as an effective method of internalization of externalities. The famous article by the Nobel laureate Vickrey (1963) has

[19]VKT—Vehicle Kilometers Traveled; VMT—Vehicle Traveled Miles; car mileage calculated respectively in kilometers or miles.

[20]Such instruments as "Vignettes", "Distance-Based Network Charges", "Congestion Charge", "Pollution Charge" "Link Tolls", "Cordon Tolls", "Point Tolls", etc.

even been translated into Russian. However, it should be taken into account that in 1963, the level of motorization in the cities of Russia did not exceed 20–30 cars per 1000 inhabitants, and the tax system in a conventional sense of this term did not exist in the Soviet Union at all.

So, the practical interest in the ideas of "Road Pricing" arose in Russia much later, after the publication of the Russian translation of Phil Goodwin's inaugural lecture he gave in 1997 before accepting a position of a transport policy professor in the University College of London (UCL). In this lecture, Professor Goodwin paid tribute to his predecessor at the UCL, Professor Reuben Smeed: "An important contribution by Smeed to transport policy was his chairmanship of the committee on road pricing, which in 1964 gave a sort of official endorsement to Alan Walters' proposition that economic welfare would be increased if road use, perceived as a "free" good, were charged for".

The perspectives of basing the taxation of the Russian road users on the "Pay-as-You-Go" tax formula were first outlined in the "Strategy 2020" program developed in 2011–2012 by a group of leading scientists and economists commissioned by the Russian Government.

Here are the main arguments in favor of such a step that can be heard in the discussions held in Russia today and are, of course, largely consonant with the arguments presented by the foreign colleagues.

1. VKT-payments can provide a stable road funding revenue, as opposed to the fuel taxes. The point is that the volume of budget revenues from VKT-payments is tied to the actual mileage covered by the vehicle fleet but not to the consumption of motor fuels such as petrochemicals. Accordingly, they are not dependent on technological changes associated with the introduction of electric vehicles and alternative fuels and with the increase of engine efficiency. In Russian context, it is particularly important that the volume of budget revenues from VKT-payments is dependent on the oil and petrochemicals taxation (from production and processing to retail) which undergoes periodical changes in Russia, dictated by the situation on the global energy market.
2. VKT-payments with rates that can be differentiated based on the time and place of a trip should be able to distribute the tax burden among the road users in a way much more equitable than under the current tax system. In the Russian context, this argument is particularly important. The point is that the fuel consumption on the roads of lower categories and roads being in the worst condition is certainly higher than on the roads of higher technical categories with a high level of maintenance. In these conditions, the taxes incorporated in the price of motor fuels automatically increase the fiscal burden per one kilometer of a trip for the Russian mainland and provide clear baseless benefits for the residents of Moscow and the Moscow oblast where the main body of high quality road network is concentrated. In other words, in the capital, the motorists enjoy the best driving conditions and therefore pay significantly less for 1 km of a trip than their provincial counterparts.

3. VKT-payments with rates that can be differentiated based not only on the time and place of a trip but also on a vehicle type, weight, and emissions should be able to integrate all the diverse formats of payment that is currently collected from vehicle owners. This circumstance is also very significant in Russia where toll roads and taxing 12-tons trucks were only introduced in the last few years.

In addition, VKT-payments, if adequate rates per 1 km are introduced in urban centers, may contribute to an effective implementation of the idea of congestion charges in city centers which is now being actively discussed in a number of Russian cities and outlined in the Federal law draft "On traffic organization".

In conclusion, it should be noted that in consideration of the realities of the "Digital Age Transportation", the transition to VKT-payment system seems almost inevitable. Thus, the widespread introduction of this system may have a historical significance for the further developments in the field of reforming the institutional framework and mechanisms of financing highways and city streets.

In particular, the role of the government as the administrator of road taxes and road funds can be radically overhauled. In fact, if one particular part of road network will generate a stable stream of revenues which can be administrated by any IT company, the idea of the direct governmental participation in this process will become completely archaic. Moreover, if each specific section of the road network will be able to provide its own income, there will be no obstacles for transferring the maintenance, repair, and reconstruction duties to private businesses.

In this sense, Russia has a chance of moving away from the scheme of a "catch-up development" in the designated sphere and work proactively on the implementation of the modern formats of "Electronic Road Pricing".

References

AASHTO (1993) The American association of state highway and transportation officials. Design guide. http://www.pavementinteractive.org/article/equivalent-single-axle-load/. Accessed 31 Aug 2016

Glavnaya Doroga (2016) http://m-road.ru. Accessed 31 Aug 2016 (In Russian)

Santos G (ed) (2004) Road pricing, vol 9: theory and evidence (Research in Transportation Economics). JAI Press

Severo-zapadnaya koncessionnaya kompaniya (2016) http://15-58m11.ru. Accessed 31 Aug 2016 (In Russian)

Smeed RJ (1964). Road pricing: the economic and technical possibilities (Smeed Report). Ministry of Transport

The City budget (2014) The law on city budget of Moscow for 2015 and planning period 2016 and 2017. http://budget.mos.ru/project_directions. Accessed 31 Aug 2016 (In Russian)

Vickrey W (1963) Pricing in urban and suburban transport. Am Econ Rev 52(2)

Walters AA (1968) The economics of road user charges. World Bank staff occasional papers; no. OCP 5. The Johns Hopkins University Press, Baltimore, MD. http://documents.worldbank.org/curated/en/694271468766162565/The-economics-of-road-user-charges. Accessed 31 Aug 2016

Western High Speed Diameter (2016) http://www.whsd.ru. Accessed 31 Aug 2016 (In Russian)

World Economic Forum (2012) The global competitiveness report 2011–2012. https://www.weforum.org/reports/global-competitiveness-report-2011-2012/. Accessed 31 Aug 2016
World Economic Forum (2013) The global competitiveness report. 2012–2013 http://reports.weforum.org/global-competitiveness-report-2012-2013/. Accessed 31 Aug 2016
World Economic Forum (2014) The global competitiveness report 2013–2014. https://www.weforum.org/reports/global-competitiveness-report-2013-2014. Accessed 31 Aug 2016
World Economic Forum (2015) The global competitiveness report 2014–2015. https://www.weforum.org/reports/global-competitiveness-report-2014-2015/. Accessed 31 Aug 2016

Chapter 8
The Peculiarities of Road Safety Practices in Russian Cities

Ekaterina Reshetova and Nikita Krupenskiy

Abstract International road safety priorities are defined by the slogan "towards zero", which describes the aim to reduce the road mortality rates to a zero. The Russian Federation (along with China, India, and Brazil) is one of the «Risky States–10», which are the 10 countries with the high level of road fatalities and where more than half of the world traffic-related deaths occur. Even though the level of the transport risk has fallen over a decade (2005–2015) from 11.13 to 4.3, Russia still hasn't reached the mark 3, which had been the target for the developed countries in the middle of the 20th century. If the mortality rate in Russia was as low as it is in the UK, Sweden or Netherlands, the number of road deaths per year would be about 3.5–4 thousand instead of 27 thousand. Motorization rates in Russian cities are essentially higher than country average. That's why cities have much more road accidents and higher mortality rates. The situation is aggravated by the widespread practice of the combination of the road and street functions in the one technical object, which is aimed to let through the large volumes of traffic and is traced along the usual urban street. This leads to the larger amount of accidents with pedestrians due to the insufficiency and inconvenience of the underpasses and overpasses. The problem of the road safety receives much attention: the official meetings involving the first persons of the state are held regularly; world best practices in road safety are surveyed; fines for the traffic regulations violation are getting higher; the concepts of "value of life" and "aggressive driving" are introduced; quick response emergency systems are implemented. Some measures are taken to prioritize the pedestrians: pedestrian streets, zones of restricted access and speed limits within the urban territory are created. Still the crucial changes in the levels of road safety can't be seen yet. The chapter provides a comparison of the main engineering tools and methods as well as institutional conditions for ensuring road safety in urban areas in Russia and in a number of developed regions of the world. Some recommendations for achieving further progress towards a civilized level of road safety in Russia are made.

E. Reshetova · N. Krupenskiy (✉)
Institute for Transport Economics and Transport Policy Studies,
National Research University Higher School of Economics, Moscow, Russia
e-mail: nkrupensky@hse.ru

M. Blinkin and E. Koncheva (eds.), *Transport Systems of Russian Cities*, Transportation Research, Economics and Policy,
DOI 10.1007/978-3-319-47800-5_8

8.1 The Analysis of Traffic Safety Indicators Analysis in Russian Cities

The level of road safety in the cities of Russia is based on the evaluation of this problem at the scale of the entire country. In this regard, it should be noted that Russia (along with China, India, Brazil, Cambodia, Egypt, Kenya, Mexico, Turkey, and Vietnam) is included in the group of "Risky States–10": the ten countries with the high level of transport-related risks (road fatalities per 10,000 cars), which accounts for more than half of the total number of road fatalities in the world.

The severity of the problem is appreciated by the experts as well as the authorities; accordingly, respective measures are being implemented in Russia in order to deal with it. In recent years, given the rapid growth of the vehicle fleet, the country managed to achieve a downward dynamics, both in transport and social risks (road fatalities per 100,000 residents).

The systematic improvement of the key accident rates in recent years in Russia corresponds to the standard world trend adapting to the situation of the global motorization level growth. So, along with the motorization level growth in Russia, transport risks have been reduced in conformance with the classical Smeed's Law (Smeed 1949) (Fig. 8.1).

However, the Russian indicators lag behind the global average, and critically, behind the OECD member countries. The indicator of transport risks in Russia during 2015 reached 4.3 units, while the socially acceptable standard in the developed countries by the middle of the 20th century was at about 3 units and the comprehensive indicators for countries leading in the OECD rankings ranged from 0.4 to 1.0 units. The backlog related to reaching the milestones achieved in the OECD countries has not been shortening, but rather growing in recent years. So, in 2000, the gap in terms of the social risks level against the average values was 5 %, in 2013, 8 %; against the average values among the countries presented in the IRTAD[1] database (OECD 2016), in 2000, 1.6 times, in 2015, 2.7 times (Figs. 8.2 and 8.3).

In 2015, the record decrease in road fatalities was achieved since the beginning of the period of mass motorization: the number of deaths decreased by 14.7 %, compared to 2014, and amounted to 23,114 persons, with the number of victims being 10 times as much. It is important to note, however, that if the level of transport risks (road fatalities per 10,000 cars) in Russia corresponded to that in the countries such as Great Britain, Sweden, or Japan, our losses would not exceed 3500–4000 deaths per year. Tables 5.1 and 5.2 present the data on the site of the road accident registration and its severity.

[1]In 1988, the OECD Road Transport Research Program established the International Road Traffic and Accident Database (IRTAD) as a mechanism for providing an aggregated database where the international data on traffic accidents and their victims as well as the exposure data are collected on a continuous basis. The IRTAD database includes accident and traffic data and other safety indicators for 29 countries.

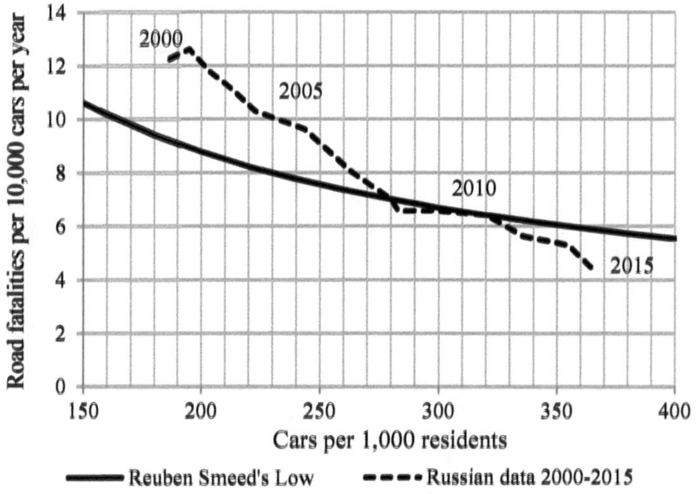

Fig. 8.1 Road fatalities per 10,000 cars versus Motorisation Level *Source* data from OECD (2016)

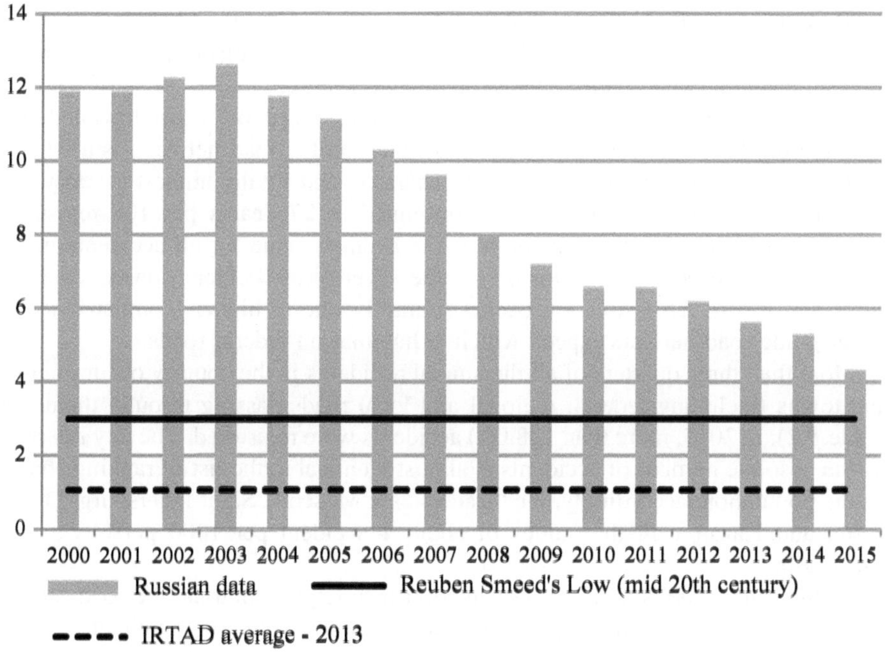

Fig. 8.2 Road fatalities per 10,000 cars *Source* data from OECD (2016)

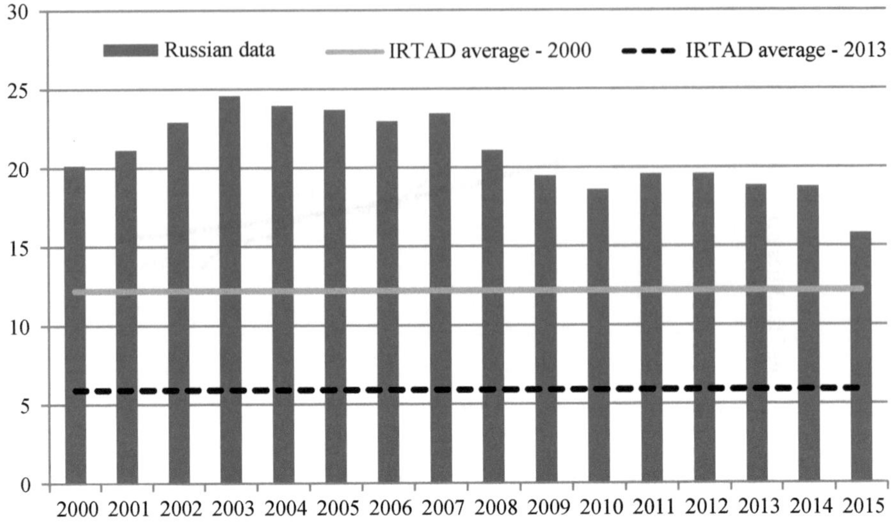

Fig. 8.3 Road fatalities per 100,000 residents *Source* data from OECD (2016)

More than one third (34 %) of all accidents registered on the highways occur on the federal roads (Table 8.1). It should be noted for comparison that according to the estimates by experts, the federal road network makes up for 40 % of the total VKT (Vehicle Kilometers of Travel[2]), therefore, based on 1 km travelled, the situation on these roads is somewhat better than on the road network as a whole.

Traffic accidents on federal roads are characterized by the highest severity and consequences of injuries. In 2015, it amounted to 27 deaths per 100 registered accidents, which is 2.1 times as high as the average value for all accidents in the country and exceeds by a factor of 3.8 the severity of accident consequences in cities and towns. This situation occurs primarily due to higher (compared to the lower grade road network) speed which is habitual on federal roads.

More than three quarters of all the annual accidents in the country occur in cities and towns (including federal, regional and local roads passing through them, see Table 8.2); in 2015, more than 128,000 accidents were registered. The city rankings by the absolute number of accidents is almost identical to the list of rankings based on the population; accordingly, the "leaders" are Moscow, Saint-Petersburg, Omsk, Perm, and Kazan with the values of about 1 accident per 1000 persons a year (Appendix).

At the same time, similar to the situation in foreign countries, the road fatality values in cities are considerably lower than the average value for the country. In the cities where the major share of the vehicle fleet is concentrated the number of road traffic-related deaths does not exceed one third of the total number of road fatalities;

[2]Currently, there is no official statistics on VKT in Russia.

Table 8.1 Structure and severity of road accidents by road status

Sites where road accidents were recorded (road status)	Registered road accidents (% share from the total number of accidents in cities and towns) (%)	Road fatalities (% share from the total number of accidents in cities and towns) (%)	Injured in road accidents (% share from the total number of accidents in cities and towns) (%)	Severity = 100 * (Road fatalities/Registered road accidents)
Federal roads	34.0	38.7	34.8	27
Regional or inter-municipal roads	47.0	45.8	47.6	23
Local roads	12.3	11.1	11.9	21
Other roads	19.0	15.5	17.6	16

Source data from OECD International Transport Forum (2015), World Health Organization (2015), European Commission (2015)

Table 8.2 Structure and severity of road accidents by place (in or outside cities and towns)

Sites where road accidents were recorded (in or outside cities and towns)	Registered road accidents (% share from the total number of accidents in cities and towns) (%)	Road fatalities(% share from the total number of accidents in cities and towns) (%)	Injured in road accidents (% share from the total number of accidents in cities and towns) (%)	Severity = 100 * (road fatalities/registered road accidents)
In cities and towns (i)	69.9	37.9	66.2	7
On roads outside cities and towns (ii)	30.1	62.1	33.8	15
On road sections, routes inside cities and towns (i ∩ ii)	6.7	8.3	6.8	26
In all places where traffic accidents were recorded (i + ii)	100.0	100.0	100.0	13

Source data from OECD International Transport Forum (2015), World Health Organization (2015), European Commission (2015)

in 2015, 8752 persons were killed in road accidents in Russian cities. At the same time, urban roads and streets account for more than two thirds of injured or disabled in road accidents; in 2015, the total number of injured was 153,089 persons. In general, the severity of the consequences of urban road accidents is almost twice as low as in the country as a whole: 7 deaths versus 13 per 100 registered accidents. This situation can be explained by the more scrupulous registration of accidents, usual for urban conditions, as well as by extremely low speeds in large cities.

It should be noted that the level of motorization in many large Russian cities has exceeded the value of 300 cars per 1000 residents and continues to grow. At the same time, the planning parameters related to road and street network in Russian cities were established based on 150–180 cars per 1000 residents.

Due to the issues described above, the road and street network in Russian cities operates in the conditions of chronical overload. The average speed of urban traffic decreased by 40 % during the last 5 years: in the largest cities it is now 15–30 km/h, in big cities, 20–40 km/h, and during the rush hours the average speed falls to 5–10 km/h. It is obvious that this factor, though quite negative in terms of mobility, has a positive influence on the severity and consequences of injuries.

At the same time, the following pattern can be noticed: the larger is the city in terms of the total number of residents, the lower is the level of transport and social risks as well as the severity and consequences of injuries (Fig. 8.4).

There are several reasons for that. Firstly, as it was mentioned earlier, in large cities where the motorization level is significantly higher than the average in Russia, the speed in the conditions of chronic "heavy traffic" is usually quite low.

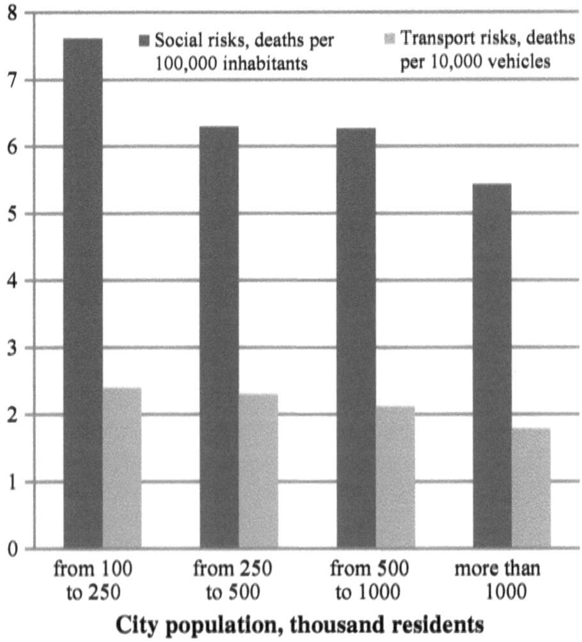

Fig. 8.4 Relationship between transport and social risks and city population (created by authors using the data provided by Main Directorate for Road Traffic Safety)

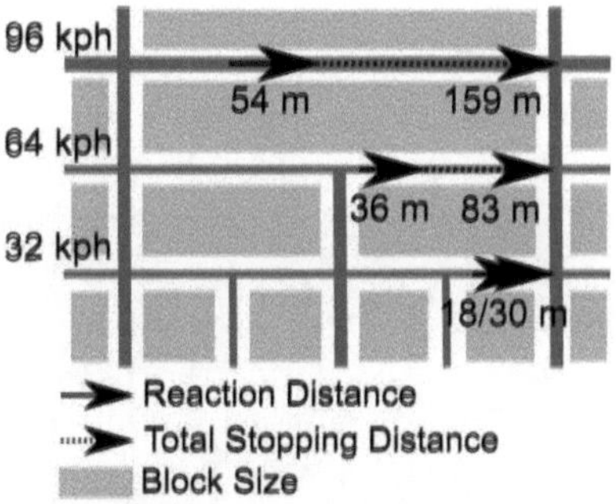

Fig. 8.5 Block size impact on traffic safety [*Source* created by author basing on (Welle et al. 2015)]

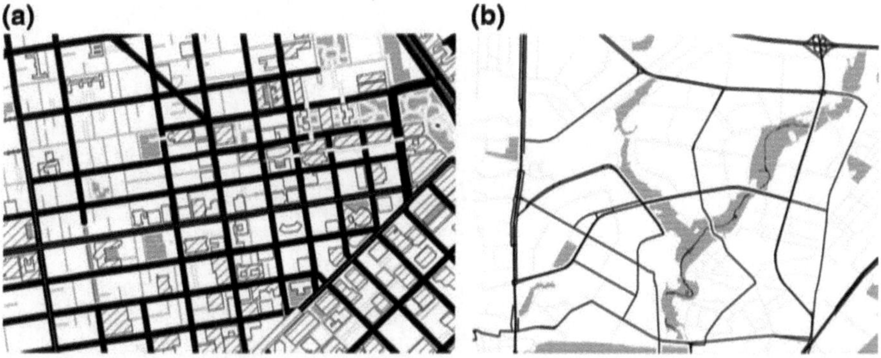

Fig. 8.6 High (**a**) and low (**b**) connectivity of road and street network [*Source* created by author using images by Stamen Design (2016)]

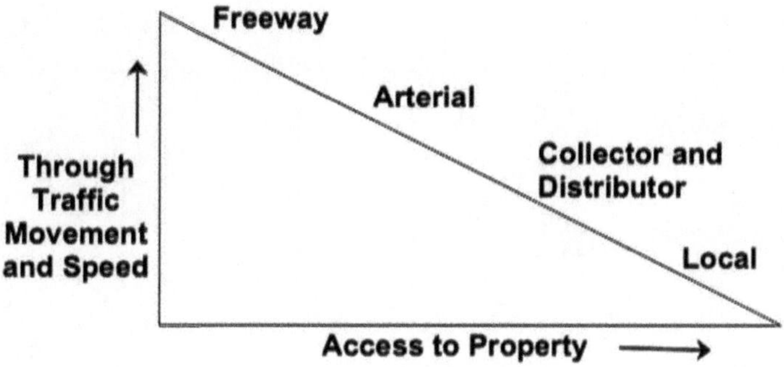

Fig. 8.7 Hierarchical transport network [*Source* created by Levinson (2004)]

(a)

(b)

Fig. 8.8 **a** Median island with a raised mid-block pedestrian crossing [*Source* created by Drdul (2008)]. **b** Absence of medians, dividing strips, green areas, etc. Leninsky Avenue, Saint-Petersburg

Secondly, in large cities (500,000 residents and more) there are bypass roads diverting rapid and transit traffic away from the city. As to many small towns, the federal (and also regional and intermunicipal) roads pass directly through their territory and play the role of major urban streets. The lack of bypass roads around these towns is explainable not only by insufficient financing of road development, which is traditional for Russia but also by the direct economic interests of local communities. In most small towns, more than 50 % of the local businesses are characterized by high dependence on the transit traffic, so the construction of bypass roads may have a negative impact on the local economic situation.

Thirdly, the largest cities (especially Moscow, St. Petersburg, and the leading regional administrative centers) possess ample financial means to implement measures ensuring road safety. In such cities, all the standard techniques of traffic engineering are being used: high quality road signs and traffic signals are installed; road marking is renewed regularly; traffic and crash barriers are placed as necessary; traffic control systems to match the highest foreign standards are applied. In recent years, traffic surveillance cameras have also gained a widespread recognition.

The budgets of the cities that are not so large, and especially those of small towns, are considerably smaller. Often their funds are insufficient even for the acquisition of new traffic lights, renovation of traffic signs, and road marking.

In 2015, on the sections of highways passing through cities and towns there were more than 12,000 accidents in which 1907 people were killed. The severity and consequences of injuries on these sections amount to 26 units, which is twice as high as the average in the country.

It should also be noted that accidents involving pedestrians are as typical for the sections of highways passing through cities and towns as for the urban roads in general. In this, socially very sensitive aspect there have been some positive achievements: the share of pedestrians in the overall number of people killed in traffic-related accidents decreased from 42 to 27 %; nevertheless, it remains abnormally high by international standards: in the member countries of IRTAD, this indicator does not exceed 12–18 %.

While the number of traffic accidents involving pedestrians has been reduced significantly, the number of such accidents on the marked pedestrian crossings that account for 4–5 % of the total number of road fatalities has not changed. It is clear that in contrast to developed countries, in Russia, death of a pedestrian on a pedestrian crossing is not a rare event triggering a widespread public response.

The problem of extremely ineffective protection of pedestrians is caused largely by the Russian practice of building roads combining the functions a street and a highway.

The largest cities (including Moscow) are characterized by "hybrid streets" (arterials) and expressways. In Russia, the category of a "high street of continuous/non-stop traffic flow" combining the features of an urban street (including intensive transit traffic) with those of a multilane highway with multi-level interchanges. It should be noted that such roads are not included in the design standards of developed countries, or in the World Road Association-PIARC recommendations.

For small cities and other communities, the similar negative role is played by the said practice of routing sections of federal and regional roads through urban streets.

In these conditions, the only instrument able to ensure the safety of pedestrians is constructing overpasses and underpasses, as recommended by the Russian design standards.[3]

Unfortunately, the reality does not reflect these standards: even in Moscow, only 25–30 % of pedestrian crossings conform to the design standards; in other cities, the situation is even worse.

Another systemic problem with road safety in the cities of Russia is the lack of adequate institutional environment that would ensure the assimilation of the most advanced scientific accomplishments and recognized humanitarian values based on the total equity of rights, obligations, and responsibilities of all participants of road traffic. In Russia, numerous categories of road users continue to enjoy formal or informal privileges that would be unimaginable in developed countries even in the case of the highest ranking persons. These privileges apply to all law enforcement agencies, senior officials of a number of governmental and corporate structures, and even ordinary but wealthy citizens.

According to a survey of traffic police employees conducted by VCIOM[4] in March 2013, more than 20 % of road users in Moscow and up to 10 % in other regions are listed in the category of "untouchables". At the same time, the respondents commented that the practice of ignoring traffic violations committed by the "untouchable persons" is fraught with troubles for an average policeman, including a possibility of dismissal. Meanwhile, it should be obvious that any law allowing for such a number of exceptions is condemned to be almost entirely ineffective.

Despite the unresolved systemic problems, similar to the described above, the situation with traffic safety in Russian cities has improved significantly in recent years: for example, the number of road fatalities per 100,000 residents has decreased noticeably during the last 10 years in most of the big cities in the country, including the largest. These improvements became possible due to purely objective factors: the increase of the motorization level in the cities and adaptation of citizens to a new transport reality, the increase in the share of new cars equipped with the latest active and passive safety technology, and the increase in the density of traffic police control, including the use of traffic surveillance cameras.

[3]GOST R 52,766—2007 "Vehicle roads for public usage. Arrangement of elements".

[4]Russian Public Opinion Research Center.

8.2 The Main Directions and Methods of Ensuring Road Safety in Urban Areas

It is not by chance that the problem of road safety belongs to the category of global ones: the transport system is not in a state of conflict-free prosperity in any large city in the world. However, the experience of the most successful cities and metropolitan areas shows that it is vital to solve the problem of the "clash of the cars and the cities» through a thought-out strategy of consistent development of private and public transport and its co-existence with non-motorized road users.

It is obvious that any measures described in a planning, engineering or fiscal document aimed at curbing the demand for daily car trips always cause a public debate. Groups lobbying certain interests, as well as some of the media, are opposed to any restrictions of car owners' interests. This is why we should look back at the best known examples of planning for road safety in urban centers across the globe, where the common feature is the clear "anti-automobile" tendency: priority is given to pedestrians, public transport, cyclists and, obviously, emergency vehicles. Drivers of private cars get their place on urban street only in case of compliance with all these priorities, which includes giving the right of way to all the road users listed above.

We will analyze the main measures taken to ensure the safety of the urban environment in foreign cities and metropolitan areas and compare them with the Russian practice (Tables 8.3, 8.4, 8.5 and 8.6).

The main trend in the transport policy enforced in the cities of Western Europe and a number of other developed regions of the world (as we can see basing on the Russian example) includes two sets of policies:

- Assistance (facilitation, promotion) in the use of public transport by improving the quality of mass transportation, in particular, by engineered isolation of rail lines and separation of bus-only lanes to ensure the independence of public transport from the traffic load in the road street network;
- Discouraging (decrease of attractiveness, anti-stimulation) car traffic through regulatory, pricing and planning measures.

Consistent application of these measures leads to positive results in terms of improved comfort of urban environment, reduced environmental externalities and, what is particularly important in the context of the topic, reduced VKT and number of accidents.

Unfortunately, the transport policy in many Russian cities lags behind these progressive trends. In the cities with limited budget (and those are a majority), the laissez-faire strategy is maintained either openly or latently: those who have private cars drive them, while those who have none are left to use jitneys. The major advantage of this strategy is minimization (or even abolition) of the budget expenses on regular public transport. The major disadvantage—considering the design of Russian cities—is the guaranteed perspective of chronic congestion and high accident rate.

Table 8.3 Key urban design elements

Safety activities	Foreign experience	Russian practice
1. Block size The larger the block size is the higher speed vehicles develop, exposing pedestrians to a higher risk (Fig. 8.5)	Reducing the size of the blocks: the more there are street intersections, the more points where cars are forced to stop and let pedestrians pass	The modern block development is characterized by psychologically uncomfortable sizes and proportions of public and residential spaces. Development is provided by multi-story (17–22 floors), multiple-section buildings At the same time, there is an interesting pattern: the older the city is, the smaller are the blocks, and vice versa. As a rule, in the central part of the city, the blocks are smaller than in suburban areas
2. Road and street network connectivity With the increase in connectivity, the length of trips diminishes and the number of possible routes from point A to point B increases, which increases also the convenience of access to public transport (Victoria Transport Policy Institute 2012)	High connectivity of road and street network affects the choice of transport mode and often increases the attractiveness of walking and cycling (Fig. 8.6a)	Low connectivity of road and street network entails excessive detours for road transport, increasing travel time and producing traffic jams (Fig. 8.6b)
3. Vehicle/travel lane width The width of a carriageway strongly influences the ability to organize cycling and pedestrian paths, parking pockets or landscaping elements	The width of a carriageway is reduced by lowering the number of traffic lanes for vehicles and simultaneously arranging pedestrian paths and cycling trails, parking pockets, green spaces, and widened sidewalks	The width of a carriageway in most cases is being increased due to the construction of additional traffic and parking lanes, additional side tracks, turning pockets etc. This widening is often made at the expense of lawns and greenery
4. Access to destinations and convenient pedestrian environment Key points of citizen attraction (residential areas, schools, shopping areas, bus stops, stations and places of work and recreation), as a rule, must be provided with a quick and hassle-free pedestrian access	High	Low

Table 8.4 Traffic calming measures

Safety activities	Foreign experience	Russian practice
1. Installation of speed bumps and speed cushions in front of pedestrian crossings	+	+
2. The artificial chicanes on the roadway to slow traffic down	+	+/− The practice is mainly used in urban centers and not everywhere
3. Narrowing of roadways due to the expansion of chokers, including curb extensions at intersections and raised intersections	+ Narrowing of roadways reduces the speed of vehicles. Expansion of sidewalks at intersections and raised intersections improve the visibility of pedestrians for drivers and reduce the distance for pedestrians to cover	+/− The practice is mainly used in urban centers and not everywhere
4. Introduction of speed limits in consideration of the features of street and road network, traffic volumes, and climatic conditions	+	+
5. Traffic circles/roundabouts	+	+/− In urban areas with heavy traffic and high housing density construction of roundabouts is uncommon; intersections with traffic lights are usually used. If the traffic is not heavy and there is enough space, then traffic circles are constructed (for example, in Togliatti)

In a few cities with high budget resources the transport policy is very controversial. The most typical example is Moscow.

On the one hand, the priority of public and non-motorized types of transport is declared. This priority is implemented by constructing new subway lines and stations, constructing additional railway lines on suburban railways and the Moscow Central Ring Railway, the first of its kind in Russia; renovating the land passenger transport rolling stock; establishing bus-only lanes for buses and trolleybuses; and introducing modern IT services for payment and passenger information.

Table 8.5 Arterial corridors and junctions

Safety activities	Foreign experience	Russian practice
1. Subdivision of the road and street network into arterials and local streets (Fig. 8.7)	Hierarchical transport network, divided into roads and streets: a clear separation of the road and street network into streets (streets, avenues, driveways, arterial roads) for pedestrians, public transport, and local traffic between neighboring areas, and motorways (freeways, expressways, urban motorways) that are only intended for high-speed traffic transit	The hybrid category "high street of continuous/non stop traffic flow", which combines functions such as a multilane roadway with interchanges at different levels and, at the same time, direct access to houses and numerous public transport stops. Among these hybrids are almost all the highways constructed in recent years and partially routed through urban streets The practice of reconstructing streets with their transformation into sections of "high street of continuous/non stop traffic flow" directly affects the local traffic conditions as well as pedestrian and vehicle connectivity between residential areas
2. Separation of traffic flows.	+ Isolation of expressway designed for speed transit and collecting local traffic	+/− Separation of traffic flows is used very seldom
3. Separation of carriageways going in different directions by special medians, dividing strips, green areas, boulevards, sidewalks and etc. (Fig. 8.8)	+	− Elimination of medians in favor of additional lanes and reverse traffic, resulting in accidents with high severity and consequences of injuries
4. Median refuge islands	+	+/−
5. Signal control of pedestrian and bicycle flows	+	+
6. Introduction of smart transport systems and, in particular, automated traffic control systems	+	+
7. The ideology of a "forgiving" road and roadside infrastructure is based on the fact that the road design should compensate for human	Promotion of traditional design solutions in terms of geometric parameters of roads and the arrangement of their engineering elements, as well as smart roads. The	The Russian road design standards were developed in the last century based on the dynamic characteristics of vehicles produced at that time and do not consider all

(continued)

Table 8.5 (continued)

Safety activities	Foreign experience	Russian practice
shortcomings, including the tendency to make mistakes and physical fragility	idea of smart roads implies the use of technological innovations that assist in increasing the information coverage of what is happening on the roads and helping predict the road conditions: lightning system of road markings and/or roadway, activated when a vehicle is approaching; various indicators of weather conditions and conditions of the road surface and etc.	the modern criteria of road safety, which have also changed It is necessary to improve the design standards for geometrical parameters of roads in accordance with their functionality and modern traffic conditions
8. Provision of lane balance of inbound and outbound directions of road network individual elements, forming transport nodes and territorial network as a whole	Design of road networks is carried out with a consideration to ensure the lane balance of incoming and outgoing directions of individual elements of road network	In many cases, there is lane imbalance of incoming and outgoing directions of individual elements of road network

The infrastructure for non-motorized traffic is being developed (cycling roads, pedestrian streets, expanded pedestrian walkways) and vast stretches of road and street network are converted into public spaces.

On the other hand, far greater financial resources are directed to the road construction projects, many of which are based on the counterproductive idea of converting main city streets into a semblance of city highways.

The allocations from the 2015–2017 Moscow budget on transport needs amounts to 13.6 billion euro.[5] These allocations were distributed between the infrastructure for road traffic and the needs of public transport in the ratio of 47–53 %. At the same time, the needs of the land public transport account for less than 5 % of total allocations.

One can understand the political motives behind such a ratio, that is, the simultaneous implementation of two competing transport policies: in Moscow, there are 4 million car owners, and that means 4 million electoral votes.

It is clear, however, that such a policy cannot ensure the radical lowering of the total VKT which is crucial for the city in terms of sustainable mobility as well as road safety improvement.

[5]By exchange rate 1€ = 75 rubles, at the moment this book is being printed.

Table 8.6 Pedestrian spaces and access to public space, bicycle infrastructure

Safety activities	Foreign experience	Russian practice
1. Arrangement of sidewalks and pavements (the foundation of safe sidewalks)	+	+
2. Road network sections with the priority for pedestrians and non-motorized road users (shared streets)	+ Used by all road users. Motorists are beginning to pay more attention to non-motorized traffic	+
3. Pedestrian streets and zones	+	+
4. Additional safety measures for places with higher concentration of children: playgrounds, parks, schools and etc. (Safe places to learn and play)	+ Additional measures to slow down the traffic, speed limits, signs installation and etc. (For example, school zone—slow down—30 km/h)	+
5. Temporarily open streets These streets are temporarily (usually during weekends) used exclusively for walking, jogging, rollers, cycling and other kinds of non-motorized traffic	+	+
6. Sections of road street network converted into public places (street plazas)	+	+
7. Limited access or congestion charge zones for vehicles	+	+/− The ban on truck traffic in the city center. Currently there are no congestion charge zones in Russia
8. Bike lanes and cycle tracks	+ Bike lanes are often physically separated from roadways, which guarantees the safety of cycling	+/− Bike lanes are usually not separated from roadways
9. The off-street trails passing through parks, squares, embankments	+	+
10. Shared bicycle street The streets with low intensity and low vehicle traffic speed, adapted for cycling	+	−

(continued)

Table 8.6 (continued)

Safety activities	Foreign experience	Russian practice
11. Additional measures for bicycle safety at intersections, bike safety at bus stops	+ May include such elements as dedicated lane for cyclists, special markings warning drivers about the approach of cyclists, special green phase of traffic lights for cyclists crossing the street. Special design of bicycle lanes should ensure the safety for both pedestrians getting on/off public transport and cyclists	+/– In large cities (Moscow, Kazan and etc.)
12. Bicycle network organization Well-organized cycling network consisting of cycling lanes connected to the common network of cycling streets, open streets, cycling priority streets, off-street routes, and special measures for convenient intersection of crossroads	+	– The absence of a common urban bicycle network on a city scale or bicycle routes of limited length, mainly off-street routes

8.3 Institutional Conditions for Ensuring Road Safety

In addition to the activities of engineering nature, experts formed a consensus on the necessary institutional conditions for the radical reduction of transport and social risks, including the urban areas (Table 8.7)

The experience of various countries around the world shows that the task of ensure the road safety can be solved solely with the means of consistent development of the road traffic institutional environment according to the standards typical for the developed countries. The first steps to be taken in order to form this kind of environment should not be limited to the introduction of practices that are currently absent but, first of all, have to lead to the change of the existing formal and informal "rules".

In conclusion, it should be noted that any further progress towards a civilized level of road safety is only possible when the public consensus is achieved on these issues, involving the authorities, the expert community, and the motorists.

Table 8.7 The major institutional measures to ensure the safety of the urban environment

Safety activities	Foreign experience	Russian practice
1. Respect for the principle of rights equity (egalitarianism), duties and responsibilities for all road users, regardless of the official status of a driver or vehicle passenger	An exception is made, usually, only for the first persons of a country	Benefits depend on the departmental affiliation of the vehicles, financial and social position of the car owner According to the sociological survey conducted by VCIOM in March 2013, the license plates evincing "high" departmental affiliation of their owners account for 20–40 % of all road users in Moscow and up to 10 % in regions. The use of legal sanctions against the drivers of these vehicles is impossible: any traffic violation is argued by service necessity and is accompanied by an urgent call to the higher level of management The benefits of this kind lead to spontaneous formation of a vast informal hierarchy of road users, creating extremely dangerous archetypes of transport behavior
2. Implementation of the "3D" concept (Dangerous, Drunk and Drugged Driving) intended to identify and eradicate dangerous driving and driving under the influence of alcohol and/or drugs	The priority is given to identifying dangerous driving, signs of which are subject to active control. In addition, the categories of dangerous driving and careless driving are clearly separated. In this case, there are increased penalties for dangerous (aggressive) driving, including driving in the process of committing a crime	The legal definition for "dangerous driving" is currently missing, although it has been actively discussed for more than a year. It is widely believed that it would be difficult to provide a legal definition of this sort and even more difficult to identify this sort of driving in practice Drunk driving (more than 0.16 ppm) is considered one of the most serious offences that can be penalized not only by license suspension but also by a large fine. In addition, if this kind of offence is repeated, it will be classified as criminal rather than administrative In general, repeated drunk driving may have the following consequences:

(continued)

Table 8.7 (continued)

Safety activities	Foreign experience	Russian practice
		• Driving license suspension for 3 years • Fine of up to 4000 Euro[a] • Incarceration for up to 2 years
3. The focus on the driver's skills, abilities, and psychology, as formulated in the concept of "3 E"—Enforcement, Education, Engineering. Within the framework of this concept, the classification of accidents, especially the fatal accidents related to poor road conditions, is recognized as unproductive	A car owner as a driver who is properly trained (Education) and properly motivated (Enforcement) is obliged to adapt to the actual road conditions, that is, the choice of driving behavior that would allow for avoiding accidents with severe consequences in all cases. This idea is reflected in a characteristic slogan "The road doesn't kill"	The majority of the accidents are usually attributed to the "poor road conditions" category
4. Following the standards of "a competent, responsible and friendly behavior" for car owners, having a universal character of social and cultural stereotypes	Some of these standards include: constant readiness to give way, inappropriateness of aggressive driving style or random maneuvering which is inconvenient for other road users; strict adherence to the sequence when merging with traffic, etc.	The reason for the most accidents are the drivers, more specifically, their incorrect, rude and sometimes boorish behavior on the road
5. In order to monitor the drivers' compliance with the rules and proper behavior on the road, "driving histories" are maintained, regularly renewing files, charting registered facts of traffic violations and, accordingly, progressively increasing the tariff for liability and damages based on the accumulation of penalty points	In the USA, in some states, the point system does not exist at all; in others, the number of points collected during a year significantly affects the cost of insurance policy. In Germany, the drivers having a small number of points are sent to special theoretical courses. In Italy, not only the worst offenders are punished but also the honest drivers are encouraged. For every two years without charges, a driver gets 2 bonus points, thus postponing the chance of license suspension further. In Great Britain, things are quite simple: drivers who have scored 12 penalty points lose their right to drive a car	Points system was used in the USSR and brought tangible results in terms of disciplining drivers. Currently, there is no such practice. From time to time, there have been discussions in academic and political circles as to whether it should be reintroduced

(continued)

Table 8.7 (continued)

Safety activities	Foreign experience	Russian practice
6. The existence of an informal positive road coalition, based on a high level of mutual trust between all road users: drivers are decent to each other, driver community respecting the police, police respecting the drivers abiding by rules	"Positive road coalition" is the interaction of law-abiding road users and police against substandard road users, aimed to protect lives, health, and psychological comfort of mutually polite road users	"Negative road coalition" is the confrontation of road users with road police and privileged participants
7. Drivers' respect of the value of life reaches the level that determines the transition to the humanitarian categories ("suffering", "social and psychological costs", etc.)	In international practice, the value of life is a value accepted for both insurance settlements and the purposes of the economic analysis of national programs and specific measures to ensure traffic safety. The category "value of life" is translated into real planning and engineering practice. In particular, the specific numerical estimates for value of life are used in the calculation of the elements of engineering arrangement of roads, for example, barrier fences or pedestrian crossings	A number of methods for assessing the value of economic damage are currently used by various departments. These methods are based on different approaches and methods of calculation and do not allow performing the calculations correctly in the regional context, considering the outcomes of death, disability and injury, gender and age characteristics of the victims, direct costs, loss of profits and consequential losses in the form of non-pecuniary damages According to calculations made by the National Research University Higher School of Economics, the economic cost of deaths and injuries resulting from traffic accidents in 2014 was 3.5 billion euro, in 2013—4 billion euro, including 3.3 billion euro as a result of the loss of life The developed method is not currently approved
8. Safe car ideology results from the need to compensate for the human imperfections through implementing modern means of passive and active safety, with emphasis on the informational and communicational technology	The idea of complete automation of driving, including the organization of autonomous communication between moving vehicles becomes very popular. The indicated conditions are formed by market competition, automobile	About half of the cars in Russia are domestically manufactured, 37 % of which are produced by AvtoVAZ (according to the research agency "Autostat"). Today the cars produced in Russia and in the CIS countries are not tested by foreign

(continued)

Table 8.7 (continued)

Safety activities	Foreign experience	Russian practice
(ICT) systems and various gadgets	manufacturers and the introduction of certain regulatory standards on national and international level	independent organizations (European EuroNCAP, American IIHS or Japanese JNCAP) which execute crash tests on cars around the world At the same time, periodical testing of the cars produced in Russia are executed in the laboratories of AvtoVAZ and Dmitrov proving ground. The resulting rankings are not very consoling in comparison with international safety ratings, and for the Russian car industry, there is something to strive for
9. Creation of the infrastructure of rapid response and emergency medical care, ensuring mitigation of accidents	Almost all countries that have made progress in reducing the death toll in road accidents have introduced emergency response system and solved the problem of providing emergency communications and calls, transport support medical teams and also the solution of medical issues of methodical, technological and pharmacological character on a high level	Large cities have corresponding financial resources to establish the appropriate infrastructure (modern hospitals, equipped with helipad, uniformly distributed first aid stations). In small communities, there is insufficient amount of funds for that
10. Maintaining a regularly updated database of standardized data on the number, structure and use of the vehicle park, as well as the accident rate by country	To implement this priority, the institute of data exchange and best practices has been founded—International Road Traffic and Accident Database (IRTAD). Regular analytical reports by IRTAD have become a working tool to be used by the community of specialists in the field of traffic safety, as well as the standard and the bar for professional discussion of problems in the field of road safety	Applied analysis method used in Russia today compares the indicators of traffic safety for a current year with the indicators of the previous year. Such indicators do not allow for taking into account the difference in population and motorization levels in different regions of Russian Federation, ranking Russian regions by the level of traffic safety, or assessing the levels of active, passive and post-accident traffic safety In addition, Russia has not yet become an IRTAD member The Ministry of Internal Affairs of Russian Federation holds a monopoly in terms of

(continued)

Table 8.7 (continued)

Safety activities	Foreign experience	Russian practice
		possessing the information on the statistics of road traffic accidents (RTA). In addition, the Ministry of Internal Affairs is responsible for the evaluation of the success of its own activities in the field of road safety. For this reason, the Ministry of Internal Affairs is interested in providing the most positive information on the results generated by the activities and concealing the negative results
11. Following the principle of "zero tolerance", according to which it is not possible to relate to the traffic related deaths as inevitability associated with global motorization	The most radical is the Swedish concept "Vision Zero"	Russia is far from implementing the concept "Vision Zero" To date, the official forecasts of transport risks level are formed in the medium term: According to official forecasts[b]: • The indicator of transport risks should be reduced by 36.1 % from the base of 2012, that is 3.9 units • Indicator of social risks should be reduced by 30.5 % from the base of 2012, that is 13.9 units In fact, the official (quite realistic) forecast is equivalent to acceptance of catch-up scenario in the field of road safety with a lag behind the best international practices of about 40 years

[a]Hereinafter 1 Euro = 76 rubles
[b]The Federal target program "Increase of traffic safety in 2013–2020", approved by a resolution of Russian Government dated in October 3, 2013 N 863 g-PP

References

Drdul R (2008) Midblock median island https://en.wikipedia.org/wiki/Traffic_calming#/media/File:Midblock_median_island.jpg. Accessed 31 Aug 2016

European Commission (2015) Traffic safety basic facts 2015—Roads outside urban areas. http://ec.europa.eu/transport/road_safety/pdf/statistics/dacota/bfs2015_urbanareas.pdf. Accessed 31 Aug 2016

Levinson D (2004) Hierarchy of roads. https://en.wikipedia.org/wiki/Hierarchy_of_roads#/media/File:HierarchyOfRoads.png. Accessed 31 Aug 2016
OECD (2016) The International Road Traffic and Accident Database (IRTAD) http://stats.oecd.org/Index.aspx?DataSetCode=IRTAD_CASUAL_BY_AGE. Accessed 31 Aug 2016
OECD International Transport Forum (2015) Road Safety Annual Report. http://www.oecd-ilibrary.org/transport/road-safety-annual-report-2015_irtad-2015-en. Accessed 31 Aug 2016
Smeed RJ (1949) Some statistical aspects of road safety research. J Royal Stat A(I):1–34
Stamen Design (2016) http://maps.stamen.com/. Accessed 31 Aug 2016
Welle B, Li Wei, Adriazola C et al (2015) Cities safer by design. Urban Design Recommendations for Healthier Cities, Fewer Traffic Fatalities.World resources institute. http://www.wrirosscities.org/research/publication/cities-safer-design. Accessed 15 Mar 2016
World Health Organization (2009) Global status report on road safety. http://whqlibdoc.who.int/publications/2009/9789241563840_eng.pdf. Accessed 31 Aug 2016
World Health Organization (2015) Global Status Report on Road Safety. http://www.who.int/violence_injury_prevention/road_safety_status/2015/en/. Accessed 31 Aug 2016

Chapter 9
Russian Cities Mobility Culture: International Comparison

Mikhail Blinkin and Egor Muleev

Abstract The travel behavior surveys are not generally required by the decision-makers in Russian cities. However, the academic community had initiated the survey which was conducted in 2014 and became the first comprehensive household travel behavior survey in Russia. This allowed stepping aside from the applied nature of the majority of the similar surveys and to concentrate on the sociological characteristics of mobility. The survey was largely an experimental one as some methods employed had unlikely been used before. Even though the project was considered to be successful and some results turned out to be surprising. For example, no significant difference has been revealed between the drivers' travel patterns and public transport users' travel behavior. The results of the survey enable the comparison of mobility patterns in Russian cities and abroad, which can be interesting both for Russian and international readers as there has been no such analysis before.

9.1 Introduction

There are two hypotheses for travel behavior analysis:

- Citizens are free to choose place of residence, travel mode, and route. Consequently, travel demand is influenced by value of time and generalized cost of travel. However, travel behavior cannot be reduced to two factors and has a very complicated structure.
- City government must be familiar with travel patterns of citizens within the city limits in order to manage infrastructure projects and travel demand. Arranging periodical surveys of traffic flows and travel patterns allows companies to help improving their strategies.

M. Blinkin · E. Muleev (✉)
Institute for Transport Economics and Transport Policy Studies, National Research University Higher School of Economics, Moscow, Russia
e-mail: emuleev@hse.ru

M. Blinkin and E. Koncheva (eds.), *Transport Systems of Russian Cities*, Transportation Research, Economics and Policy,
DOI 10.1007/978-3-319-47800-5_9

These hypotheses seem quite feasible for cities with developed labor and real estate markets and a longstanding tradition of local democracy.

By 1917, Russian cities were just beginning to establish this type of administration. The level of population mobility was low, and a number of trips limited. Joseph Brodskiy, a Nobel laureate, in literature reflected on this level of mobility in his poem *The Classical Ballet*: "A small plot gave you birth, the nest you lived and died in".[1] In 1900, a Russian engineer Heinrich Girshson made a comparative study of a number of trips per citizen in various cities across the world. According to him, in 1893, there were 290 trips per citizen in New York, 228 in Chicago, 128 in Brookline, 172 in Boston, 145 in London, 140 in Berlin, 100 in Paris, 87 in Budapest, 70 in Vienna, 60 in Saint-Petersburg, and 50 in Moscow" (Girshson 1901).

The quality of public transportation in pre-revolutionary Russia did not encourage people to travel. One of the first Russian urbanists wrote: "I blanch at the prospect of using the Moscow public transport. In Berlin, however, it is a pleasure" (Ozerov 1906).

A bit later, a Russian engineer Greg Dubelir improved the thesis of Ozerov's book regarding the need for travel behavior analysis (Dubelir 1910). Dubelir visited USA and was inspired by the mass development of private automobile ownership. He linked the ability to travel faster and farther with the oncoming century of motorization. Dubelir believed that the tendency of "traffic volumes increasing due to private automobile use" should be a prime factor for consideration in the contemporary town planning.

Between 1917 and 1932, planning issues were discussed by some of the most talented engineers in USSR. They believed that the socialist idea was worth combining with rational and humanist approach to town planning. For example, A. Zilbertal emphasized that transportation planning issues cannot be reduced to a mathematical problem. The solution is dependent on the society's demand for comfort as well as on the value of time (Zilbertal 1932).

Nevertheless, while sociological surveys were allowed in the mid 1960s (Firsov 2012), travel surveys have been conducted very rarely. One Estonian researcher did complete such a survey in the 1980s due to liberalization of the communist regime and relatively high level of motorization in the republic (Arrak 1982).

The methodology of transportation planning in USSR (see Chap. 1) has been different from the Western approach. Low motorization level along with the absence of housing market and labor market compelled people to spend most of their lives in the same place. As a result, travel behavior was influenced by public transportation supply. Moreover, urban planning suggests that industrial areas and residential districts should be built in close proximity. As a result, the commute options available to the working class were limited to either boarding a tram or walking.

[1]Joseph Brodsky's "The classical ballet..." is dedicated to Mikhail Baryshnikov.

As cities grew bigger, the interest in transportation surveys increased. In the 1960s–1980s, cities in developed countries tried to manage public transport in an effective way. The methodology of transportation studies developed accordingly to the new challenges.

The approach to transportation planning in USSR was based on two principles. Firstly, it was the planning process supported by monitoring surveys of ridership and passenger turnover. Ridership statistics was the central indicator used by land electric transport planners until 1992. Passenger turnover was the central category for city bus companies. This information was necessary not only for transportation planning but also for transportation industry and financial support.

The average length of a trip is the data required for both indicators. For trolleybus and tram lines, these numbers were constant. New lines were rarely built, so the trip distance was a fixed value. In contrast, bus lines developed quickly to emulate the urbanization process.

However, the statistical committee concentrated on collecting data on ridership while average distances were not required to be documented. As a result, the methodology of transportation studies developed accordingly to applied tasks.

The second principle was compiling a trip matrix, which became necessary due to the growing sophistication of travel behavior. In a big city, destination points could vary, as could departure points in residential areas. The theoretical groundwork of the Soviet transportation studies was based on technocratic vision and suggested quantitative methods and statistical analysis. Travel behavior analysis was not to be conducted at an individual level. All trips were analyzed as depersonalized flows of passengers, and its fluctuation would premeditate either adding or deleting one train route or another.

The approach to travel behavior analysis in USSR was reduced to the so-called "route-mobility". A number of trips per specified duration served as a main point of reference for transportation planners and officials. It is important to emphasize that human mobility was tantamount to the intensity of public transport usage. Consequently, field methods propose different approaches: "(1) active continuous control, (2) automatic occupancy control, (3) visual estimation of the number of passengers, (4) survey methods" (Efremov et al. 1980). The last one is also subdivided into the questionnaire, table, and talon[2]-based methods of counting. Of those, the questionnaire-based survey is the most difficult to implement. "This kind of surveys pursues the goal of evaluating the performance of public transportation service, pedestrian activities, time budgeting, and so on. Travel diaries allow collecting info on travel frequency, travel modes, and other characteristics. The results can be used in general layout development, public transportation administration, and transportation planning" (Lobanov 1990). All the Soviet sources emphasize that proper application of survey methods are extremely important. However, sociological sources that might help address methodological problems were

[2]Talon-based method is used for ridership distribution survey. Passengers take a talon on stop (or station) and drop it off at destination point.

neglected. Technical austerity provokes adoption of a statistical apparatus rather than sociological. So, the overall situation with the implementation of methodology seems quite similar to the respective experience in the developed countries abroad.

In the recent history of Russia, the sociologically oriented travel surveys have never been conducted. The legacy of USSR, the statistics-based approach is being preserved. Many of the specialists in the field have only basic understanding of the scale and structure of human mobility in Russia. As a result, there is no scientific validation of the forces assumed to drive it. Operational data like reasons for travel, travel modes, distances, travel duration etc. have not been collected, either. It should be mentioned that travel surveys were occasionally conducted on a citywide scale. However, the results of such surveys were concealed from public, and researchers are not allowed to share the figures and methodology in academic literature.

The decision to conduct a travel behavior survey would not usually be based on an intention to solve a specific problem. The establishment of priorities in the transportation policy in Russia is not guided by rational calculation. In fact, this process may be suggested as another promising topic for research.

In the case of a travel survey, a team of professional sociologists focus on the lack of data and methodological gaps. We tried to shift the attention from intensity of vehicle usage to human mobility and formulate some of the circumstantial parameters.

9.2 Methodology

The preparation of the first nationwide travel survey in Russia included the suggestion to conduct a critical review of the methods used abroad and subsequently adapt their main components to the local specifics.

A number of interviews had been conducted prior to the initial formulation of a questionnaire. The questionnaire was revised 5 times before the final version was approved. All stages suggested testing procedures on three randomly selected respondents. Nevertheless, while the diary method is quite popular in foreign countries, it is hard to implement in Russia. The common practice is asking the respondent to fill in the blank spaces in the diary after the general questions have been posed by an interviewer. However, a 600-respondents pilot survey shows that only 10 % of all the respondents were eager to complete the travel diary. The only way to advance was to incorporate the "yesterday travel" section into the main questionnaire. As a result, the final version of the questionnaire comprised 8 blocks and 57 questions. Despite a money reward offered for completing a 4-days travel diary, only 10 % of all the respondents who took part in the final survey actually completed the travel diary via the Internet.

The GPS technology can provide interesting outcomes, and it was necessary to find a suitable opportunity to use it. A test group in Moscow used to track their trips with a special application on a smartphone for a couple of days. However, the

Table 9.1 Response rates of the first nationwide travel survey

COOP1	RR3	REF2	CON2
37.3	16.2	23.7	43.6

results were extremely inaccurate, and testers complained at the interface and battery demand, so the idea was rejected.

Eventually, the CATI (Computer Assisted Telephone Interview) technology was selected as the main instrument of surveying. The probability stratified sample included mobile and landline phones at a ratio of 75 to 25 % respectively. 8028 respondents completed the questionnaire, and 1041 respondent filled the travel diary. However, the sample of the internet survey was not random. The survey was conducted in September 2014. AAPOR's indexes are shown in Table 9.1 and demonstrate good results in comparison to the country-wide surveys.

9.3 General Observations

The first question in the survey suggests identifying three "types" of mobility in the long-term perspective. The difference between them is based on the characteristics of occupation and leisure activities. The first type is called the "sedentarists" group. It refers to the people who rarely leave their place of residence except on occasions such as a vacation. The second group is called "migrants" and includes active people who frequently go on business trips, spend weekends in the countryside, or work in the transportation sector. Finally, the third type is named "rotation workers". This term refers to the seasonal workers or those who spend a lot of time in other cities or even countries. Figure 9.1 shows the distribution of these three groups among the respondents.

The greatest share of the respondents identified themselves as "sedentarists". This is a characteristic feature of human mobility in Russia: people are not interested in travelling frequently and far away. According to the research conducted for "Strategy-2020",[3] the travel behavior of 20 % of Russians shows the same figures as the average in advanced countries. 10 % have no access to transportation at all. Moreover, 81 % of the population never leaves the Russian Federation (Strategiya 2013: 46).

This case of low mobility seems true for short-term and long-term perspectives alike. According to the travel survey, the share of the respondents who prefer to spend their free time outside their home is similar to the opposite—45 % in both cases. Less than one third (27 %) of the respondents changed their place of

[3]"Strategy 2020" is the concept for socio-economic development in the Russian Federation till 2020. The concept was developed by academics in 2010. It comprises 21 promising directions including transportation.

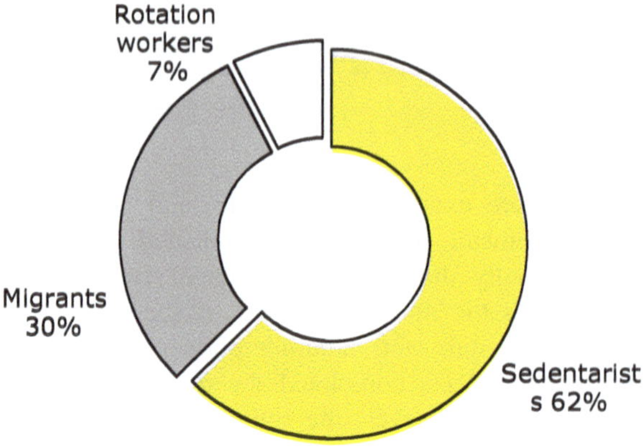

Fig. 9.1 The distribution by long-term mobility type (*Source* Created by authors)

residence in the 5 years preceding the survey. Furthermore, half of the people who changed the place of residence did not feel any changes in their personal travel behavior. The distribution of the three mobility types among the communities does not show any significant variations except in millionaire cities and rural areas.

The millionaire cities have the smallest share in the "sedentarist" group—54 %, while 72 % of people identified with this group live in villages.

In the Soviet academic literature, a lot of authors noted the correlation between the number of trips taken by public transportation in communities and the size of population. The relationship is quite simple: residents of big cities take more trips in comparison to people living in rural areas.

The same result is shown by the travel survey, although the difference between the biggest and the smallest communities is minor (Fig. 9.2).

The average number of trips per day is 3.1. For the "migrants" group, it is 3.4, and for the "sedentarists", it is slightly smaller (3.0). The "rotation workers" occupy the middle position and take 3.1 trips per day.

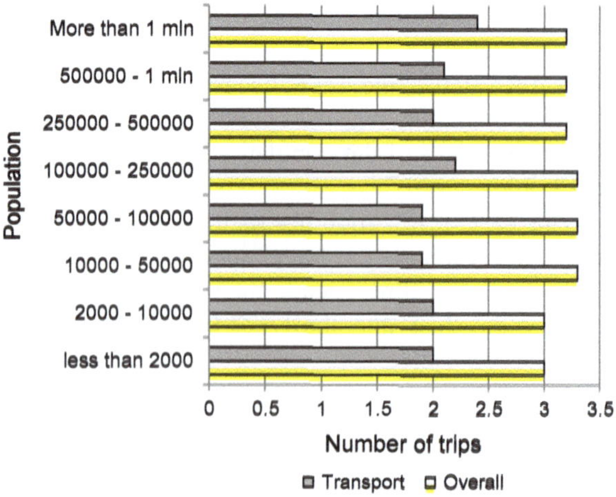

Fig. 9.2 Average number of trips by type of settlement (*Source* Created by authors)

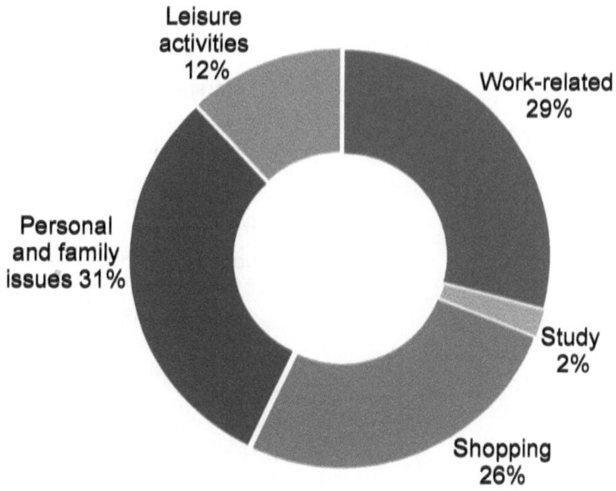

Fig. 9.3 Distribution of trips by purpose (*Source* Created by authors)

The most popular reason for travel is "private and family issues", comprising 31 % of all the responses. The second place is taken by work-related trips, including commuting—29 % (Fig. 9.3).

Likewise, there is no significant difference in the distribution of trips with relation to the size of a city. However, it seems like a fairly unstable interrelation between factors such as working, shopping, and personal and family issues, and it tends to change with time.

We assume that the travel behavior of the respondents could have been conditioned by the kind of activities at their destination point. These activities could have affected the choice of travel mode, route selection (if it involved alternatives) and time budgeting. The questionnaire included several questions regarding the route choice and the necessity to arrive somewhere right on time. 90 % of the respondents had to get to work at the specific time. It is quite clear that 90 % should have also used the same route to get to it. The situation is the same for those commuting to their place of study. Other reasons for travel vary by time and route, and those appear constant for about 10 % of the respondents.

9.4 Private and Public Transport

73 % of all public transportation trips are made by various kinds of buses. *Marshrutka* means a privately owned minibus (fewer than 20 passengers) which operates on a preset route and sometimes runs on schedule. Revenue is collected by the driver. They became very popular since USSR fell apart and local governments denied financial support for "classic" public transportation. For a lot of average-scale cities, "marshrutka" serves as a main mode of public transportation.

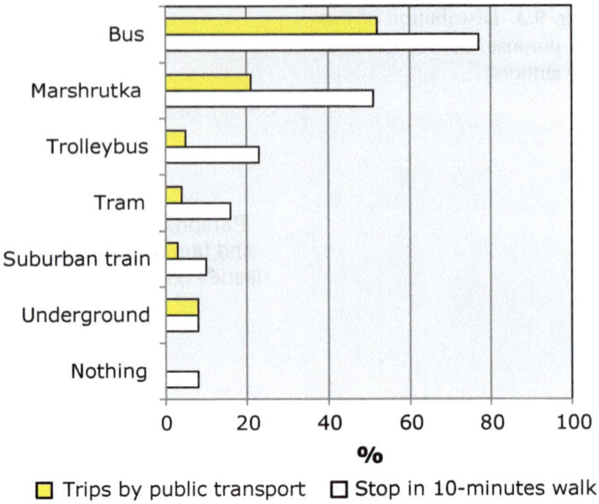

Fig. 9.4 Ridership distribution by public transport mode and availability of infrastructure (*Source* Created by authors)

In spite of the fact that 77 % of the respondents reported living at a 10 min walk from a bus stop, the use of public transportation decreases year after year. The Soviet urban planning approach included development strategies that were transit oriented to an outstanding degree.

However, it did little to improve the performance of Russian public transportation. Moreover, no public transportation at all is referred to in 8 % of the responses. The largest group of these respondents (65 %) lives in the countryside (Fig. 9.4).

Despite the disparity between the availability and the demand for electrically powered public transportation, 59 % of all the respondents are either "totally satisfied" or "satisfied" by the performance of the public transport. 23 % use public transport every day, while 26 % do not ride it at all.

To some extent, such a situation might be explained by the rise of private automobile ownership. Automobility in Russia is a relatively new subject for research. Between 1993 and 2013, it increased by the factor of 3.6: from 75 to 273 vehicles per 1000 people. Traffic jams and parking problems are the everyday reality in the big cities. Urban and transportation planning have changed slightly to provide traffic capacity for automobiles. Roads are being widened, huge parking lots are being built, tram and trolleybus lines are being discontinued, and suburbs without public transportation are being developed. Financial incentives such as affordable credit, fuel prices, low taxes, and insurance costs encourage car ownership. Moreover, rural areas become isolated because some of them have extremely limited or no public transportation service at all.

One of the questions that the survey posed to the automobile owners was, 'There are two types of people. For the first group, the car is an item of daily necessity while the second consists of those car owners who admit the possibility of getting away without it. What type is more descriptive of you?' The results show that 64.4 % of the automobile owners cannot imagine their everyday life without a car,

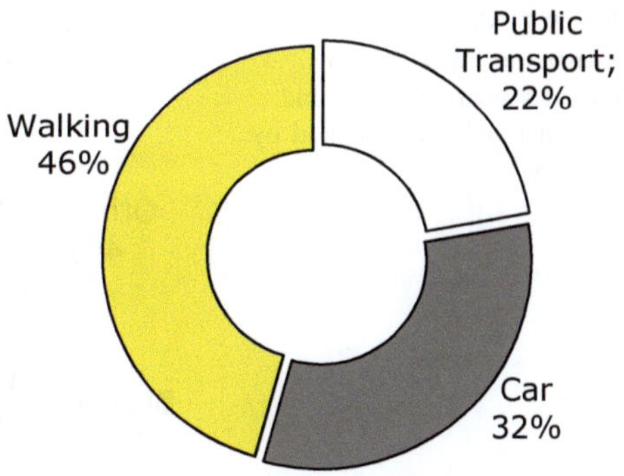

Fig. 9.5 Modal split (*Source* Created by authors)

and just 34.4 % chose the second answer. Russians love their cars, but only 32 % of the car owners who participated in the survey admitted not riding public transport at all. Nonetheless, the modal split shows the dominance of the walking mode (Fig. 9.5).

The nature of such distribution has obvious reasons related to the size of population. Public transportation is popular in the biggest cities while its share tends to decrease along with the size of the population in a community. The share of the walking mode is higher than 40 % everywhere, while that of the private car usage remains stable (Fig. 9.6).

The surprisingly high share of the walking mode might be explained by the high standard of performance by local public services such as shops, schools, sport facilities, etc. In 79 % of cases, either "totally satisfied" or "satisfied" were chosen for an answer.

All in all, the distribution by vehicle ownership shows the prevalence of automobile owners. It is also important to mention that the motorization rate among the surveyed was 328 cars per 1000 people, while the average rate for the country in 2014 was 317 cars per 1000 people (Fig. 9.7).

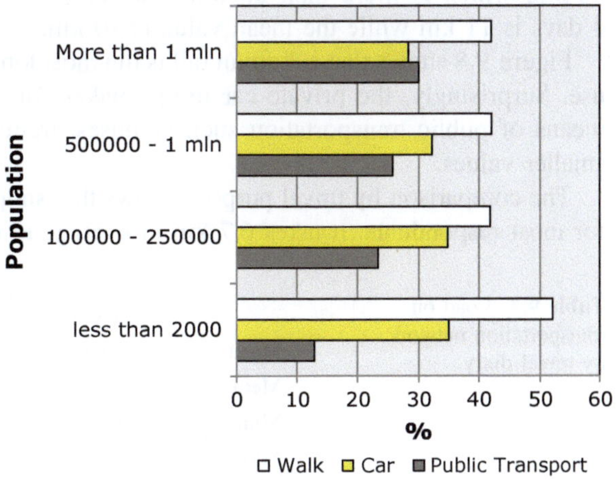

Fig. 9.6 Modal split by the size of population (*Source* Created by authors)

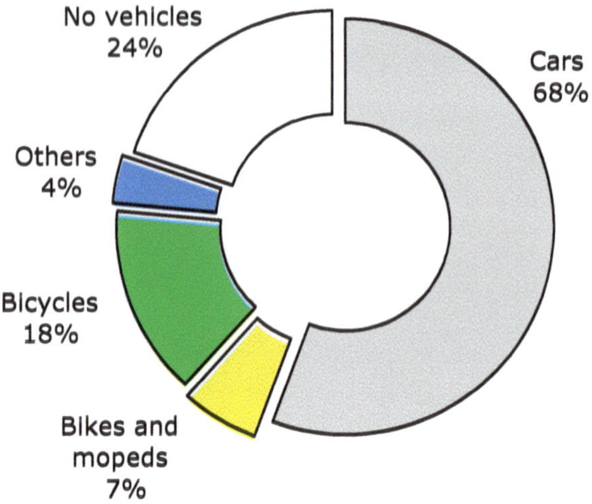

Fig. 9.7 Vehicle ownership. *Note* "Other" includes boats, tractors and rototillers (*Source* Created by authors)

9.5 Distances

The accuracy of distance calculation supplied by the respondents is doubtful. A number of qualitative surveys clarifies that most people find it hard to estimate distance correctly. Therefore the question about distances was only included in the travel diary survey. Table 9.2 shows results for 4 days of survey. However, it addresses the load on the transportation system rather than the average distances travelled by people because the maximum values appear extremely high.

A special statistical algorithm helps dealing with outliers (extremely high values). Table 9.3 shows average distances travelled by people and suggests a typical situation for everyday corporeal mobility.

The percentiles show that 10 % of the less mobile population takes 2 km trips per day in total. On the other hand, 10 % of the hyper-mobile respondents cover 56 km, which is more than 28 times as long. Finally, the median value for all 4 days is 11 km while the mean value is 20 km.

Figure 9.8 shows that suburban rail is the most long-range transport for everyday use. Surprisingly, the private car usage makes for only a half of that. However, means of public transportation such as buses, trams, and trolleybuses have even smaller values.

The comparison by travel purpose shows that shopping is the closest destination for most respondents. It takes 6.7 km in average to reach a store (Fig. 9.9).

Table 9.2 Load on transportation network by travel diary

	1 day	2 day	3 day	4 day
Mean	31.5	27.5	28.2	32.4
Median	12.0	11.6	12.0	13.0
Minimum	0.01	0.30	0.05	0.02
Maximum	1285.0	1267.0	1179.0	1100.0
Dispersion	5436.6	3826.1	3512.7	4334.0

Source Created by authors

Table 9.3 An average distances of corporeal mobility

		1 day	2 day	3 day	4 day
Mean		21.1	19.9	20.2	20.8
Median		11.0	11.0	11.0	11.9
Minimum		0.01	0.30	0.05	0.02
Maximum		115	105	110	100
Percentile	10	2.0	2.0	2.0	2.0
	20	3.5	3.0	3.0	4.0
	30	5.0	5.0	5.0	5.2
	40	8.0	8.0	8.0	8.0
	50	11.0	11.0	11.0	11.9
	60	16.0	15.0	15.0	16.0
	70	25.0	22.0	22.0	24.0
	80	36.0	33.0	34.3	37.0
	90	56.0	55.3	52.0	56.0

Source Created by authors

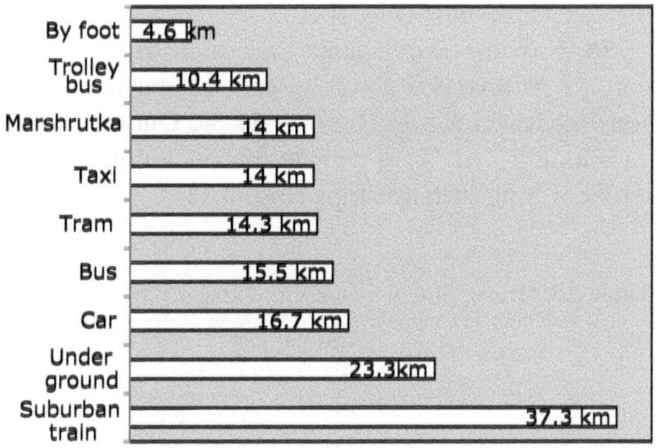

Fig. 9.8 Average distances by mode of transportation (*Source* Created by authors)

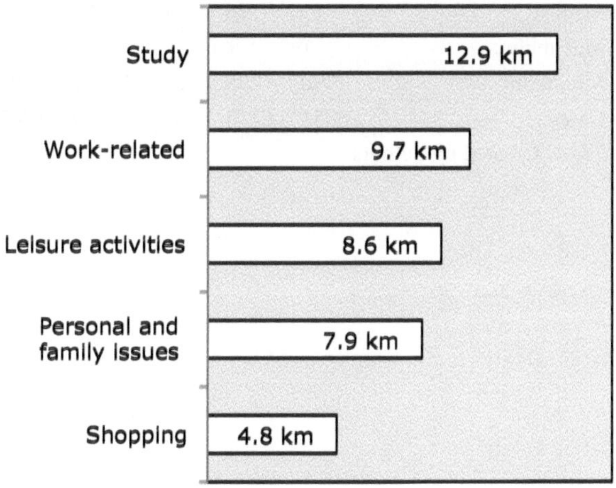

Fig. 9.9 Average distance per travel purpose (*Source* Created by authors)

9.6 Travel Time

There is a disparity between the travel diary results and the questionnaire survey. The average travel time, based on the telephone interview, is 20.7 min, while the travel diary shows 32 min. Median value has similar variation disparity: 15 and 25 min respectively. Probability sample allows us to expand boundaries of analysis and take the confidence interval into consideration (Table 9.4).

The available data allows concluding that people spend more than 8 but less than 28 min on each trip.

9.7 Long-Distance Trips

Long-distance trip is a journey that is more than 200 km long and implies overnight accommodation outside the place of residence. One third of such trips are taken for the purpose of vacationing. 37 % of all the respondents had not taken such trips in the last 12 months (Fig. 9.10).

86 % of the respondents took such trips within the Russian Federation while only 13 % travelled abroad. Personal and family issues are the most popular reasons for travel across Russia—27 %. Out-of-country trips are made primarily for vacationing—61 %. A private car (including taxi) has a surprisingly big share in terms of long-distance trips (Fig. 9.11).

Table 9.4 Travel time for each trip during a day

		Trips per day									
		1st	2nd	3d	4th	5th	6th	7th	8th	9th	10th
Mean		27	25	23	21	21	21	22	21	14	12
95 % Conf. lev.	Low	27	24	22	20	19	19	19	18	10	8
	Up	28	25	24	22	22	23	25	24	18	15
Median		15	15	15	15	15	15	15	15	15	10
Maximum		1	1	1	1	1	1	1	1	1	1
Minimum		290	25	250	240	200	240	180	150	60	60
Cases		6853	6113	3645	2430	1267	729	370	217	111	72

Source Created by authors

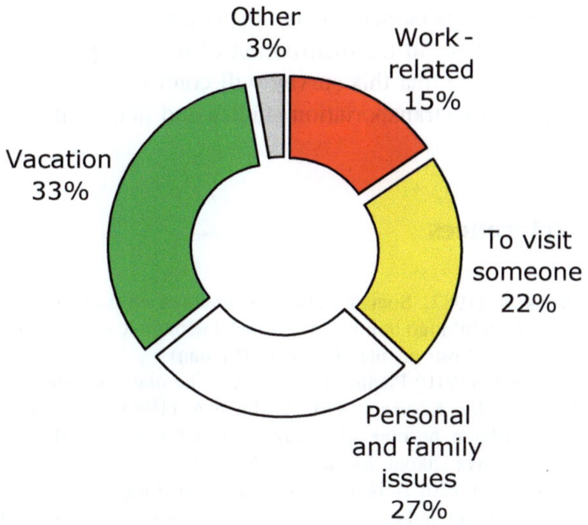

Fig. 9.10 Purposes of long-distance trips. *Note* "Other" includes medical trips and leisure activities such as fishing (*Source* Created by authors)

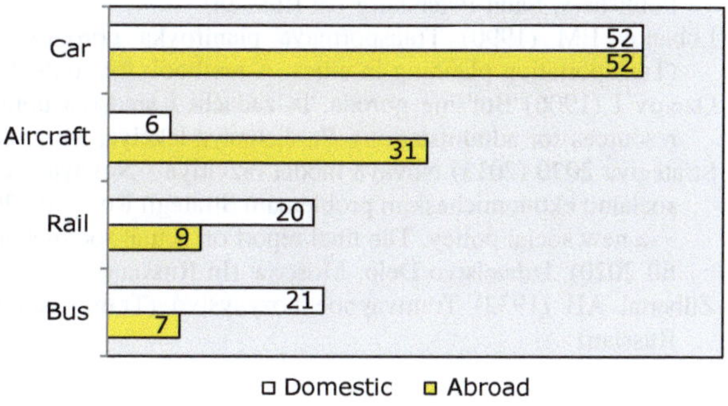

Fig. 9.11 Modal split for long-distance trips (*Source* Created by authors)

9.8 Conclusion

The first nationwide travel survey aimed to achieve three main goals.

First of all, the theoretical basis for this kind of research was examined, basing on the respective experience in the advanced countries. Decision-making process should also be included in the analysis in order to outline the consequences of this kind of research.

Secondly, methodological experience could be used for similar research in Russia. Although the regional and federal administrations do not call for collecting this kind of data, it is required as a starting point for future research.

Finally, the data gap has been filled. The average distance is 20 km per day, an average trip takes approximately 21 min, walking is the most popular transportation mode, and personal and family issues along with work-related trips comprise the biggest share in the distribution of trips by purpose.

We hope that this survey will contribute to developing the specifically Russian approach to transportation studies and help improve the quality of life.

References

Arrak A (1982) Socialno-ekonomicheskaya effektivnost passazhirskih perevozok (na primere avtomobilnogo transporta) (A social-economic effectiveness of public transport (an example of buses). Eesti raamat, Tallin (In Russian)

Dubelir G (1910) Planirovka gorodov (The plan for cities). Slovo, Sankt-Peterburg (In Russian)

Efremov IS, Kobozev VM, Yudin VA (1980) Teoriya gorodskih passazhirskih perevozok: Uchebnoe posobie dlya vuzov (The theory of public transport: a textbook for students). Vysshaya shkola, Moskva (In Russian)

Firsov BM (2012) Istoriya sovetskoy sociologii: 1950–1980-e gody (The history of Soviet sociology) Izdatelstvo Evropeyskogo universiteta v Sankt-Peterburge, Saint-Petersburg (In Russian)

Girshson G (1901) Gorodskie dorogi bolshoj skorosti: Sooruzhenie, ekspluataciya i finansovoe polozhenie (The metropolitans: construction, exploitation, finance situation). Yablonskiy Publishers, Saint-Petersburg (In Russian)

Lobanov EM (1990) Transportnaya planirovka gorodov: Uchebnik dlya studentov vuzov. (Transportation planning in cities. A textbook for students) Transport, Moscow (In Russian)

Ozerov I (1906) Bol'shie goroda, ix zadachi i sredstva upravleniya (Big cities, objectives and resources for administarion). Publichnaya lekciya, Moscva (In Russian)

Strategiya 2020 (2013) Novaya model razvitiya - Novaya socialnaya politika. Finalnyj otchet po socialno ekonomicheskim problemam Strategii Rossii do 2020 (A new model for development —a new social policy. The final report on actual socio-economic problems of Russian strategy till 2020). Izdatelstvo Delo, Moscow (In Russian)

Zilbertal AH (1932) Tramvaynoe hozyaystvo (Tram economy). Gostransizdat, Leningrad (In Russian)

Chapter 10
A Forecast for Transport System Development in Russia

Mikhail Blinkin and Nikolay Zalesskiy

Abstract The forecast for transportation systems in Russian cities is based on the hypothesis of catching-up development, according to which Russia's urban transport evolution repeats the succession of events in western cities with a several decade lag. Thus, the analysis of the previous trends of urban transportation development in countries of the West is used for the forecast of the future trends in Russian cities evaluating. The chapter describes a few cases of the interrelationship between the motorization rate and public transport usage. Such factors of transportation system development in each particular city as the size of local budget, budget system issues, quality of local management are also discussed. One of the most important features in Russia is a leading position of Moscow and St. Petersburg in terms of the possibility the best global practices implementation. Hypothesis and estimations are projected to the road map of the transport supply technological development as it is described in recent publications. The most important changes are expected to be the emergence of the digital age transportation (massive networking and user integration via user interfaces of transport systems) and wide adoption of sharing economy models and driverless cars.

10.1 Budget Restrictions

The traditional explanation of the low quality of transportation system in cities of Russia refers to the insufficiency of municipal budgets. Indeed, the current procedure of tax revenue distribution among the various levels of budget system (federal, regional, and municipal) in Russia is highly discriminatory towards the municipal level. Therefore, according to the existing laws, municipalities are obliged to organize and maintain their own urban transportation services. The success of these operations, based on the Soviet-style maintenance and sometimes, development, is

M. Blinkin · N. Zalesskiy (✉)
Institute for Transport Economics and Transport Policy Studies, National Research University Higher School of Economics, Moscow, Russia
e-mail: nzalesskiy@hse.ru

M. Blinkin and E. Koncheva (eds.), *Transport Systems of Russian Cities*, Transportation Research, Economics and Policy,
DOI 10.1007/978-3-319-47800-5_10

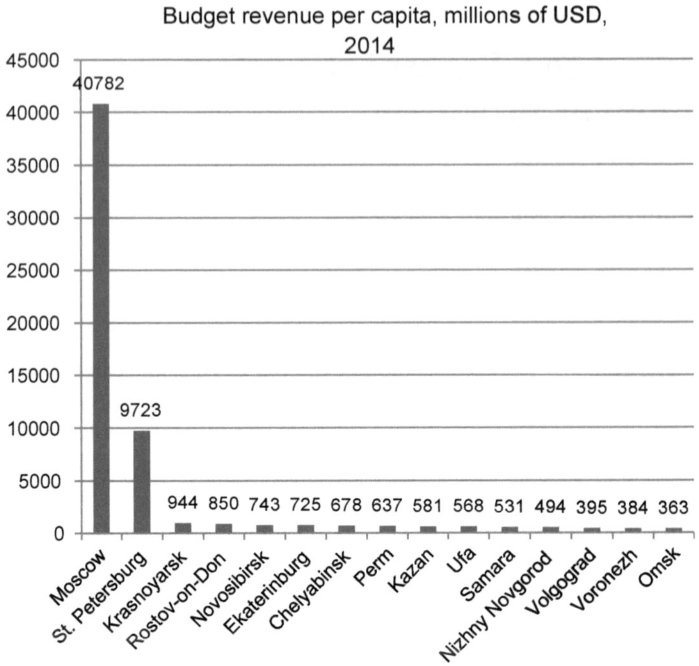

Fig. 10.1 Ranking of Russian cities with population over 1 million people in terms of municipal budget revenues (Conversion from ₽ into USD using the average exchange rate for 2014) [*Source* Created by authors using the data from Kommersant.ru (2014)]

largely dependent on transfers from a regional or, in some special cases, federal budget.

In some cases, especially in the 'national' republics, transfers from regional or federal budgets can be very significant and lead to investments that are quite large by Russian standards. The best example is Kazan which is closer to Moscow and St. Petersburg than to the provincial cities in terms of urban transport development figures.

The situation in the two Russian "capital cities", Moscow and St. Petersburg,[1] is completely different from other cities. Both capitals have a status of federal subjects and some specific advantages[2] and, at same time, are the largest economic centers with their own significant tax base.

Figures 10.1 and 10.2 show the rankings of Russian cities with population over 1 million people in terms of municipal budget revenues and budget revenues per capita. It is easy to notice that Moscow and St. Petersburg have exceptionally high

[1]St. Petersburg is perceived as a 'symbolic' capital city as part of Russian tradition.

[2]The most significant of these advantages is being the place of tax registration for the largest commodity companies. Some of them (e.g. some divisions of Gazprom) are registered in St. Petersburg, others (such as Rosneft and Lukoil), in Moscow.

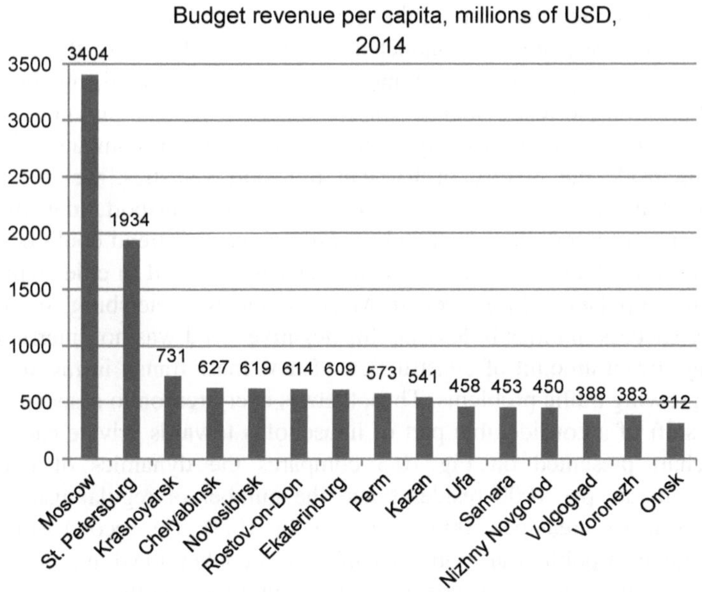

Fig. 10.2 Rankings of Russian cities in terms of distribution of income for the city budget per capita (Conversion from ₽ into USD using the average exchange rate for 2014) [*Source* Created by authors using the data from Kommersant.ru (2014)]

concentration of cumulative budget potential among the Russian cities. They account for 86.5 % of the total budget revenues of the cities presented in the chart.

Obviously, in these conditions, forecasts of transportation systems development in Moscow and St. Petersburg must be drastically different from the forecasts for provincial cities both in terms of quantity and quality of projects and institutions and in terms of implementing technological innovations.

We should also take into account the dependence of each city's development not only on the urban budget capabilities but also on the quality of local governance. In particular, there is a certain political will to find new forms of urban mobility that would improve the city's competitiveness.

10.2 Catching-up Development

As noted earlier, the motorization rate in the largest Russian cities at the moment is about 400 cars per 1000 people. The positive trend started in 1990 and can still be observed. Almost everywhere you can see the decline of urban public transport. This trend, in its turn, has changed only in a few cities, and only recently.

Unfortunately, regular household transport behavior surveys, usual in the developed countries, are not conducted in Russia.[3] Therefore, the retrospective evaluation of the impact of the past and ongoing changes on citizens' behavior can be based only on indirect evidence: for example, on the data on annual passenger volumes serviced by large (usually municipal) mass transit companies.[4]

Leaving aside the quality of Russian transport statistics, we may say with confidence that the volume of mass urban passenger transport ridership shows a stable downward trend. For small and medium cities, this trend could be attributed to population decline. However, the same trend is observed in cities with stagnant or growing population. Moreover, in Moscow and St. Petersburg, the most economically successful cities in Russia, this negative trend was not interrupted[5] even by the significant amount of construction of new mass transit infrastructure along with the growing traffic problems. Thus, there is every reason to associate this trend with the shift of a considerable part of households towards private cars.

The chart presented on Fig. 10.3 compares the dynamics of two figures: motorization rate per 1000 residents and the number of regular users of public transport in urban areas per 1000 residents (Fig. 10.3). The second value is based on the volume of public transport ridership and a conventional assumption that on average, a regular public transport user makes two trips daily.

Now we will compare the Russian case of 'motorization vs public transit' with similar cases of two leading countries of the developed world, USA and Germany. USA has one of the highest motorization rates in the world, and it has been growing since WWI. They also possess vast experience of implementing the 'car dependent cities' paradigm, while urban public transport systems have a long history of ups and downs. Germany also has one of the highest motorization rates in the world, and it has been growing since WWII. Urban development is mostly transit-oriented; urban public transport systems are among the best in the world.

For the purpose of our analysis, it would make sense to choose the historical periods when the motorization rate in both USA and Germany was growing from the level comparable to that of the late Soviet years, (i.e. dozens of vehicles per 1000 residents) to the level when this growth was stabilized (800 vehicles per 1000 residents in the USA and 500–600, in Germany).

Motorization in the USA had grown from 5 to 320 between 1910 and 1950. At the same time, the intensity of public transport usage (with some fluctuations) was showing a clear negative trend. Over those years, both the country and the world went through a series of major events that left a considerable impact on the mobility of the population and its modal split: the beginning of the Ford era, the Great

[3]The results of the first of these surveys are presented in Chap. 9.

[4]The data on passenger volumes carried by private operators, the so-called "marshrutkas" (jitneys), is not part of the official statistics. Rough estimates of these volumes are produced in some indirect way. The data on passenger volumes, attributable to the share of private cars, cannot be found in the official statistics.

[5]Moscow in 2014–2015 had seen some improvement in public transport development. However, there is no evidence so far to suggest that the beginning of long-term negative trend.

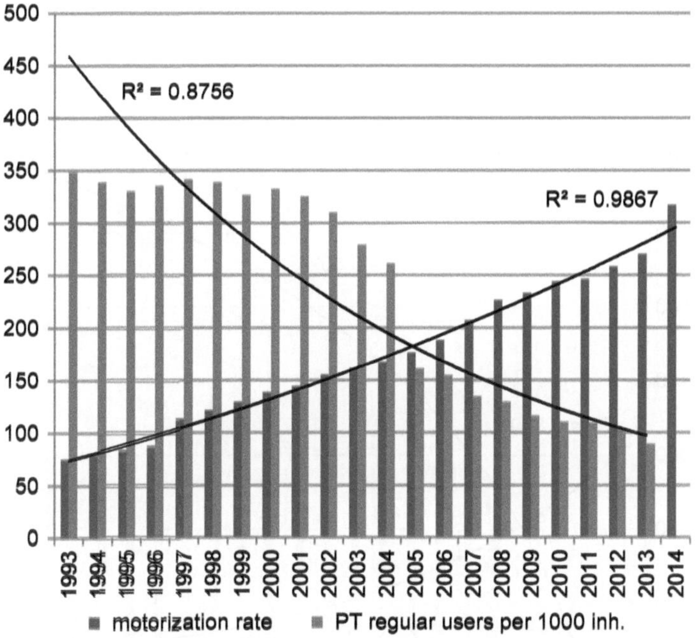

Fig. 10.3 The dynamics of motorization rate per 1000 residents, and the number of regular users of public transport in urban areas per 1000 residents [*Source* Created by authors using the data from Russian Federation Federal State Statistics Service (2016)]

Depression, WWII (and the transformation of transport behavior associated with wartime circumstances), the exodus of the middle class into suburbs (Jones 1981), etc. Given that, the respective lines representing the trends of motorization rate and public transport usage appear to intersect at the point of 250 vehicles per 1000 residents (Fig. 10.4). Afterwards, public transport usage steadied itself at a modest but relatively stable level. In recent years, the volume of public transport usage has been showing a steady increase (Public Transportation Fact Book 2016) due to the growth of urban population and fundamental changes in the national transport policy through a number of federal laws[6] introduced in the 1990–2010s.

In West Germany, the motorization rate has grown from 80 to 400 vehicles per 1000 residents over 30 years between 1960 and 1990. During the period, inter alia, the 1973 oil crisis had struck, which led to a significant increase of motor fuel prices. This, accordingly, had spurred the usage of public transit between 1975 and 1980 (Buehler and Pucher 2012). However, the overall trend remained negative. The respective lines representing the trends of motorization rate and public transport usage appear to intersect at the point of 160 vehicles per 1000 residents (Fig. 10.5). Later, the level of public transport usage showed a slight positive trend

[6]ISTEA of 1991, TEA-21 of 1998, SAFETEA-LU of 2005, MAP-21 of 2012.

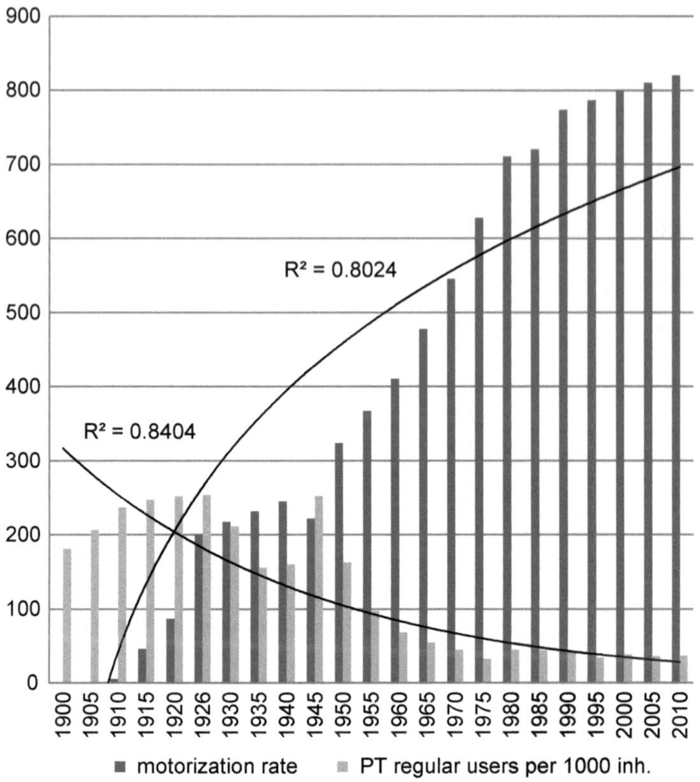

Fig. 10.4 Changes in motorization rate per 1000 residents, and the number of regular users of public transport in urban areas per 1000 residents in the USA [*Source* Created by authors using the data from Public Transportation Fact Book (2016)]

caused by the reunification with GDR with its transit-oriented policies of a socialist kind as well as by large investments in public transport infrastructure.

The chart shows that the emerging Russian motorization trends are more like those the USA used to demonstrate. The next important milestone will be reached when the dynamics of public transport usage approaches the level of the 1990s Germany or 1950s USA, i.e., when the graph will stabilize at a low level or reach the bottom and then slowly return back to higher figures.

Unfortunately, at least with regard to the provincial Russian cities, we must deal with standards rather less advanced than those applying to the US and German cities. A typical example based on lower standards is a city in one of the developing countries such as South Africa, where urban mobility is shared between private cars and jitneys (minibus-taxi), as shown in Fig. 10.6, in the situation of ongoing public transport stagnation.

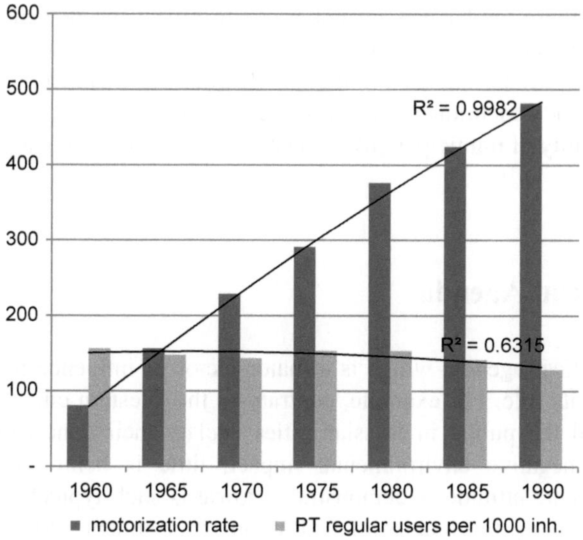

Fig. 10.5 The dynamics of motorization rate per 1000 residents, and the number of regular users of public transport in urban areas per 1000 residents in Germany [*Source* Created by authors using the data from Kraftfahrt-Bundesamt (2016) and Buehler and Pucher (2012)]

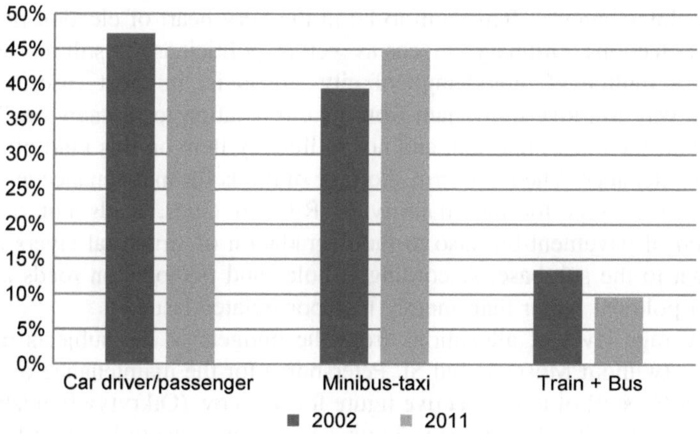

Fig. 10.6 Modal split in Johannesburg, 2002 and 2011. Non-motorized modes are excluded [*Source* Created by authors using the data from Official website of the city of Johannesburg (2013)]

A "Capital city" scenario will be discussed below. In general, the situation in the country is more likely to develop along the lines of the USA in 1950s: the stagnation of public transport offset by the dominance of private cars. At the same time, jitneys will dominate the sphere of public transportation.

Of course, these tendencies are not likely to repeat the precedents of the past decades in a precise manner. A major impact will be made by technological innovations (especially the beginning of "Digital Age Transportation") and, of course, the already mentioned specifics of Russian cities in terms of budget sufficiency and quality of municipal governance, including urban planning and transport policy.

10.3 Political Agenda

The local political agenda, which is a major factor of influence in Russian cities, will also play its role. For example, contrary to the Western cities, while both the authorities and the public in Russian cities declare their concern with regard to reducing the negative environmental impact, little is being done in practice. Roughly the same attitude is demonstrated towards such typically "Western" priorities as improving the quality of pedestrian environment and providing infrastructure for non-motorized travel.

The centerpiece of a typical urban agenda related to the transportation problems has to do with private car ownership and include characteristic issues such as "struggle against traffic jams", "parking lots", and, most importantly, "multilevel interchanges". These days, multilevel interchanges occupy a remarkable position in the collective unconscious of Russian car owners and local politicians. Promises to build new interchanges often seem to be in the very heart of election programs at municipal elections. Ordinary citizens as well as politicians are still not used to the idea that a multilevel interchange of city streets is, in most cases, "the most expensive way to move traffic jam from its old location to the new one".

However, the most important and not so illusory item on this kind of an agenda is road maintenance. The long-term shortage of funds for maintenance and repair of road network, usual for the majority of Russian cities, leads not only to the destruction of pavement but also to the degradation of structural layers of a roadway, down to the sub-base. Accordingly, holes and potholes on roads and streets become a political rather than merely transport-related issue.

The average level of allocations from the budgets of the subjects of Russian Federation (without Moscow and St. Petersburg) for the maintenance of 1 km of a road is 2.5 % (sic!) of the respective figure for Moscow (Otkrytyy byudzhet goroda Moskvy 2016).[7] Clearly, in such circumstances, any institutional and/or technological innovations are unlikely to be of any relevance for the urban agenda in Russian provincial cities.

[7]For regions other than Moscow data obtained from the summary on the road financing in the Russian Federation provided by the Russian Association of Territorial Bodies of Road Management "RADOR".

10.4 The "Capital City" Scenario

The situation in the capital cities—Moscow and St. Petersburg, as noted above, is fundamentally different from the nationwide. For example, 5.65 billion dollars out of the Moscow city budget was directed at solving transportation problems in 2015.[8]

Allocations from the Moscow city budget for the construction of subway lines allows for building up to 10 km or more of lines per year. The 54 km Moscow Ring Railway is planned to be launched in 2016, routed mainly through brownfields remaining from the Soviet era in the middle area of the city. It is expected that the facility will significantly stimulate the interest of developers in the redevelopment of the "rust belt". The total investment in this project amounted to about 1 billion dollars. Great opportunities emerge for integrating the traditional HRT systems (suburban rail, subway and Moscow Ring Railway), that is, actually including HRT in the urban transport system.

Between 1991 and 2010, more than 30 % of tram tracks were dismantled in Moscow, but in recent years, the length of the tram network began to grow, and its quality improved. At the same time, a significant renovation of all kinds of land public transport is going on; all the land passenger transport, including jitneys, is transferred to a single government gross cost contract; a new payment system and ticket menu is being introduced, similar to those used in the cities of Western Europe; a passenger is provided with online information on public transport via information boards at stops and via special apps. Systems of bike sharing and car sharing are being actively developed. The central part of the city has seen a large-scale reconstruction of streets, targeted at providing clear priority to pedestrians and cyclists rather than to cars. The real breakthrough was the abolition of free parking and, accordingly, the introduction of paid parking, which occurred in 2012.

At the same time, the structure of Moscow budget allocations for transport (Fig. 10.7) does not show an explicit priority for public transport: the sums allotted for infrastructure improvements benefitting private car users are almost similar to those allotted for public transport infrastructure.

Moreover, Moscow satisfies the classical principle of "*highway users must pay their way*"[9] in no possible way: in 2015–2016, and according to the forecast for 2017, the ratio of user payments versus road costs for Moscow is 0.25. The latent budget subsidy for car owners in Moscow which is calculated as the difference between the road costs and user payments is about 2 billion dollars annually. Thus, a powerful program of public transport development is not accompanied by strong measures to adequately discourage private car trips. In fact, we are witnessing a very controversial transport policy.

[8]Conversion from ₽ into USD using the average exchange rate for 2015.

[9]President Dwight D. Eisenhower, Feb. 22, 1955.

Fig. 10.7 The structure of the Moscow city budget allocations for transport millions USD (2015) [*Source* Created by authors using the data from Otkrytyy byudzhet goroda Moskvy (2016)]

Under this policy, the volume of public transport ridership (and therefore, the number of regular users of public transport per 1000 residents) remains at a stable level which is much more modest than the figures of the Soviet times but extremely high by international standards. At the same time, the growth of motorization is being maintained at the same pace (Fig. 10.8).

The transport policies of the Moscow government, with all their contradictions, can still be regarded as rather progressive. At the same time, urban development policies continue to be archaic. Moscow, despite some progress in the redevelopment of the industrial zones, is a place of conservation of the most non-market and archaic management practices in urban spatial development. This leads to the practice of new greenfields development which is probably not the most astute. From the standpoint of estimation and forecasting of consequences of these practices for transport development, the most critical factor is the dense multi-story development of the Moscow suburbs all along the urban perimeter.

Such development is typical in equal measure for the peripheral areas of the city within its old administrative boundaries, for the municipalities of the Moscow region within the nearest zone of the Moscow metropolitan area, and for the so-called ‘new Moscow’, i.e., the vast south-western sector of the Moscow region included in the administrative borders of Moscow in 2012.

During the last decade, the pace of housing construction in the municipalities surrounding Moscow has been much more vigorous than in Moscow itself.

The well-known poor quality of transportation services in these areas does not preclude the high demand for relatively affordable housing therein. A typical buyer who has recently arrived from the Russian province prefers heavy traffic inconvenience in Moscow to the alternative of settling in much poorer regions.

Appending new territories to Moscow was originally justified by the excessive density within the city and the need for moving administrative and business functions towards the ‘New Moscow’. The talks about transferring some of the government offices to new areas were soon phased out, and developing vast

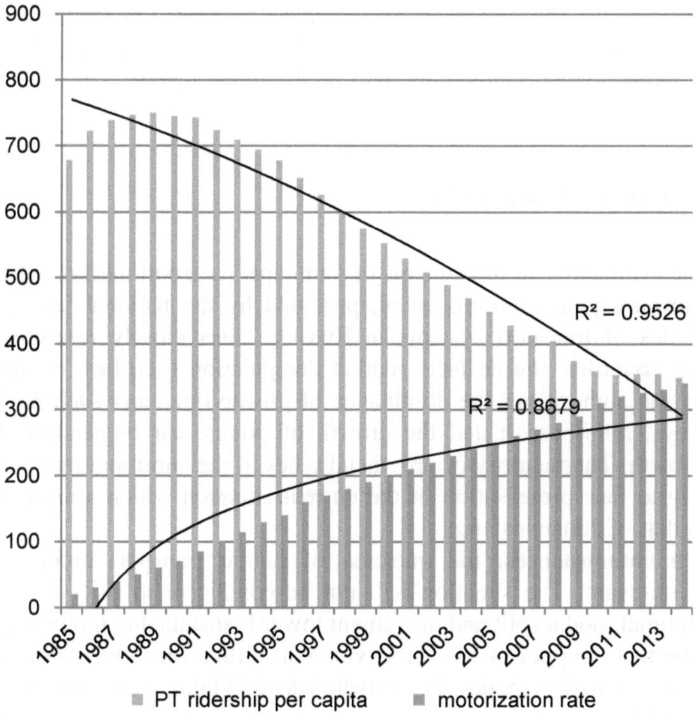

Fig. 10.8 Motorization rate per 1000 residents and number of regular users of public transport in urban areas per 1000 residents in Moscow [*Source* Created by authors using the data from Russian Federation Federal State Statistics Service (2016) and authors' calculations]

greenfields into high-rise, high-density residential areas started. The first subway stations in the 'New Moscow' opened in 2015–2016; it is assumed that they will be able to meet the transportation needs of the new areas of high-density development. In part, this hypothesis is valid; the only problem is that in the situation of the existing subway lines often operating at the limit of their freight capacity (about 60–65 passengers per hour per direction at peak hours), their extension to the far periphery is at risk of sharply degrading the quality of services provided to the residents of 'old' areas.

Today, the only factor limiting the density and number of stories of the new housing built in the Moscow suburbs is the insufficient capacity of the road network as well as that of the feeder lines of public transport that can carry passengers to a nearest subway station.

In view of the above, and for many years to come, this type of development is likely to ensure stable demand for high-performance public transport in Moscow. So, it is quite possible that in Moscow, the German scenario of the 1960–1990 will be realized. High demand for public transportation services will be combined with the continuing growth of the motorization rate among the population.

The fact that the Moscow scenario would be ‘degraded in quality’ stems from the prognostic assessment of the Moscow HRT system capacity. The possible impact of technological innovations will be discussed below.

10.5 Provincial Scenario

Despite the availability of instruments of municipal transport market regulation and public transport quality improvement, provided by the national law system, the characteristics of transport systems in Russian cities closely resemble those in developing countries: the de facto market deregulation (i.e., lack of management actions by the authorities); the dominance of private operators such as “marshrutkas” (jitneys, mini-buses); and the growth of private car ownership along with income growth. The majority of provincial cities in the short and medium term are most likely to face preservation of the current situation which almost completely replicates that of Johannesburg.

In such circumstances, the question of public transport quality obviously becomes irrelevant. The situation will not be conducive to the establishment of socially optimal modal split and movement towards sustainable mobility patterns. It is also clear that the poor quality of service will further incentivize citizens to look for alternative transport modes, i.e., private cars and taxis. Taxi fares have become quite affordable since the expansion of the illegal private taxi market in the 1990s and the introduction of ride-sharing services like GETT or Uber in recent years. The poor quality of the pedestrian environment and the spatial configuration of the urban area will hold back the growth of the share of non-motorized trips.

Despite the obvious general drawbacks of the provincial scenario, it has an extremely important advantage for the urban authorities: the scenario can be implemented without the involvement of any budget funds and does not require complex management decisions.

It should be noted that the “marshrutka” model has become widespread in the vast majority of Russian cities. Even in Moscow, this transport mode has been used on the major feeder lines to subway until very recently.

The liquidation of the electric transit lines becomes almost inevitable as part of this scenario in most cities in Russia. Tram and trolleybus systems degrade and shut down even in the cities where spatial configuration provide preconditions for their usage as trunk sections of public transport network (Fig. 10.9).

The development prospects of tram and trolleybus enterprises in those cities where they have survived until today are still not very optimistic. For accuracy, it should be noted that there is a number of very positive exceptions in this sphere, which include Kazan, Yekaterinburg, Samara, Naberezhnye Chelny, and some smaller cities.

The prospect for maintenance and development of the subway systems should be discussed separately.

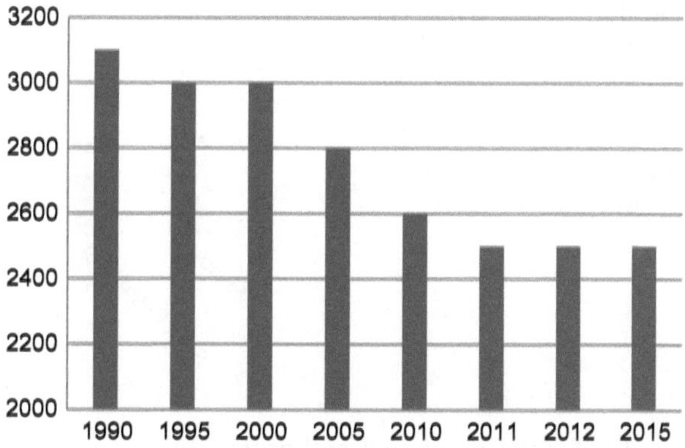

Fig. 10.9 Length of tram lines in Russian cities [*Source* Created by authors using the data from Russian Federation Federal State Statistics Service (2016)]

Constructing a subway system similar to those in Moscow and St. Petersburg was an extremely popular idea in all major Russian cities during the Soviet period. Today, the rush hour passenger flow in these cities does not exceed 15,000–20,000 passengers per hour on the major routes, which makes the efficiency of the subway system entirely redundant. However, the ideas that were popular in the Soviet times have been preserved throughout the post-Soviet period.

The most illustrative example is the case of Omsk with its population of 1.15 million residents where the subway system has been under construction since before the dissolution of the USSR. Until now, only one (sic!) station has been finished and conserved. The prospects of further construction are unclear due to the lack of funding as well as the availability of the less costly alternatives. Similarly conserved subway constructions can be found in Krasnoyarsk and Chelyabinsk. All these objects require significant spending from local budgets to cover the continuous maintenance expenses.

Subway systems currently operate in five cities with more than 1 million residents (with the exception of Moscow and St. Petersburg): Novosibirsk, Samara, Kazan, Nizhniy Novgorod, and Yekaterinburg (Figs. 10.10 and 10.11). However, they have largely turned from a panacea (as in Moscow and St. Petersburg) to a

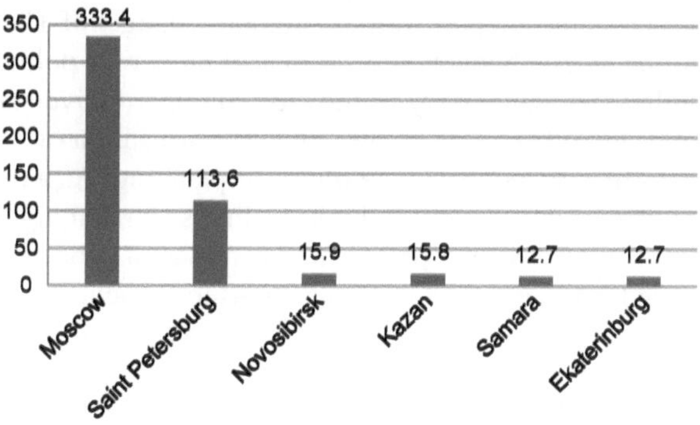

Fig. 10.10 Length of subway systems lines in Russian cities [*Source* Created by authors using the data from Russian Federation Federal State Statistics Service (2016)]

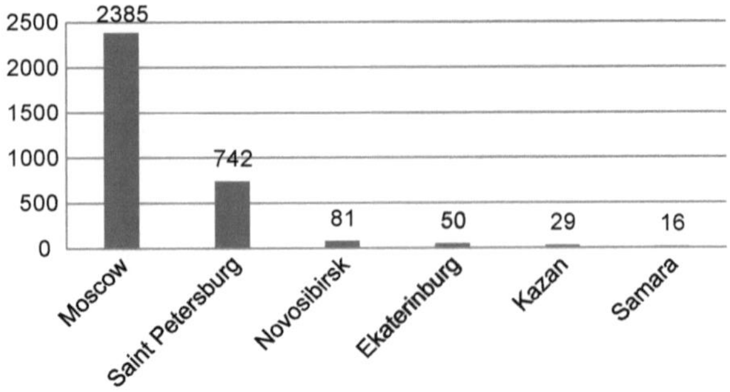

Fig. 10.11 Annual ridership of the subway systems in Russia, millions, 2015 [*Source* Created by authors using the data from Russian Federation Federal State Statistics Service (2016)]

burden for the municipal authorities. For example, in Samara, the subway system provides only 15 % of the total public transport ridership, being a very costly consumer of budget subsidies. The tram system provides more than 50 % of total public transport ridership, while its unit maintenance costs are much lower.

10.6 Mixed Strategies

We have described two polar scenarios for the development of transportation systems in Russian cities. The 'Capital city' scenario is a downgraded analog of the German scenario, while the 'peripheral' scenario is an analog of the developing countries scenario.

Of course, some mixed scenarios can also be found. The degree to which the scenario of each particular city resembles one of the radical examples is not really dependent on its own budgetary limits: budget incomes per capita differ sharply when Moscow is compared to other cities but do not differ much when 'other cities' are compared with each other.

The individual scenario for each city is much more dependent on the stability of the local budget support from regional budgets and, of course, on the quality of transportation planning and the performance of the local government on the whole.

One of the best examples is the city of Kazan.[10] Budget incomes per capita in the city are close to the average Russian level (Fig. 10.2). At the same time, the city gets a stable support from the regional budget of The Republic of Tatarstan as well as the federal budget on occasions of numerous politically significant events like the millennial anniversary of the city, Summer Universiade in 2013, 2018 FIFA world cup, and other sport events. Effective utilization of budgetary opportunities is made possible by very adequate transport planning. As a result, the city actively develops modern rail systems. Six high-quality, separated high-capacity tram lines have been introduced over the past year. There has also been a massive introduction of dedicated bus lanes. Land public transport, including jitneys, has moved towards a

[10]Some other examples have been already descripted in the previous chapters.

single municipal gross cost contract model. A project of a congestion charging zone in the city center has been started for the first time in Russia. Unfortunately, the local planning policies are largely characterized by the same flaws that we discussed in the case of Moscow. However, it is possible to expect positive changes in the modal split, i.e., the increase in the share of public transport and non-motorized trips.

Another interesting case is Volgograd. The city has an effective and high-capacity LRT line with the total length of 17.3, 7.1 km of which is an underground ROW-A section while land sections operate as ROW-B. The tram system has good chances for extension and provides service for more than 50 million passengers annually which is quite comparable to Russian subway systems except for Moscow and St. Petersburg.

We can also mention the city of Naberezhnye Chelny (527,000 residents in 2016) where the budget income per capita is lower than in any of the cities with a population over 1 million. Nevertheless, the quality of the local tram system is being maintained and the sections with the highest demand improved.

With all these positive points, even in these cities, the structure of urban transportation in general is characterized by the dominance of private "marshrutka"-type operators and prevailing orientation of households at private car use.

An important feature of the intermediate scenarios is an attempt to implement the best practices which have already been accumulated in Moscow. This dissemination of innovations can already be noticed: the most typical example is paid parking, which gradually began to make its way into some big cities (more than a dozen out of 75 cities with population over 250,000).

In many cities, usually under the influence of local 'hipster urbanism' communities, priorities such as the quality of pedestrian environment and comfortable public transport started appearing in the local urban agendas along with objections against using the scarce budgetary resources for the construction of large multi-level intersections at the central streets of the city. In some cases, these ideas are embodied in the decisions by local authorities.

Finally, it should be noted that following the generation change among the municipal officials, young people, well-educated and familiar with the best global practices, are getting the chance of taking over municipal offices. In these circumstances, the cities have a chance to significantly improve the quality of their transportation systems without calling for substantial financial resources.

There is also the reverse pattern, the examples of deteriorating management quality and returning to the lower, average Russian level. One of the examples is Perm which became an advanced city in terms of urban and transportation planning in the late 2000s and early 2010s. This process took place at the backdrop of an overall upsurge of urban culture and promotion of the city brand in other regions and abroad by experts invited by the reform-minded governor of the Perm region. However, after the change of power in the region, the level of urban and transportation planning dropped back to the national average.

10.7 Modern Trends

Firstly, in terms of the contemporary trends of development of urban transport systems, we will be guided by the consensus about the reality of the 'Digital-Age' and the impact of its concepts on the urban mobility technology which has been formulated in a number of publications in recent years. The recent Russian experience makes it reasonable to believe that the advanced IT tools will be engaged by Russian cities more or less in the same way as it happened in the developed countries but with a certain delay and errors inevitable when trying to replicate foreign experience with no consideration of local context.

Secondly, we cannot expect that the urban mobility problem in Russia will be alleviated due to the intra-urban residential mobility factor, common for the Western cities. The reason is that the real estate market in Russia has never been customer-friendly: the mortgage rates are very high, and so are the costs associated with a change of residence prior to the mortgage expiration. Therefore the replacement of transport mobility with residential mobility in Russia remains unrealistic.

Thirdly, it is also possible to notice that the need for daily commuting is unlikely to disappear in the process of IT development. The development of telecommunications has stimulated close interaction between parties to such a degree that they cannot do without meeting in person. The prediction that the need for daily migrations will be gone because of IT development seems quite improbable (Schiller et al. 2010).

10.7.1 Paradigm Changing

It's likely that cities are facing the third shift of the urban mobility paradigm in the last 350 years. The first paradigm is associated with the idea of regular public transportation which was exhaustively formulated in 1661 in the famous letter by Blaise Pascal to Duke de Roannez: "...Les carrosses devaient toujours suivre le même itinéraire et devaient respecter les horaires fixés, les départs ayant lieu tous les demi-quarts d'heures de leur terminus même si ceux-ci étaient vide. Les passagers devaient payer leur place cinq sols[11]..." The second paradigm is based on the idea of total transport self-sufficiency of households. It became a reality in 1908 with the arrival of the first affordable car, the Ford-T: '... greatest need today is a light, low-priced car with an up-to-date engine with ample horsepower'.

The third and most recent paradigm is associated with the advent of the 'Digital Age': '...The arrival of the 'information everywhere' world has opened up new

[11]According to Blaise Pascal, public transportation should not be cheap: 5 sou in 1661 were no less than the present-day €10.

opportunities to make the existing transportation network far more efficient and user-friendly' (Fishman 2012).

This fundamental paradigm shift has been made possible by the ability of IT tools to provide:

- Personalization of the originally mass-oriented 'Pascal's' mobility pattern, i.e., individualization meets the demand for public transportation;
- Inclusion of the originally individualistic 'Ford' mobility pattern into the sphere of social relations (car sharing, ride sharing, etc.);
- Alignment and optimization (based on the criteria of time, price or convenience) of series of multimodal moves, which includes utilizing all forms of public transport such as private cars, taxi, car sharing, bike sharing, etc.

Modern IT-tools such as portable gadgets, user-friendly apps, and big data can lead to fundamental changes in urban transport behavior. Today, almost everyone has access to the online information on the location of public transport and expected time of its arrival, availability of free spaces on parking lots and current price of parking (including dynamic pricing mode), and presence of a free car (when using car sharing) or bicycle (when using bike sharing) in the immediate vicinity. Everyone can also take advantage of convenient apps to pay for a trip or call for a taxi while calculating the time needed for arriving at a station prior to the train departure etc.

It is important to note that all of these 'Digital Age' features are coming to the largest Russian cities more or less at the same time or with a very slight delay in comparison to the developed world.

10.7.2 Driverless Cars

Products such as 'Self-driving Car', 'Driverless Car', and 'Autonomous Car' are either being manufactured or are will be manufactured very soon by all the leading automotive companies in the world as well as by the beginners in this field among the leading IT companies. There is an active discussion on the legal (technical, customer-related, etc.) preconditions and consequences of the potential widespread usage of these technologies in Russia. It is possible to say with sufficient confidence what kind of customers may or may not provide the demand for the 'Driverless' feature.

The most obvious and undisputed potential applications for such systems are ROW-A designed rail lines. In this case, there are neither engineering nor legal, nor labor market-related obstacles that might prevent the introduction of driverless trains. Unfortunately, even employing driverless trains will not help surpassing the record operating parameters of the most developed HRT systems introduced in Moscow since 1935. Today, there are 40–42 pairs of trains per hour on some of the lines of the Moscow subway, and this number is established by the planning parameters of subway stations and conditions of embarkation and disembarkation of passengers. Therefore the existing capacity values (about 60–65, 000 passengers per hour in one direction) will clearly not be increased by the introduction of new technologies.

Another obvious and undisputed possible use for the 'Driveless' feature is for the parking private cars. In this case, there are also no obstacles for the widespread adoption of this technology, which is especially relevant for the largest Russian cities with their shortage of residential parking places and parking lots.

A very promising way of using the 'Driverless' technology is for urban trips via car-sharing opportunities. Today, a customer can register with one of the car-sharing companies and, having installed the required application on a smartphone, use it to find a free car bearing the company's logo in the nearest vicinity, yet it is necessary to walk to its location. The nearest free car may happen to be farther away than the closest subway station, or a customer may feel too tired to walk to the car. These circumstances significantly limit the competitiveness of car sharing as compared to the private car and ride-sharing services. It may be a different situation when every car belonging to car-sharing companies gets equipped with the driverless technology. In this case, customers will be able to order the car directly to their location. Thus, it becomes a kind of a 'taxi without driver' which can be cheaper than GETT or Uber but will also come on demand. For this purpose, however, one particular legal problem should be solved first: traffic rules should allow the driverless vehicles owned by car-sharing companies to move autonomously for a limited distance (for example, 1–1.5 km), at a low speed (around 25–30 km/h), and directly to the point of client pick-up. Such legal provisions should be sufficient for the function of calling on a free car to be implemented. The risk of accidents in this particular case is quite minimal: a small car (there are only small cars within the 'car sharing' system) at a minimum speed does not present a considerable hazard in terms of road safety. Apparently, options like this will become part of everyday urban reality very soon.

In the international practice, one of the most promising applications for the 'Driverless' feature is for the cars on urban freeways. Hitherto, the critical factor limiting the capacity of such roads was a 'biological driver' usually having very slow reaction time, which facilitates high variability of speed versus traffic flow density. Today, when every single car can be equipped with an autopilot and freeways are provided with suitable ITS systems, a double or triple increase in capacity can be reached by solving a standard problem of controlling the collective behavior of automata. Technical obstacles to the realization of this scenario that was discussed by the traffic engineering experts as early as in the 1980s no longer exist today. Both the legal and the behavioral limitations exacerbate the situation, yet it is not hopeless. In view of the qualities of street and road network in Russian cities, described in the previous chapters, this issue, unfortunately, is almost irrelevant for Russia.

As to the potentially most popular application, 'a passenger car on a city street', managing of the collective behavior of automata within a road network in a city is much more difficult than on freeways. Anyway, the present-day IT tools should be able to cope with the task tomorrow, if not today. The only question is whether we need to solve this problem in Russia as well as in the more advanced countries. It seems that for the purpose of the 'Digital Age Transportation', there is no point in installing a sophisticated IT-shell on archaic transport systems inherited from the Ford's century.

Anyway, vehicle to vehicle (v2v) or vehicle to infrastructure (v2i) technologies will be implemented in Russia in the years to come.

10.7.3 Intelligent Transportation System

National or municipal programs of developing and implementing Intelligent Transportation System (ITS) proliferated around the world at the turn of the 1980s and 1990s. Initiators of this motion claimed that ITS should be able to significantly raise the level of information about traffic conditions and promote safety improvement, efficiency and reliability of transport systems through the use of modern IT and communication technologies in road traffic and public transport systems. Meanwhile, the representatives of the American transport and urban establishment claimed that a common problem of such projects is that the solutions to urban transport problems are being sought in the sphere of new technologies while these problems are rather rooted in the short-sighted policies and weak organizational procedures than in backward technologies (Vuchik 1999).

Unfortunately, V. Vuchik's words are quite befitting the experience of ITS implementation in Russia. In some segments of the transport system, this experience was successful, and the prospects for the coming years are quite favorable; in others, the attempts were quite unconvincing.

Positive examples include equipping land public transport vehicles with GPS/GLONASS, as well as further development of this idea and creation of the information systems for passengers through online boards at stops and mobile apps. Thereby positive effects have been achieved, essential for customers, enterprises, and transport planners. At the same time, the possibilities for falsifying trip statistics, very characteristic of Russian reality, were eliminated. The availability of this system allowed for the easy transfer of the public transport companies, including the "marshrutka" operators, to gross contracts.

Another positive example is the introduction of sophisticated parking payment systems using mobile apps and SMS that allowed for eliminating cash payments in Moscow and other cities.

Negative examples include the very costly ITS systems in charge of traffic control. The problem is that the efficiency of urban traffic ITS algorithms for routing orientation of drivers and coordinated management of traffic lights almost fully depends on the road network connectivity (Inose and Hamada 1975). Therefore, for the networks with high grade connectivity, those algorithms can bring really impressive benefits. Otherwise, on the networks with low or zero grade connectivity, as is the case with Moscow and many other cities in Russia, those algorithms can provide only superficial effects.

In Russian cities, there are three groups of targets in terms of implementing new technologies:

1. The targets where introducing new technologies can yield significant positive effects. A typical example is the possibility of objective evaluation of transport operator performance based on the GPS/GLONASS data and the data available due to introduction of electronic forms of payment. The passenger convenience and the transport planning quality improve due to decriminalization of private marshrutka carriers.
2. The targets where introducing new technologies creates the possibility of radical renewal of existing transport services as well as a significant increase in the quality of services.
   The most glaring example is the taxi business which has been transformed dramatically (and positively, too) due to the introduction of modern services such as GETT, Uber or their Russian analog, Yandex-taxi.
   Over the past 25 years, the era of big taxi companies and central control services that existed in all major cities of the USSR passed into the period of uncontrolled, often semi-criminal private carting. In order to overcome this negative trend in Moscow and several other cities, attempts have been made to restore the Soviet model, which means that the place of the state taxi companies should be taken over by private companies enjoying municipal support.
   The emergence of the services described above has rendered these attempts completely redundant. New technologies allow for shaping the market in a modern and user-friendly format, which also ensures complete transparency of financial flows.
   The revolutionary technology of car sharing and driverless cars can also be used for this kind of targets.
3. The targets where the introduction of new technologies is purely symbolic and does not lead to any significant positive results. A typical example, which has already been discussed above, is the attempt to introduce very costly elements of ITS in the sphere of traffic control along with the preservation of an archaic topological structure of road networks.
   Summarizing the perspectives of implementing new technologies (including those quite revolutionary) in Russian cities, it should be stated: in the most important aspect, that is, in terms of providing mass transportation carrying tens of thousands of passengers per hour, no real breakthroughs are expected in the foreseeable future. Using the most advanced high-tech systems, including those discussed above, does not necessarily facilitate major changes in mass public transportation, except a few that might improve circuit surveillance and control and allow for shifting the modal split towards the socially optimal point.

References

Buehler R, Pucher J (2012) Demand for public transport in Germany and the USA: an analysis of rider characteristics. Transp Rev 32(5):541–567

Fishman T (2012) Digital-age transportation: the future of urban mobility. http://d27n205l7rookf.cloudfront.net/wp-content/uploads/2012/12/Digital-Age-TRANSPORTATION.pdf. Accessed 31 Aug 2016

Inose H, Hamada T (1975) Road traffic control. University of Tokyo Press, Tokyo

Jones D (1981) Urban transit policy: an economic and political history. Prentice Hall, New Jersey

Kommersant.ru (2014) Luchshie goroda Rossii (The best cities of Russia). http://www.kommersant.ru/doc/2441305. Accessed 31 Aug 2016 (In Russian)

Kraftfahrt-Bundesamt (2016) Bestand in den Jahren 1960 bis 2016 nach Fahrzeugklassen. http://www.kba.de/DE/Statistik/Fahrzeuge/Bestand/FahrzeugklassenAufbauarten/b_fzkl_zeitreihe.html. Accessed 31 Aug 2016

Official website of the city of Johannesburg (2013). Strategic integrated transport plan framework for the city of Johannesburg. http://www.joburg.org.za/images/stories/2013/May/CoJ%20SITPF%20Draft%2013%20May%202013.pdf. Accessed 31 Aug 2016

Otkrytyy byudzhet goroda Moskvy (2016) http://budget.mos.ru/. Accessed 31 Aug 2016 (In Russian)

Public Transportation Fact Book (2016) http://www.apta.com/resources/statistics/Pages/transitstats.aspx. Accessed 31 Aug 2016

Russian Federation Federal State Statistics Service (2016) http://www.gks.ru/. Accessed 31 Aug 2016 (In Russian)

Schiller P et al (2010) An introduction to the sustainable transportation. Earthscan, London

Vuchik V (1999) Transportation for livable cities. CUPR Press, NJ

Appendix

M. Blinkin and E. Koncheva (eds.), *Transport Systems of Russian Cities*, Transportation Research, Economics and Policy,
DOI 10.1007/978-3-319-47800-5

Major accident rates in the major cities of Russia (2015)

City	Population, thousands	Car fleet, thousand units	Number of accidents	Number of deaths	Number of injured	The severity of the consequences, deaths per 100 victims	Social risks, deaths per 100,000 residents	Transport risks, deaths per 10,000 cars
Moscow	12,198	3,840,740	10,228	650	11,700	5.3	5.3	1.7
St. Petersburg	5192	1,648,739	5383	253	6231	3.9	4.9	1.5
Omsk	1174	326,281	2591	94	3157	2.9	8.0	2.9
Perm	1036	239,713	1988	89	2421	3.5	8.6	3.7
Kazan	1206	365,171	1962	65	2253	2.8	5.4	1.8
Chelyabinsk	1183	315,650	1946	87	2373	3.5	7.4	2.8
Krasnoyarsk	1052	307,950	1895	76	2184	3.4	7.2	2.5
Ufa	1106	313,461	1881	63	2367	2.6	5.7	2.0
Novosibirsk	1567	441,133	1761	79	2010	3.8	5.0	1.8
Rostov-on-Don	1115	321,763	1716	72	2124	3.3	6.5	2.2
Barnaul	636	171,434	1608	51	1958	2.5	8.0	3.0
Tyumen	697	237,060	1447	39	1927	2.0	5.6	1.6
Nizhny Novgorod	1268	342,832	1384	48	1674	2.8	3.8	1.4
Saratov	842	250,903	1334	53	1644	3.1	6.3	2.1
Irkutsk	620	201,004	1233	48	1470	3.2	7.7	2.4
Voronezh	1024	305,395	1027	80	1206	6.2	7.8	2.6
Vladivostok	605	420,260	992	27	1141	2.3	4.5	0.6
Penza	523	127,861	967	35	1207	2.8	6.7	2.7
Khabarovsk	607	142,340	955	46	1167	3.8	7.6	3.2
Orenburg	561	170,550	906	34	1153	2.9	6.1	2.0
Ryazan	533	166,003	875	36	1120	3.1	6.8	2.2

(continued)

(continued)

City	Population, thousands	Car fleet, thousand units	Number of accidents	Number of deaths	Number of injured	The severity of the consequences, deaths per 100 victims	Social risks, deaths per 100,000 residents	Transport risks, deaths per 10,000 cars
Ulyanovsk	619	162,325	858	28	1071	2.5	4.5	1.7
Krasnodar	830	299,690	810	54	958	5.3	6.5	1.8
Ivanovo	409	86,330	802	29	989	2.8	7.1	3.4
Astrakhan	533	138,910	796	18	1014	1.7	3.4	1.3
Tula	488	152,564	776	31	923	3.2	6.4	2.0
Yaroslavl	604	151,152	748	43	899	4.6	7.1	2.8
Lipetsk	510	146,443	744	36	894	3.9	7.1	2.5
Kirov	493	124,839	715	28	859	3.2	5.7	2.2
Ulan-Ude	427	92,186	713	21	819	2.5	4.9	2.3
Kursk	435	116,189	703	32	888	3.5	7.4	2.8
Samara	1172	400,579	679	47	807	5.5	4.0	1.2
Vladimir	353	92,160	649	50	822	5.7	14.2	5.4
Cheboksary	474	93,522	635	31	791	3.8	6.5	3.3
Volgograd	1017	251,008	634	55	731	7.0	5.4	2.2
Chita	339	99,835	616	34	749	4.3	10.0	3.4
Izhevsk	642	173,979	615	20	679	2.9	3.1	1.1
Kaluga	343	122,436	607	26	754	3.3	7.6	2.1
Kemerovo	549	137,830	603	43	761	5.3	7.8	3.1
Tver	414	132,999	588	26	695	3.6	6.3	2.0
Stavropol	426	111,520	555	34	651	5.0	8.0	3.0
Tambov	289	90,157	548	16	717	2.2	5.5	1.8

(continued)

(continued)

City	Population, thousands	Car fleet, thousand units	Number of accidents	Number of deaths	Number of injured	The severity of the consequences, deaths per 100 victims	Social risks, deaths per 100,000 residents	Transport risks, deaths per 10,000 cars
Arkhangelsk	351	81,273	526	23	634	3.5	6.6	2.8
Kurgan	326	95,331	505	39	583	6.3	12.0	4.1
Blagoveshchensk	224	59,905	503	24	637	3.6	10.7	4.0
Oryol	320	93,938	479	16	583	2.7	5.0	1.7
Bryansk	407	95,177	474	24	564	4.1	5.9	2.5
Yakutsk	299	77,678	466	27	554	4.6	9.0	3.5
Kaliningrad	453	165,218	458	21	492	4.1	4.6	1.3
Vologda	311	88,464	443	13	546	2.3	4.2	1.5
Pskov	208	71,411	435	14	540	2.5	6.7	2.0
Smolensk	330	100,501	424	18	508	3.4	5.5	1.8
Tomsk	565	147,605	396	37	476	7.2	6.5	2.5
Novgorod	222	69,248	382	10	450	2.2	4.5	1.4
Makhachkala	583	100,114	375	44	449	8.9	7.5	4.4
Kostroma	276	69,493	370	15	439	3.3	5.4	2.2
Saransk	303	81,210	365	27	437	5.8	8.9	3.3
Syktyvkar	243	74,934	358	17	456	3.6	7.0	2.3
Yoshkar-Ola	263	55,056	356	8	445	1.8	3.0	1.5
Murmansk	305	92,861	327	6	393	1.5	2.0	0.6
Petrozavodsk	275	88,230	321	12	373	3.1	4.4	1.4
Belgorod	384	116,659	312	15	391	3.7	3.9	1.3
Petropavlovsk-Kamchatsky	181	94,976	297	8	367	2.1	4.4	0.8

(continued)

(continued)

City	Population, thousands	Car fleet, thousand units	Number of accidents	Number of deaths	Number of injured	The severity of the consequences, deaths per 100 victims	Social risks, deaths per 100,000 residents	Transport risks, deaths per 10,000 cars
Abakan	176	57,276	288	12	351	3.3	6.8	2.1
Vladikavkaz	308	73,691	286	20	369	5.1	6.5	2.7
Kyzyl	114		223	26	252	9.4	22.8	
Nalchik	239	54,179	193	20	195	9.3	8.4	3.7
Maykop	144	45,418	191	10	238	4.0	6.9	2.2
Cherkessk	124	27,301	133	19	166	10.3	15.3	7.0
Grozny	284	48,372	121	22	195	10.1	7.7	4.5
Anadyr	15	1599	11	1	13	7.1	6.6	6.3

MIX
Papier aus verantwortungsvollen Quellen
Paper from responsible sources
FSC® C105338

If you have any concerns about our products,
you can contact us on
ProductSafety@springernature.com

In case Publisher is established outside the EU,
the EU authorized representative is:
Springer Nature Customer Service Center GmbH
Europaplatz 3, 69115 Heidelberg, Germany

Printed by Libri Plureos GmbH
in Hamburg, Germany